Recent Progress in Sustainability and Durability of Concrete and Mortar Composites

Recent Progress in Sustainability and Durability of Concrete and Mortar Composites

Editors

Ofelia-Cornelia Corbu
Ionut Ovidiu Toma

Basel • Beijing • Wuhan • Barcelona • Belgrade • Novi Sad • Cluj • Manchester

Editors
Ofelia-Cornelia Corbu
Technical University of
Cluj-Napoca
Cluj-Napoca
Romania

Ionut Ovidiu Toma
The "Gheorghe Asachi"
Technical University of Iasi
Iasi
Romania

Editorial Office
MDPI
St. Alban-Anlage 66
4052 Basel, Switzerland

This is a reprint of articles from the Special Issue published online in the open access journal *Coatings* (ISSN 2079-6412) (available at: https://www.mdpi.com/journal/coatings/special_issues/concrete_mortar_composites).

For citation purposes, cite each article independently as indicated on the article page online and as indicated below:

Lastname, A.A.; Lastname, B.B. Article Title. *Journal Name* **Year**, *Volume Number*, Page Range.

ISBN 978-3-7258-0715-4 (Hbk)
ISBN 978-3-7258-0716-1 (PDF)
doi.org/10.3390/books978-3-7258-0716-1

© 2024 by the authors. Articles in this book are Open Access and distributed under the Creative Commons Attribution (CC BY) license. The book as a whole is distributed by MDPI under the terms and conditions of the Creative Commons Attribution-NonCommercial-NoDerivs (CC BY-NC-ND) license.

Contents

About the Editors vii

Preface ix

Ofelia Corbu and Ionut-Ovidiu Toma
Progress in Sustainability and Durability of Concrete and Mortar Composites
Reprinted from: *Coatings* 2022, 12, 1024, doi:10.3390/coatings12071024 1

Xingyu Wang, Fengkun Cui, Long Cui and Di Jiang
Research on a Multi-Objective Optimization Design for the Durability of High-Performance Fiber-Reinforced Concrete Based on a Hybrid Algorithm
Reprinted from: *Coatings* 2023, 13, 2054, doi:10.3390/coatings13122054 5

Fengyu Song, Didi Huo, Yanmin Wang, Dunlei Su and Xiaocun Liu
The Effect of Different Particle Sizes of SiO_2 in Sintering on the Formation of Ternesite
Reprinted from: *Coatings* 2023, 13, 1826, doi:10.3390/coatings13111826 24

Ofelia Corbu, Attila Puskas, Mihai-Liviu Dragomir, Nicolae Har and Ionuț-Ovidiu Toma
Eco-Innovative Concrete for Infrastructure Obtained with Alternative Aggregates and a Supplementary Cementitious Material (SCM)
Reprinted from: *Coatings* 2023, 13, 1710, doi:10.3390/coatings13101710 36

Xianglong Zuo, Shen Zuo, Jin Li, Ning Hou, Haoyu Zuo and Tiancheng Zhou
Study of the Design and Mechanical Properties of the Mix Proportion for Desulfurization Gypsum–Fly Ash Flowable Lightweight Soil
Reprinted from: *Coatings* 2023, 13, 1591, doi:10.3390/coatings13091591 70

Liliana Maria Nicula, Daniela Lucia Manea, Dorina Simedru, Oana Cadar, Ioan Ardelean and Mihai Liviu Dragomir
The Advantages on Using GGBS and ACBFS Aggregate to Obtain an Ecological Road Concrete
Reprinted from: *Coatings* 2023, 13, 1368, doi:10.3390/coatings13081368 91

Chunqin Tan, Mu Wang, Rongyao Chen and Fuchang You
Study on the Oil Well Cement-Based Composites to Prevent Corrosion by Carbon Dioxide and Hydrogen Sulfide at High 2Temperature
Reprinted from: *Coatings* 2023, 13, 729, doi:10.3390/coatings13040729 118

Sergiu-Mihai Alexa-Stratulat, George Stoian, Iulian-Adrian Ghemeș, Ana-Maria Toma, Daniel Covatariu and Ionut-Ovidiu Toma
Effect of a New Multi-Walled CNT (MWCNT) Type on the Strength and Elastic Properties of Cement-Based Mortar
Reprinted from: *Coatings* 2023, 13, 492, doi:10.3390/coatings13030492 135

Jin Li, Yongshu Cui, Dalu Xiong, Zhongmei Lu, Xu Dong, Hongguang Zhang, et al.
Experimental Study on the Bending Resistance of Hollow Slab Beams Strengthened with Prestressed Steel Strand Polyurethane Cement Composite
Reprinted from: *Coatings* 2023, 13, 458, doi:10.3390/coatings13020458 149

Brăduț Alexandru Ionescu, Alexandra-Marina Barbu, Adrian-Victor Lăzărescu, Simona Rada, Timea Gabor and Carmen Florean
The Influence of Substitution of Fly Ash with Marble Dust or Blast Furnace Slag on the Properties of the Alkali-Activated Geopolymer Paste
Reprinted from: *Coatings* 2023, 13, 403, doi:10.3390/coatings13020403 165

Haoyu Zuo, Jin Li, Li Zhu, Degang Cheng and De Chang
Identification Fluidity Method to Determine Suitability of Weathered and River Sand for Constructions Purposes
Reprinted from: *Coatings* **2023**, *13*, 327, doi:10.3390/coatings13020327 . . . . **186**

Manpreet Singh, Priyankar Choudhary, Anterpreet Kaur Bedi, Saurav Yadav and Rishi Singh Chhabra
Compressive Strength Estimation of Waste Marble Powder Incorporated Concrete Using Regression Modelling
Reprinted from: *Coatings* **2023**, *13*, 66, doi:10.3390/coatings13010066 . . . . **200**

Li Li, Tianlai Yu, Yuxuan Wu, Yifan Wang, Chunming Guo and Jun Li
Research on the Properties of a New Type of Polyurethane Concrete for Steel Bridge Deck in Seasonally Frozen Areas
Reprinted from: *Coatings* **2022**, *12*, 1732, doi:10.3390/coatings12111732 . . . . **217**

Yunkai Zhang, Fei Liu, Yuhan Bao and Haiyan Yuan
Research on Dynamic Stress–Strain Change Rules of Rubber-Particle-Mixed Sand
Reprinted from: *Coatings* **2022**, *12*, 1470, doi:10.3390/coatings12101470 . . . . **237**

Qiaoyi Li, Yonghai He, Guangqing Yang, Penghui Su and Biao Li
The Cracking Resistance Behavior of Geosynthetics-Reinforced Asphalt Concrete under Lower Temperatures Using Bending Test
Reprinted from: *Coatings* **2022**, *12*, 812, doi:10.3390/coatings12060812 . . . . **256**

Georgiana Bunea, Sergiu-Mihai Alexa-Stratulat, Petru Mihai and Ionuț-Ovidiu Toma
Use of Clay and Titanium Dioxide Nanoparticles in Mortar and Concrete—A State-of-the-Art Analysis
Reprinted from: *Coatings* **2023**, *13*, 506, doi:10.3390/coatings13030506 . . . . **267**

About the Editors

Ofelia-Cornelia Corbu

Dr. Ofelia-Cornelia Corbu was the Research Laboratory Director (2007–2019) within the Faculty of Construction, Technical University of Cluj-Napoca. She obtained her Ph.D. in the field of Civil Engineering at the Technical University of Cluj-Napoca in 2011. She obtained the title of Senior Scientific Researcher in 2018, recognized at the university and research center level. She is a scientific researcher with a Ph.D. in Civil Engineering with experience in practical and didactic activity, with achievements in the field of ecological construction materials (having 4 patents) that lead to the reduction of CO_2 by using recycled waste with different roles in concrete/ mortar/ geopolymeric concrete. She has a post-doctorate and two internships at UK universities. Her research interests are in the field of advanced recycling, namely the valorization of recycled waste in ecological composites with applications in the field of construction.

Ionut Ovidiu Toma

Ionut Ovidiu Toma is an Associate Professor and Head of the Department of Structural Mechanics at the Faculty of Civil Engineering and Building Services, "Gheorghe Asachi" Technical University of Iasi, Romania. He obtained his Ph.D. in the field of Civil Engineering in Reinforced Concrete Structures in 2007 from the Tokyo Institute of Technology, where he also worked as a researcher for the Center of Urban Earthquake Engineering until March 2008. His research interests are the use of recycled materials in concrete structures and the safety and reliability of engineering structures in seismically active areas.

Preface

The current Special Issue contains the works of authors from all over the world and provides other academics with insights from their own research experiences. A large part of the works refers to the creation of the material itself with a novelty character, developed in various projects with the aim of creating sustainable materials with better characteristics than conventional materials. The wide use of alternative materials to natural raw materials, such as recycled wastes for a cleaner environment, was the focus of some of the research works included in this Special Issue. Supplementary cementitious materials with excellent pozzolanic qualities that can partially or totally replace the amount of cement in the mixture for a reduction of the CO_2 footprint were also present in research papers submitted and accepted for publication in the current volume. Other authors reveal the influence of material optimized by various methods in construction products.

This Special Issue covers a wide range of scientific interests related to concrete/mortar, and the editors of this volume are grateful for the contribution of the authors.

Ofelia-Cornelia Corbu and Ionut Ovidiu Toma
Editors

Editorial

Progress in Sustainability and Durability of Concrete and Mortar Composites

Ofelia Corbu [1] and Ionut-Ovidiu Toma [2,*]

[1] Faculty of Civil Engineering, Technical University of Cluj-Napoca, 400027 Cluj-Napoca, Romania; ofelia.corbu@staff.utcluj.ro
[2] Faculty of Civil Engineering and Building Services, The "Gheorghe Asachi" Technical University of Iasi, 700050 Iasi, Romania
* Correspondence: ionut.ovidiu.toma@tuiasi.ro; Tel.: +40-232-701455

Citation: Corbu, O.; Toma, I.-O. Progress in Sustainability and Durability of Concrete and Mortar Composites. *Coatings* **2022**, *12*, 1024. https://doi.org/10.3390/coatings12071024

Received: 14 July 2022
Accepted: 17 July 2022
Published: 19 July 2022

Publisher's Note: MDPI stays neutral with regard to jurisdictional claims in published maps and institutional affiliations.

Copyright: © 2022 by the authors. Licensee MDPI, Basel, Switzerland. This article is an open access article distributed under the terms and conditions of the Creative Commons Attribution (CC BY) license (https://creativecommons.org/licenses/by/4.0/).

The origins of concrete as a construction material date back more than 2000 years ago, but the origins of the term itself are still under debate due to its many different interpretations throughout history. Concrete, in its generally accepted form by today's definition, is a hard rock-like man-made construction material that consists of aggregates, sand, water, and cement. The term *"concrete"* comes from the Latin word *"concretus"*, meaning *compact* or *grown together*, and it was used for characterizing the materials obtained from the combination of two or more different materials. From this perspective, concrete can be considered to be among the first composite materials to be used by mankind.

As opposed to plastic, the facility with which concrete can be used to fill formworks or molds with any shape is one of its key advantages. At the same time, most of the constituent materials, with the exception of cement, can be found at very low cost, locally or at a very short distance from the construction site. Concrete compressive strength is high, which makes it suitable for members primarily subjected to compression such as columns and arches.

As with any other construction material, concrete has its disadvantages, too. The rheological, mechanical, and durability properties of concrete is highly dependent on the water/cement ratio. While higher values of the water/cement ratio results in a more workable concrete, the risk of segregation is also high. On the other hand, a too low value of the water/cement ratio, without the use of proper water reducing admixtures, may result in an improper binding of all aggregates within the cement paste [1,2]. Either way, the end result is a material that cannot fulfill its role for structural applications.

While concrete is very good in compression, it is also inherently weak in tension. The solution came from using materials that have a good behavior in tension, e.g., steel, to take over tensile stresses in concrete elements. The distribution of such materials could be in the form of individual reinforcing bar or as distributed reinforcement. Both reinforcing methods have their benefits, and sometimes they are used in combination. While "traditional" reinforcement is made almost entirely of steel, distributed fibers may have a variety of sources from natural [3] to man-made [4]. The advantage of using fibers in mortar or concrete resides in their ability to bridge small cracks and arrest their development. The scientific literature also highlights their contribution to increasing the toughness of mortar and concrete to repeated impact loads [5].

The European directive 2003/87/CE establishing a scheme for greenhouse emission allowance trading within the Community was the first international program of its kind in the world. The main objective was the reduction by 20% of the greenhouse emissions, by 2020, below the levels recorded in 1990. Despite this bold step towards reducing the environmental impact of modern society the UN Sustainable Development Goals (UN-SDG) project accounts that, globally, the greenhouse gas (GHG) emissions are 50% higher than in 1990. One major reason for this increase in the GHG emissions is the continuously increasing demand for new buildings due to a rapid expansion of urban areas. This put a tremendous pressure on the production of building materials, among which concrete is the leading material. According to the latest CEMBUREAU report, the production of cement in the European Union has risen from 175.1 million tons in 2017 to 182.1 million tons in 2019. Taking into account that cement represents 10% ~ 15% of the concrete volume one could only imagine the environmental burden the construction industry creates.

The depletion of natural resources is another factor of concern worldwide. Natural aggregates, constituting between 60% and 75% of concrete volume and which are recognized as the second source of CO_2 in concrete production, are quickly becoming a scarce "commodity".

However, one of the most important characteristics of cement based materials, e.g. mortar and concrete, is their ability to incorporate large amounts of wastes as partial or total replacement for cement and/or natural aggregates. Extensive research works were conducted to assess the suitability of various industrial by-products as supplementary cementitious materials (SCM) in mortar and concrete and their influence on the rheological, elastic, and mechanical properties of cement based materials [6,7]. Fly-ash (FA), a by-product of the coal-fired power plants, consisting mostly of glassy spherical particles has been long recognized as one of the SCMs with significant beneficial influence on the durability and long term mechanical properties of mortar and concrete when used in lower concentrations (less than 20% by weight of cement). On the other hand, one of the drawbacks of using fly ash resides in slow early strength development of cement based materials [8]. However, this effect could be mitigated by means of chemical activation or addition of nanomaterials. Another frequently used SCM is the ground granulated blast furnace slag (GGBS). The use of GGBS as partial replacement of cement in mortar or concrete leads to increased values of the mechanical properties, increased abrasion resistance, and lower permeability [9]. This has direct influence on the long-term durability of concrete which necessitates less maintenance works. Silica fume (SF), a by-product from the production process of silicon or alloys containing silicon, is another frequently used SCM with net benefits in terms of durability of cement based products [10]. Recent studies acknowledge the potential use of gypsum based by-products as SCMs in mortar and concrete [11,12]. Although less investigated compared to other SCMs, mortar and concrete made with gypsum-based SCM exhibited higher values of mechanical properties.

Thorough investigations were also conducted in the direction of natural aggregates substitution by recycled aggregates from various sources [1]. The benefits of using waste glass in mortar or concrete has been acknowledged not only from an environmental point of view but also from its contribution to the overall strength and durability properties [13] of the cementitious composites acting as a pozzolanic material [14]. At the same time, there is a growing concern related to the continuous increase in plastic wastes all around the world. It is estimated that plastic wastes exceed 25 million tons per year, worldwide. Studies revealed that the use of plastic waste in construction materials resulted in a decrease in mechanical properties, but the density and the thermal conductivity of the new material decreased [15,16]. This would make the use of plastic wastes more attractive for the construction industry provided more research effort is invested in this direction. Construction industry is not only one of the largest consumer of natural resources; it also generates a lot of wastes from the rehabilitation and demolition processes. The feasibility of using recycled aggregates from demolition wastes, in the context of the circular economy paradigm, resulted in many research works being published in the scientific literature. Their effect on the mechanical properties of mortar or concrete is hard to fully understand and quantify, given the large variety in the characteristics of the recycled aggregates depending mostly on their origin and initial composition [2,17,18]. The general consensus is that the mechanical and durability properties of mortar and concrete incorporating construction demolition wastes decrease, and new methodologies are investigated to improve the quality of recycled aggregates from these sources. Another pressing issue of the modern society is related to the increasing number of discarded worn tires. In view of ever stricter regulations in terms of landfills, tire components found their way in mortar and concrete mixes either in terms of distributed textile and steel fibers as reinforcement or as aggregates [19,20].

Water, the third component of a concrete mix, represents 15%–20% of the total mix. It also represents an invaluable resource for supporting all forms of life on earth. Research works were dedicated to using water from sources other than the fresh water. Studies on the use of wastewater from ready-mix concrete plants revealed that the solid content from the mixing water coupled with total amount of dissolved solids represent the main influencing factors on the mechanical properties of the resulting concrete [21]. The use of sea water, as an alternative to fresh water, in concrete mixes may have both negative and positive effects on alkali-activated slag cement and blast-furnace cement concrete. The former was identified in the form of lower initial compressive strength, increased open capillary porosity, and increased corrosion of steel reinforcement while the latter consists in higher values of the mechanical properties due to the combined effect of alkali metal compounds and salts [22].

A more recent alternative to cement based mortar and concrete is the geopolymer concrete. It is based on aluminosilicate precursors from either natural (metakaolin) or industrial by-products (fly ash, ground granulated blast furnace slag) sources. From the point of view of GHG emissions, geopolymer concrete has net benefits over Portland cement concrete, as it produces 80% less CO_2 during its production phase and requires 60% less energy to be produced [23–25].

We trust that the issues raised during this short editorial will help acknowledge the need for further research works in making mortar and concrete the construction materials of choice for future generations, especially in a way that significantly reduces these materials' environmental impact.

Author Contributions: Conceptualization, O.C. and I.-O.T.; methodology, O.C. and I.-O.T.; writing—original draft preparation, I.-O.T.; writing—review and editing, O.C. All authors have read and agreed to the published version of the manuscript.

Conflicts of Interest: The authors declare no conflict of interest.

References

1. Martínez-García, R.; Jagadesh, P.; Fraile-Fernández, F.; Morán-del Pozo, J.; Juan-Valdés, A. Influence of Design Parameters on Fresh Properties of Self-Compacting Concrete with Recycled Aggregate—A Review. *Materials* **2020**, *13*, 5749. [CrossRef] [PubMed]
2. Sosa, M.E.; Villagrán Zaccardi, Y.A.; Zega, C.J. A critical review of the resulting effective water-to-cement ratio of fine recycled aggregate concrete. *Constr. Build. Mater.* **2021**, *313*, 125536. [CrossRef]
3. Ahmad, J.; Majdi, A.; Al-Fakih, A.; Deifalla, A.F.; Althoey, F.; El Ouni, M.H.; El-Shorbagy, M.A. Mechanical and Durability Performance of Coconut Fiber Reinforced Concrete: A State-of-the-Art Review. *Materials* **2022**, *15*, 3601. [CrossRef] [PubMed]
4. Blazy, J.; Blazy, R.; Drobiec, Ł. Glass Fiber Reinforced Concrete as a Durable and Enhanced Material for Structural and Architectural Elements in Smart City—A Review. *Materials* **2022**, *15*, 2754. [CrossRef] [PubMed]
5. Abid, S.R.; Murali, G.; Ahmad, J.; Al-Ghasham, T.S.; Vatin, N.I. Repeated Drop-Weight Impact Testing of Fibrous Concrete: State-Of-The-Art Literature Review, Analysis of Results Variation and Test Improvement Suggestions. *Materials* **2022**, *15*, 3948. [CrossRef]
6. Naqi, A.; Jang, J. Recent Progress in Green Cement Technology Utilizing Low-Carbon Emission Fuels and Raw Materials: A Review. *Sustainability* **2019**, *11*, 537. [CrossRef]
7. Corbu, O.; Ioani, A.M.; Al Bakri Abdullah, M.M.; Meiță, V.; Szilagyi, H.; Sandu, A.V. The Pozzoolanic Activity Level of Powder Waste Glass in Comparisons with other Powders. *Key Eng. Mater.* **2015**, *660*, 237–243. [CrossRef]
8. Li, G.; Zhou, C.; Ahmad, W.; Usanova, K.I.; Karelina, M.; Mohamed, A.M.; Khallaf, R. Fly Ash Application as Supplementary Cementitious Material: A Review. *Materials* **2022**, *15*, 2664. [CrossRef]
9. Nicula, L.M.; Corbu, O.; Ardelean, I.; Sandu, A.V.; Iliescu, M.; Simedru, D. Freeze–Thaw Effect on Road Concrete Containing Blast Furnace Slag: NMR Relaxometry Investigations. *Materials* **2021**, *14*, 3288. [CrossRef]
10. Du, Y.; Gao, P.; Yang, J.; Shi, F. Research on the Chloride Ion Penetration Resistance of Magnesium Phosphate Cement (MPC) Material as Coating for Reinforced Concrete Structures. *Coatings* **2020**, *10*, 1145. [CrossRef]
11. Toma, I.-O.; Covatariu, D.; Toma, A.-M.; Taranu, G.; Budescu, M. Strength and elastic properties of mortars with various percentages of environmentally sustainable mineral binder. *Constr. Build. Mater.* **2013**, *43*, 348–361. [CrossRef]
12. Shi, C.; Qu, B.; Provis, J.L. Recent progress in low-carbon binders. *Cem. Concr. Res.* **2019**, *122*, 227–250. [CrossRef]

13. Corbu, O.-C.; Bompa, D.V.; Szilagyi, H. Eco-efficient cementitious composites with large amounts of waste glass and plastic. *Proc. Inst. Civ. Eng.-Eng. Sustain.* **2022**, *175*, 64–74. [CrossRef]
14. Ahmad, J.; Zhou, Z.; Usanova, K.I.; Vatin, N.I.; El-Shorbagy, M.A. A Step towards Concrete with Partial Substitution of Waste Glass (WG) in Concrete: A Review. *Materials* **2022**, *15*, 2525. [CrossRef]
15. Al-Sinan, M.A.; Bubshait, A.A. Using Plastic Sand as a Construction Material toward a Circular Economy: A Review. *Sustainability* **2022**, *14*, 6446. [CrossRef]
16. Babafemi, A.J.; Sirba, N.; Paul, S.C.; Miah, M.J. Mechanical and Durability Assessment of Recycled Waste Plastic (Resin8 & PET) Eco-Aggregate Concrete. *Sustainability* **2022**, *14*, 5725. [CrossRef]
17. Vitale, F.; Nicolella, M. Mortars with Recycled Aggregates from Building-Related Processes: A 'Four-Step' Methodological Proposal for a Review. *Sustainability* **2021**, *13*, 2756. [CrossRef]
18. Da Silva, S.R.; Andrade, J.J.d.O. A Review on the Effect of Mechanical Properties and Durability of Concrete with Construction and Demolition Waste (CDW) and Fly Ash in the Production of New Cement Concrete. *Sustainability* **2022**, *14*, 6740. [CrossRef]
19. Abdelmonim, A.; Bompa, D.V. Mechanical and Fresh Properties of Multi-Binder Geopolymer Mortars Incorporating Recycled Rubber Particles. *Infrastructures* **2021**, *6*, 146. [CrossRef]
20. Oprişan, G.; Entuc, I.-S.; Mihai, P.; Toma, I.-O.; Taranu, N.; Budescu, M.; Munteanu, V. Behaviour of rubberized concrete short columns confined by aramid fibre reinforced polymer jackets subjected to compression. *Adv. Civ. Eng.* **2019**, *2019*, 1360620. [CrossRef]
21. Yao, X.; Xi, J.; Guan, J.; Liu, L.; Shangguan, L.; Xu, Z. A Review of Research on Mechanical Properties and Durability of Concrete Mixed with Wastewater from Ready-Mixed Concrete Plant. *Materials* **2022**, *15*, 1386. [CrossRef]
22. Krivenko, P.; Rudenko, I.; Konstantynovskyi, O.; Vaičiukynienė, D. Mitigation of Corrosion Initiated by Cl− and SO_4^{2-}-ions in Blast Furnace Cement Concrete Mixed with Sea Water. *Materials* **2022**, *15*, 3003. [CrossRef]
23. Elzeadani, M.; Bompa, D.V.; Elghazouli, A.Y. Preparation and properties of rubberised geopolymer concrete: A review. *Constr. Build. Mater.* **2021**, *313*, 125504. [CrossRef]
24. Abdullah, M.M.A.B.; Faris, M.A.; Tahir, M.F.M.; Kadir, A.A.; Sandu, A.V.; Mat Isa, N.A.A.; Corbu, O. Performance and Characterization of Geopolymer Concrete Reinforced with Short Steel Fiber. *IOP Conf. Ser. Mater. Sci. Eng.* **2017**, *209*, 012038. [CrossRef]
25. Burduhos Nergis, D.D.; Vizureanu, P.; Ardelean, I.; Sandu, A.V.; Corbu, O.C.; Matei, E. Revealing the Influence of Microparticles on Geopolymers' Synthesis and Porosity. *Materials* **2020**, *13*, 3211. [CrossRef] [PubMed]

Article

Research on a Multi-Objective Optimization Design for the Durability of High-Performance Fiber-Reinforced Concrete Based on a Hybrid Algorithm

Xingyu Wang [1], Fengkun Cui [1,*], Long Cui [2] and Di Jiang [3]

[1] School of Civil Engineering, Shandong Jiaotong University, 5 Jiaoxiao Road, Jinan 250357, China; 21107030@stu.sdjtu.edu.cn

[2] Shandong Provincial Academy of Building Research Co., Ltd., 29 Wuyingshan Road, Jinan 250031, China; 15550025256@163.com

[3] Shandong Huiyou Municipal Landscape Group Co., Ltd., 29 East Automobile Factory Road, Jinan 250031, China

* Correspondence: 204118@sdjtu.edu.cn

Abstract: To achieve durable high-performance fiber-reinforced concrete that meets economic requirements, this paper introduces a hybrid intelligent framework based on the Latin hypercube experimental design, response surface methodology (RSM), and the NSGA-III algorithm for optimizing the mix design of high-performance fiber-reinforced concrete. The developed framework allows for the prediction of concrete performance and obtains a series of Pareto optimal solutions through multi-objective optimization, ultimately identifying the best mix proportion. The decision variables in this optimization are the proportions of various materials in the concrete mix, with concrete's frost resistance, chloride ion permeability resistance, and cost as the objectives. The feasibility of this framework was subsequently validated. The results indicate the following: (1) The RSM model exhibits a high level of predictive accuracy, with coefficient of determination (R-squared) values of 0.9657 for concrete frost resistance and 0.9803 for chloride ion permeability resistance. The RSM model can be employed to construct the fitness function for the optimization algorithm, enhancing the efficiency of multi-objective optimization. (2) The NSGA-III algorithm effectively balances durability and cost considerations to determine the optimal mix proportion for the concrete. After multi-objective optimization, the chloride ion permeability resistance and frost resistance of the high-performance fiber-reinforced concrete improved by 38.1% and 6.45%, respectively, compared to the experimental averages, while the cost decreased by 2.53%. The multi-objective optimization method proposed in this paper can be applied to mix design for practical engineering projects, improving the efficiency of concrete mix design.

Keywords: high-performance fiber-reinforced concrete; durability; multi-objective optimization; Latin hypercube experimental design; response surface methodology; NSGA-III

1. Introduction

High-performance fiber-reinforced concrete (HPFRC) exhibits superior ductility and toughness compared to ordinary concrete [1–4]. As a result, HPFRC has found extensive applications in practical engineering in recent years [5–7]. Concrete structures in the northern coastal regions of China are not only subject to the detrimental effects of chloride salt intrusion but also the unique freeze−thaw cycles in the northern sea areas. The nonlinear coupled effects arising from the interaction of these two factors accelerate material degradation and performance deterioration of concrete components, highlighting the prominent issue of durability [8–11]. Besides adopting measures from a structural design perspective, such as implementing protective coatings and increasing the thickness of the concrete protective layer, it is imperative to conduct research on enhancing the concrete's inherent durability.

The durability of concrete mainly considers frost resistance and chloride ion permeability, as these two factors directly affect the long-term performance and safety of concrete structures. The frost resistance of concrete refers to its performance under low temperatures and freeze−thaw cycling conditions. The volume expansion of water in concrete during freezing may lead to the generation and expansion of microcracks, thereby reducing the structural integrity and load-bearing capacity of the concrete. In cold winter regions, frost resistance is a key factor in ensuring the integrity and safety of concrete structures; the impermeability of chloride ions involves the permeability resistance of concrete to chloride ions. The penetration of chloride ions (mainly from salt water or seawater) is one of the main reasons for the corrosion of steel bars in concrete. Corrosion of steel bars can seriously affect the integrity and durability of concrete structures. Therefore, improving the impermeability of concrete to chloride ions can effectively prevent steel corrosion and prolong the service life of concrete structures. By improving these two properties, the durability and service life of concrete structures can be significantly improved [12–16]. Numerous scholars have studied the parameters that affect the durability of concrete, and the research results indicate that the frost resistance and chloride ion permeability of concrete are mainly influenced by the mix ratio of raw materials such as cement, water, aggregates, and additives [17–21]. These factors determine the durability and service life of concrete structures. Currently, most people refer to the "General Concrete Mix Design Code" [22] for mix proportion design and employ orthogonal experiments to seek the optimal mix. However, when using orthogonal experiments to find the best mix, there are drawbacks, such as a substantial workload, low predictive accuracy, and suboptimal results [23–25]. Additionally, it cannot establish a clear functional relationship between factors and response values in a specified region [26,27].

To unravel the complex relationship between concrete mix proportions and resistance to freezing and chloride ion penetration, statistical models are often introduced in relevant experiments and analyses. The response surface method (RSM) is commonly used to predict the durability of concrete [28,29]. RSM is a product of the fusion of mathematics and statistics, capable of establishing mathematical models between multiple factors and one or more response values with minimal experimental data [30,31]. It evaluates the impact of interaction among factors on response values, determines the optimal response values, and offers advantages over orthogonal experiments, such as requiring fewer trials, lower costs, and higher predictive accuracy [32]. Naraindas Bheel et al. [33] employed RSM's central composite design (CCD) to establish a relationship between 13 different raw material contents and eight target values in engineered cementitious composites (ECC). They validated the predicted values through experiments, and the results showed a strong correlation between the predicted values and the experimental data. Wang et al. [34] used RSM's central composite design to perform experimental design on basalt fiber foam concrete and achieved multi-objective optimization by incorporating utility functions. Zhang et al. [35] utilized RSM with a Box−Behnken design (BBD) to obtain the optimal aggregate grading and admixture dosage for permeable concrete made with recycled aggregates. The aforementioned studies demonstrate that the application of RSM in optimizing construction material mixtures offers significant advantages. However, research on the application of RSM for optimizing the mix proportions of high-performance fiber-reinforced concrete is relatively scarce.

In addition, when designing the mix proportion of concrete, the economic cost requirements of the engineering application must be taken into consideration [36]. However, there exists a conflict between the durability of concrete and the economic cost [37]. In recent years, the nondominated sorting genetic algorithm (NSGA) has been applied to concrete mix proportion design, providing a new solution for multi-objective optimization problems (MaOPs) [38,39]. The basic NSGA proposed by Srinivas and Deb [40] has been widely used to solve MaOPs, but it comes with high computational complexity. Therefore, Deb et al. [41]

proposed the NSGA-II algorithm, which incorporates elite preservation, fast nondominated sorting, and crowding distance selection operators. NSGA-II has advantages such as fast operation speed and good convergence. However, the crowding distance selection in three-dimensional and higher dimensional objective spaces may not be effective, leading to a reduction in the diversity of solutions. Reducing the complexity of the dataset may potentially improve the accuracy of deep learning models. Simplifying the process of the dataset can help deep learning models learn key features related to problems more effectively, thereby improving their performance and generalization ability. In practice, finding the appropriate level of dataset complexity often requires adjustments based on domain knowledge and experimental results [42,43].

Hence, Deb and Jain [44] introduced NSGA-III. In comparison, NSGA-III directly searches for the Pareto optimal solutions in the space, eliminating issues such as transformation parameters and information loss, making the search process simple and intuitive. Furthermore, the inherent characteristics of genetic algorithms make NSGA-III widely adaptable; the combination of continuous and discrete variable inputs does not significantly affect the algorithm's performance. NSGA-III guides the selection of non-dominated solutions using uniformly distributed reference points in space, effectively ensuring the widespread distribution and diversity of nondominated solutions in high-dimensional objective spaces. In fact, NSGA-III is currently recognized as the best algorithm for MaOPs [45–47]. At present, NSGA-III has been applied and demonstrated effective in multi-objective optimization in various fields such as automation technology, water supply, and aerospace [48–50]. However, there is a notable scarcity of reported applications of NSGA-III in the domain of concrete mix proportion design.

This study commences with the utilization of a Latin hypercube experimental design methodology for mix proportion development. Subsequently, upon obtaining specimen samples, concrete specimens are fabricated, and frost resistance, as well as chloride ion permeability tests, are conducted. This facilitates the acquisition of the relative dynamic modulus of elasticity and chloride ion migration coefficient for concrete specimens corresponding to various mix proportions. A response surface model is then established. Subsequently, the constructed response surface model is integrated with the NSGA-III algorithm, thereby achieving multi-objective optimization for high-performance fiber-reinforced concrete.

2. Preliminary Information

2.1. Latin Hypercube Design

Before designing and optimizing the mix proportion of high fiber reinforced concrete, it is necessary first to use certain experimental design methods to sample the design space and generate a certain number of sample points. The commonly used experimental design methods include orthogonal design, uniform design, Latin hypercube sampling, etc. The Latin hypercube design (LHD) is a method used for experimental design and sampling design space, and its core idea is to ensure that each level value is evenly and randomly paired with other levels in each dimension. This design approach helps to achieve wide coverage in the design space while reducing the number of samples, which improves sampling efficiency compared to completely random sampling methods.

The key elements to ensure that the results of Latin hypercube sampling are unbiased and effective are as follows:

(1) Uniformity: The core goal of LHD is to ensure a uniform distribution of sample points in each dimension, ensuring comprehensive coverage of the design space.
(2) Randomness: By randomly selecting sample points on each dimension, LHD ensures that sufficient randomness is introduced during the sampling process so that the results are not affected by specific points.
(3) Reduce sample size: Compared to comprehensive sampling, LHD reduces the required sample size by effectively selecting sample points, improving sampling efficiency.

When applying LHD to design space sampling: In a multidimensional design space, LHD divides each dimension into equal intervals and selects a sample point within each interval to ensure a uniform selection of sample points throughout the entire design space. This helps to capture representative features of the design space rather than just sampling in certain local areas. The basic theory is as follows:

Assuming the probability distribution function of each element of the K-dimensional random variable x is F_i (I = 1, 2, ..., K). The elements of vector x are independent of each other, and each element is sampled N times, which is the value of the jth (j = 1, 2, ..., N) sampling of the k (k = 1, 2, ..., K) th element. Define $N \times K$-dimensional matrix P. Each column of P is composed of a random arrangement of elements in the sequence {1, 2, ..., N}. If the random variable ξ_{jk} follows a uniform distribution on the interval [0,1], the result obtained after sampling is:

$$xjk = F_k^{-1}[(pjk - 1 + \xi jk)/N] \qquad (1)$$

In the equation, p_{jk} is $N \times$ The j row and k column elements of the K-dimensional matrix P.

Assuming the existence of function $h(x)$, the unbiased estimate of the mean $E(h(x))$ of function $h(x)$ is defined as:

$$\hat{h} = \sum_{j=1}^{N} h(x_j)/N \qquad (2)$$

The variance of the unbiased estimate $\hat{h}$ for simple random sampling is:

$$D(\hat{h}) = D(h(x))/N \qquad (3)$$

The variance of the unbiased estimation of Latin hypercube is:

$$D(\hat{h}) = D(h(x))/N + (N-1)cov(h(x_{1n}), h(x_{2n}))/N \qquad (4)$$

It can be proven that the probability of $(N-1)cov(h(x_{1n}), h(x_{2n}))/N$ approaches a negative value. Therefore, Latin hypercube sampling is easier to converge than random sampling.

The key factor in ensuring unbiased and efficient results when dividing the experimental domain of LHD lies in its design method, which ensures the representativeness of the samples by uniformly and randomly selecting sample points. This helps to explore the design space more effectively in tasks such as experimental design and parameter optimization, reducing the number of required experiments and improving the efficiency and cost-effectiveness of experiments. Numerous scholars have further verified the above viewpoint through theoretical research [51,52]. Therefore, we select LHD to determine the sample points required for concrete mix design.

2.2. Response Surface Model

Response surface methodology is a method of optimizing experimental conditions suitable for fitting the complex nonlinear response relationship between optimization objectives and experimental factors. The multivariate second-order response surface model is generally represented by the following equation.

$$y(x) = \beta_0 + \sum_{i=1}^{m} \beta_i x_i + \sum_{i=1}^{m} \beta_{ii} x_i^2 + \sum_{i<j}^{m} \beta_{ij} x_i x_j \qquad (5)$$

In the equation, $y(x)$ represents the response objective function; x_i, x_j represents the i-th and jth experimental factors; β_0 represents a constant term, β_i, β_{ii}, β_{ij} represents various coefficients; m represents the number of parameters to be optimized.

2.3. NSGA-III Algorithm

Nondominated sorting genetic algorithm III (NSGA-III) is a widely used multi-objective optimization (MO) algorithm designed to solve two types of problems: maintaining good solution diversity and optimizing solution convergence. This is an improved version that compensates for the shortcomings of its predecessor, NSGA-II, in losing solution diversity and accuracy when dealing with high-dimensional problems.

The core operations of NSGA-III include nondominated sorting, calculation of crowding distance, evolutionary operations (selection, crossover, and mutation), and environmental selection. Its unique features and mechanisms are mainly reflected in the following points:

Reference point mechanism: NSGA-III introduces the concept of reference points to improve the diversity of solutions. During the initialization phase, the algorithm generates a set of reference points. These reference points are used in each generation to select solutions and create the next generation. The solutions are selected to minimize their distance from the reference point. This ensures the distribution and coverage of the understanding.

Multiple nondominated levels: NSGA-III implements multiple nondominated sorting of solutions. The solution is divided into several nondominated layers, each layer being superior to its lower layer. In each generation, the algorithm prioritizes solutions from higher levels.

Crowding distance: In order to maintain population diversity, NSGA-III uses a crowding distance mechanism. Among solutions with the same level, solutions with lower crowding (i.e., solutions with more "space" around them) will be preferred. This helps to prevent the algorithm from overly focusing on a small portion of the search space, thereby achieving diversity of understanding.

Additional parents: When selecting solutions to create the next generation, NSGA-III not only considers the current parents (so-called P population) but also considers new possible solutions generated through offspring (so-called Q population). This is also known as a "joint population", and this design can increase the diversity of solutions and accelerate the speed of evolution.

Special environment selection strategy: When a new P population needs to be selected, NSGA-III will first select nondominated solutions and add excess solutions to the population according to the reference point allocation strategy, which ensures the convergence of the solution in multi-objective optimization problems.

Overall, NSGA-III effectively addresses multi-objective optimization problems through these mechanisms, overcomes weaknesses in the diversity of solutions, and provides uniformly distributed solutions at the Pareto frontier, thereby enhancing the convergence of the algorithm. This characteristic makes NSGA-III perform well in handling practical engineering problems such as high-performance fiber-reinforced concrete. Meanwhile, to balance the relationships between objective functions, an adaptive normalization technique is introduced. The ideal point for the population, $S_t = F_1 \cup F_2 \cup \ldots \cup F_l$, is defined as the minimum point attained by the population S_t on each respective objective. When normalizing multiple objectives, it is necessary to construct hyperplanes by seeking limit points to determine intercepts. Subsequently, the obtained intercepts are utilized to normalize the objectives individually. Considering that the mixed NSGA-III produces a Pareto solution set that closely approximates the actual optimal solution set of the problem, the obtained Pareto solution set after multi-objective optimization can be considered the final optimal solution. Therefore, the corresponding maximum value of the i-th objective in the corresponding population can be used to replace the intercept of the corresponding objective.

$$f_i^n(x) = \frac{f_i(x) - z_i^{min}}{z_i^{max} - z_i^{min}}, \text{for } i = 1, 2, \cdots, M \tag{6}$$

In the formula, M represents the number of targets; x represents the decision variable; $f_i(x)$ represents the target value of x on the corresponding i-th target; z_i^{min} and z_i^{max} represent the minimum and maximum values of the population on the i-th target, respectively; $f_i^n(x)$ represents the normalized target value of the i-th target.

3. Method

This paper presents a smart hybrid system designed for simultaneously optimizing both the durability and cost-effectiveness of high-performance fiber-reinforced concrete, achieving multi-objective enhancement. Figure 1 shows the flowchart of the model. The overall framework of this article is divided as follows.

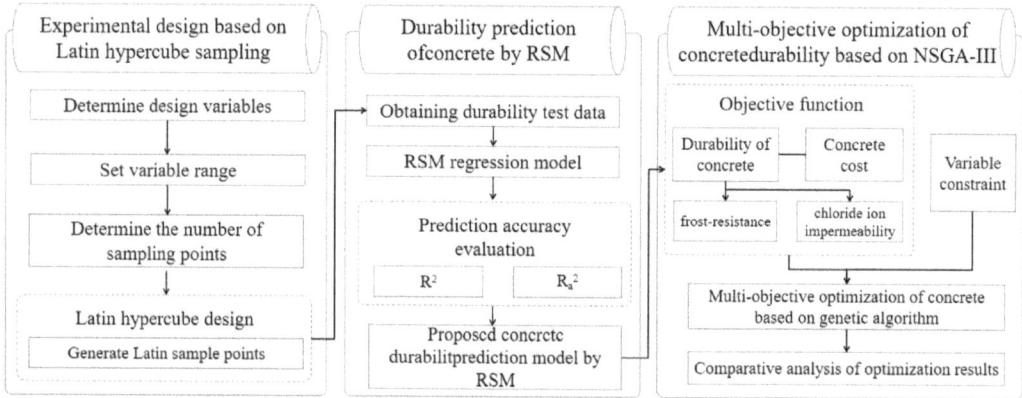

Figure 1. Flow chart of the proposed model.

3.1. Latin Hypercube Experimental Design

(1) Determine design variables

Determine the key design variables that affect the mix design of high-performance fiber-reinforced concrete, such as water content, cement content, fly ash content, fine aggregate content, coarse aggregate content, water-reducing agent content, and fiber content. The design variables determined in this article are all independent variables, further ensuring that the Latin hypercube sampling results are unbiased.

(2) Set variable range

To ensure the rationality of the mix proportion of high-performance fiber-reinforced concrete, a suitable range of raw material content is set through consulting relevant literature and preliminary mix proportion tests [53–55].

(3) Determine the number of sampling points

Determine the number of Latin hypercube sampling points to generate. This depends on the complexity of the problem and the sampling requirements for the design space.

(4) Generate Latin hypercube sampling

Generate uniformly distributed sampling points within the design variable range using the Latin hypercube sampling method. This article uses the pyDOE library in Python for Latin hypercube sampling.

(5) Durability test and data preprocessing

Using the generated Latin hypercube sampling points as input parameters for concrete mix proportions, prepare concrete and conduct corresponding frost resistance and chloride ion permeability tests to obtain the relative dynamic elastic modulus and chloride ion migration coefficient corresponding to different sample points. The dimensions and attribute ranges of various input variables representing the proportion of raw materials in concrete are not the same and cannot be directly compared. Therefore, formula (7) is used to unify input variables and output energy consumption into intervals $[-1,1]$ to achieve data normalization and unify the dimensions of variables so that each feature plays a role in the prediction process.

$$y = (y_{max} - y_{min}) \times \frac{x - x_{min}}{x_{max} - x_{min}} + y_{min} \qquad (7)$$

3.2. Establishing an RSM Model

Based on the durability test results, a response surface model is constructed to obtain the nonlinear relationship between the durability of high-performance fiber-reinforced concrete and the amount of raw materials added. The reliability of the RSM model is evaluated using correlation coefficient R^2 and adjustment coefficient R_a^2. Generally, $R^2 \in [0,1]$, and the closer R^2 is to 1, the higher the fitting accuracy of the response surface model, usually requiring $R^2 > 0.9$. The calculation formula is shown in the following equation.

$$R^2 = 1 - \frac{S_r}{S_m + S_r} \qquad (8)$$

$$R_a^2 = \frac{S_r/D_r}{(S_m + S_r)/(D_m + D_r)} \qquad (9)$$

In the formula, S_r is the sum of squares of the residuals; S_m is the sum of regression squares; D_r is the residual degree of freedom; D_m is the degree of freedom of regression.

3.3. Multi-Objective Optimization Based on NSGA-III

3.3.1. Concrete Durability Objective Function

Based on the response surface model, construct a chloride ion impermeability model for high-performance fiber-reinforced concrete, represented by f_1 and the frost resistance model, represented by f_2.

$$f_1 = \max[RSM(x_1, x_2, \ldots, x_n)] \qquad (10)$$

$$f_2 = \min[RSM(x_1', x_2', \ldots, x_n')] \qquad (11)$$

Among them, $x_1, x_2, \ldots, x_n$ are the input variables of the response surface model used for prediction.

3.3.2. Economic Cost Function

In practical engineering, concrete structure needs to control the economic cost of concrete while meeting the durability requirements. The objective function f_3 of optimizing the economic cost of concrete is expressed as:

In practical engineering applications, it is necessary to balance the cost and durability of high-performance fiber-reinforced concrete. The function f_3 with cost as the optimization objective is represented as follows:

$$f_3 = \min \sum_{i=1}^{n} v_i x_i \qquad (12)$$

Among them, x_i represents the i-th raw material that constitutes high-performance fiber-reinforced concrete, and v_i represents the cost of the i-th raw material.

3.3.3. Constraint Condition Setting

To guarantee an effective and practical composition of high-performance fiber-reinforced concrete, establishing an appropriate range for the content of raw materials and setting suitable constraints is essential. The general form of constraints is:

$$b_{\min} \leq x_i \leq b_{\max} \tag{13}$$

In the formula, x_i represents the raw material of the i-th high-performance fiber-reinforced concrete, while $b_{\min}$ and $b_{\max}$ represent the minimum amount of the i-th raw material, respectively.

3.3.4. Multi-Objective Optimization Based on NSGA-III

Using the MATLAB platform, implement the NSGA-III algorithm with the aim of enhancing the durability of concrete while concurrently minimizing its cost. The result of this algorithm will be the set of Pareto optimal solutions for concrete mix proportions. The fundamental steps for acquiring the Pareto optimal solution set through the NSGA-III algorithm include:

(1) Initialize population: Randomly generate an initial population, where each individual contains the variables of the problem and the values of the objective function.
(2) Set algorithm parameters: Determine the parameters of the algorithm, such as population size, crossover probability, mutation probability, maximum number of iterations, etc.
(3) Execute the NSGA-III algorithm: Use the core steps of the NSGA-III algorithm, including nondominated sorting, crowding allocation, genetic operations (crossover and mutation), etc. These steps will gradually optimize the individuals in the population, generating a set of approximate Pareto frontier solutions.
(4) Termination condition: Define the stopping condition, such as reaching the maximum number of iterations, Pareto frontier convergence, etc.
(5) Obtaining results: After the algorithm runs, the final Pareto frontier solution set is obtained, which represents the nondominated solution set of the problem.
(6) Analysis results: For each Pareto frontier solution, analyze its performance on various objective functions and select the solution that best meets the requirements of the problem.

4. Case Analysis

4.1. Engineering Background

The Xinan River Grand Bridge is situated on Binhai East Road, Laishan District, Yantai City, Shandong Province, China. It was completed in 2003 and serves as a major transportation artery connecting the Laishan and Muping districts. Given its substantial traffic volume, the project environment is depicted in Figure 2. This sea-crossing bridge is located in the frozen waters of northern China. Concrete components within the fluctuating water levels are subjected not only to chloride erosion but also to the unique freeze−thaw cycles prevalent in northern maritime regions. The coupling effect arising from the interaction of these two factors accelerates the material degradation and performance deterioration of the concrete elements. Consequently, the durability issues of the bridge are notably prominent.

To address these challenges, this study focuses on the development of high-performance fiber-reinforced concrete to enhance the structural durability of the Xinan River Grand Bridge.

According to the "Durability Design Standard for Concrete Structures" [56], the specimens are placed in a rapid freeze−thaw machine for 300 freeze−thaw cycles, and the relative dynamic elastic modulus of the specimens is measured to represent the frost resistance of the concrete. The chloride ion migration coefficient of the concrete after 28 days of curing is measured using the RCM method to represent the chloride ion permeability of the concrete. This study focuses on the C50 concrete used in the aforementioned projects. Figure 3a,b shows photos of the relative dynamic elastic modulus test and chloride ion migration coefficient of the tested concrete specimens. The raw materials used in this experiment include cement produced by Shandong Shanshui Group (Jinan, China), first-class fly ash produced by Hengyuan New Materials Co., Ltd., (Dongying, China), polycarboxylic acid high-efficiency water-reducing agent produced by Kaili Chemical, and polyacrylonitrile fiber produced by Huixiang Fiber, among others.

Figure 2. Photograph of the project.

Figure 3. Concrete durability tests. (**a**) Frost resistance test; (**b**) RCM test.

4.2. Proportion Design of High-Performance Fiber-Reinforced Concrete Based on LHD

This study mainly considers the influence of seven factors on the two durability indicators of high-performance fiber-reinforced concrete. These seven factors are water content (x_1), cement content (x_2), fly ash content (x_3), fine aggregate content (x_4), coarse aggregate content (x_5), water reducer content (x_6), and fiber content (x_7). This article uses the pyDOE library in Python to conduct Latin hypercube sampling and obtain 36 sets of Latin experimental samples. The design variables determined in this article are all independent variables, further ensuring that the Latin hypercube sampling results are unbiased.

The response surface model contains $1 + 2m + m(m-1)/2$ coefficients to be solved. When m = 7, the test should include at least 36 sets of test sample points, and the test arrangement in Table 1 meets the requirements.

Table 1. Initial Data Information.

Item	Units	Parameter Type	Date (36)		
			Min	Max	Ave
x_1	kg/m^3	Input	135	165	150
x_2	kg/m^3	Input	385	435	410
x_3	kg/m^3	Input	33	126	79.5
x_4	kg/m^3	Input	680	700	690
x_5	kg/m^3	Input	1116	1142	1129
x_6	kg/m^3	Input	4.16	5.77	4.965
x_7	kg/m^3	Input	24.36	73.08	48.72

4.3. Prediction of Durability Utilizing a Response Surface Model

4.3.1. Collection of Sample Data

Using the Latin hypercube experimental design, 36 sets of Latin test samples were obtained and subjected to rapid freeze–thaw test and RCM test, respectively. The results of freeze–thaw resistance and chloride ion permeability of 36 sets of Latin test samples were obtained, as shown in Table 2.

Table 2. Sample Dataset and Related Information.

Item	Units	Parameter Type	Date (36)				
			Min	Max	Ave	Median	SD
x_1	kg/m^3	Input	135	165	150	150.00	9.64
x_2	kg/m^3	Input	385	435	410	410.00	16.08
x_3	kg/m^3	Input	33	126	79.5	79.50	29.91
x_4	kg/m^3	Input	680	700	690	690.00	6.43
x_5	kg/m^3	Input	1116	1142	1129	1129.00	8.36
x_6	kg/m^3	Input	4.16	5.77	4.965	4.96	0.51
x_7	kg/m^3	Input	24.36	73.08	48.72	48.72	15.66
RD	%	Output	85.1	91.2	88.15	87.78	1.62
CP	10^{-12} m^2/s	Output	1.6	3.5	2.55	2.58	0.53
CO	yuan	Output	638.54	1068.39	853.46	853.47	123.38

4.3.2. Evaluation of Forecast Results

(1) Frost resistance model of concrete based on response surface.

According to the results of 36 groups of data, a concrete frost resistance model based on response surface is constructed, and a multiple regression model with water content, cement content, fly ash content, fine aggregate content, coarse aggregate content, water reducer content, and fiber content as response values is obtained. See Figure 4 and Table 3 for verification results.

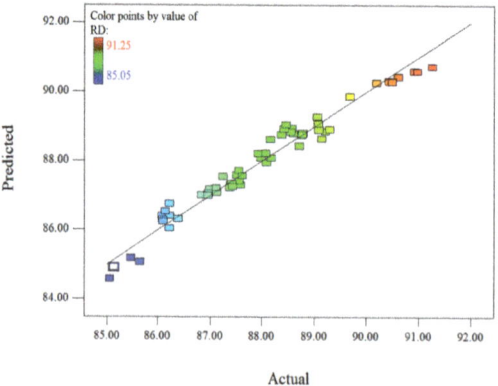

Figure 4. Accuracy verification of approximate model for frost resistance response surface.

Table 3. Error Analysis of Frost Resistance Regression Model.

Items	Std. Dev.	C.V.%	R^2	R_a^2
value	0.45	0.51	0.9657	0.9111

As can be seen from Figure 4 and Table 3, the determination coefficient R^2 of the regression model is 0.9657, which is close to 1, indicating that the predicted relative dynamic elastic modulus of high-performance fiber-reinforced concrete is highly correlated with the actual value. The adjustment coefficient R_a^2 is 0.9111, which is greater than 0.8; the coefficient of variation is 0.51%, less than 10%. It shows that the second-order response surface model has a good fitting degree and can effectively and accurately predict the relative dynamic elastic modulus of high-performance fiber-reinforced concrete under different mix proportions.

(2) Response surface-based model for chloride ion permeability resistance

Based on 36 sets of data results, a response surface-based chloride ion permeability resistance model was constructed to obtain a multiple regression model with response values of water content, cement content, fly ash content, fine aggregate content, coarse aggregate content, water-reducing agent content, and fiber content. The validation results are shown in Figure 5 and Table 4.

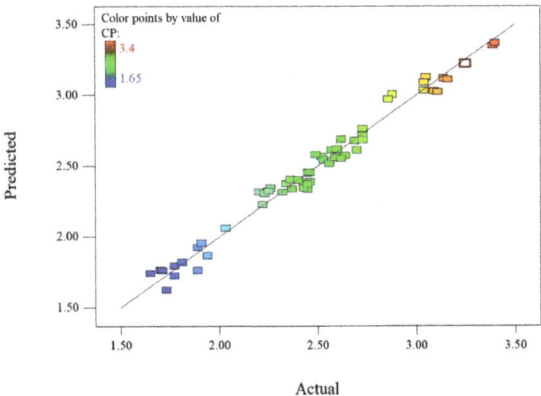

Figure 5. Accuracy verification of chloride ion impermeability response surface approximation model.

Table 4. Error analysis of regression model for chloride ion impermeability.

Items	Std. Dev.	C.V.%	R^2	R_a^2
value	0.10	4.08	0.9803	0.9490

From Figure 5 and Table 4, it can be seen that the determination coefficient R^2 of the regression model is 0.9803, which is close to 1, indicating that the predicted value of the chloride ion migration coefficient of high-performance fiber-reinforced concrete is highly correlated with the actual value. The adjustment coefficient R_a^2 is 0.9490, greater than 0.8; the coefficient of variation is 4.08%, less than 10%. This indicates that the second-order response surface model has a good fitting degree and can effectively and accurately predict the chloride ion migration coefficient of high-performance fiber-reinforced concrete under different mix ratios.

4.4. Multi-Objective Optimization Utilizing NSGA-III

4.4.1. Formulation of the Objective Function

In engineering endeavors, enhancing the durability of concrete typically accompanies increased costs. To strike a balance between cost-effectiveness and ensuring optimal durability, a multi-objective optimization approach is employed.

(1) Optimization objective function of concrete frost resistance based on response surface model.

The response surface model serves the purpose of predicting the relative dynamic elastic modulus of concrete. Subsequently, the objective function for optimizing the frost resistance of concrete is formulated as follows:

$$\max \cdot f_1 = \max[RSM(x_1, x_2, \ldots, x_7)] = \beta_0 + \sum_{i=1}^{7} \beta_i x_i + \sum_{i=1}^{7} \beta_{ii} x_i^2 + \sum_{i<j}^{7} \beta_{ij} x_i x_j \quad (14)$$

In the formula, x_1, x_2, x_3, x_4, x_5, x_6, and x_7 respectively represent the water content, cement content, fly ash content, fine aggregate content, coarse aggregate content, water-reducing agent content, and fiber content.

(2) Optimization objective function of chloride ion impermeability of concrete based on response surface model.

Based on the response surface model, the chloride ion migration coefficient of concrete is predicted, and the objective function of optimizing the chloride ion impermeability of concrete is established. The objective function for optimizing the chloride ion permeability of concrete is as follows:

$$\max \cdot f_2 = \max[RSM(x_1, x_2, \ldots, x_7)] = \beta'_0 + \sum_{i=1}^{7} \beta'_i x_i + \sum_{i=1}^{7} \beta'_{ii} x_i^2 + \sum_{i<j}^{7} \beta'_{ij} x_i x_j \quad (15)$$

(3) The objective function of concrete economic cost optimization

The costs of the raw materials employed in the concrete for the specific project under examination are presented in Table 5.

Table 5. Pricing information for the raw materials used in concrete.

Component	Units	Cost (Yuan)
Water	kg	0.0018
Cement	kg	0.4
Fly ash	kg	0.51
Fine aggregate	kg	0.12
Coarse aggregate	kg	0.14
Superplasticizer	kg	5.6
Polymer fiber	kg	7.8

Based on the above price information, the economic objective function of high-performance fiber-reinforced concrete is as follows:

$$\min \cdot f_3 = 0.0018x_1 + 0.4x_2 + 0.51x_3 + 0.12x_4 + 0.14x_5 + 5.6x_6 + 7.8x_7 \quad (16)$$

4.4.2. Using NSGA-III Algorithm for Durability and Economic Optimization

(1) Acquiring the Pareto optimal solution set for concrete mix proportions

In this investigation, the NSGA-III algorithm was characterized by a crossover rate of 0.7, a mutation rate of 0.01, a population size of 40, and a maximum generation limit of 80. Over the course of 80 iterations, the NSGA-III algorithm was applied to fulfill the durability specifications of high-performance fiber-reinforced concrete, concurrently aiming to achieve a noteworthy reduction in economic costs. Ultimately, the Pareto solution set for optimizing the mixture proportions of high-performance fiber-reinforced concrete was acquired, as depicted in Figure 6.

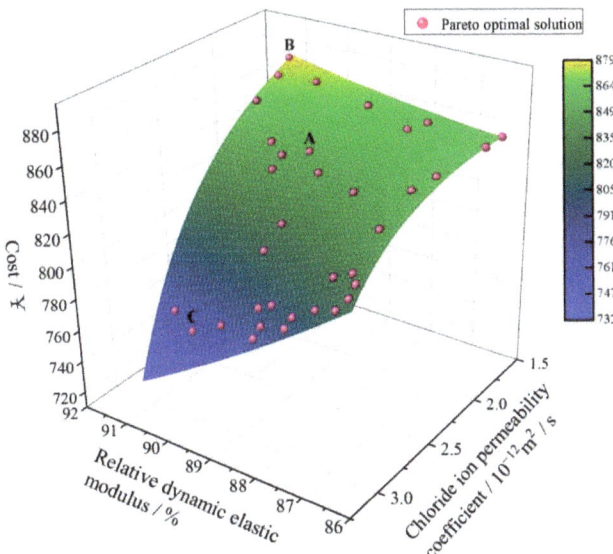

Figure 6. 3D View of the Pareto optimal solution.

Figure 6 illustrates the Pareto solution set obtained through the optimization of the mix ratio for high-performance fiber-reinforced concrete using the NSGA-III algorithm. On the horizontal axis of the figure, the relative dynamic elastic modulus and chloride ion permeability of high-performance concrete are depicted, serving as parameters to assess the durability of the concrete. Meanwhile, the vertical axis represents the cost associated with high-performance fiber-reinforced concrete.

By scrutinizing the variations in surface color, it becomes evident that, with an escalation in concrete cost, the surface color undergoes a discernible shift from deep blue to light green. The Pareto points concentrated in the blue region of the curved surface exhibit diminished values in both cost and durability indicators. As durability indicators advance, there is a concurrent increase in cost, signifying a positive correlation between the durability and cost of high-performance fiber-reinforced concrete. To a certain degree, augmenting the economic outlay of concrete can proficiently enhance its durability.

Figure 7 depicts the projections of Figure 6 on the freeze−thaw resistance and chloride permeability planes.

After applying optimization with NSGA-III, the frost resistance index of high-performance fiber-reinforced concrete lies within the 86% to 92% range. The optimized chloride ion migration coefficient varies within the range of 1.7 to 3.1×10^{-12} m^2/s, while the cost falls between 740 and 871 yuan. In accordance with the overarching trend observed in the Pareto optimization solution set, a trade-off is evident between the chloride ion permeability and the economic cost of high-performance fiber-reinforced concrete. Increasing the chloride ion migration coefficient leads to a proportional rise in economic expenses. In other words, enhancing the chloride ion resistance of concrete necessitates an increase in economic expenditure. In contrast, the relationship between the relative dynamic modulus of elasticity and economic costs for high-performance fiber-reinforced concrete remains inconclusive.

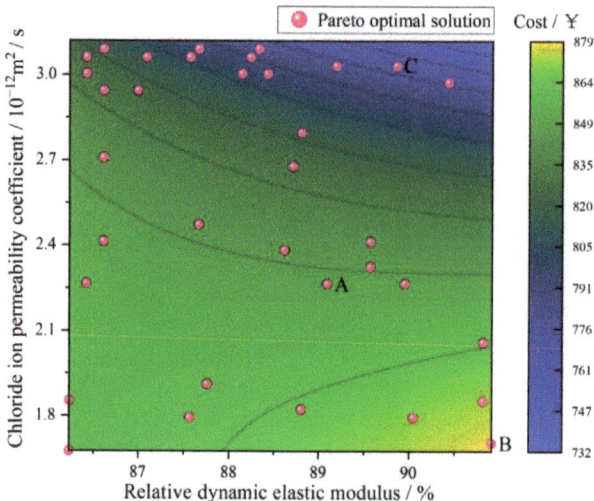

Figure 7. Projection of the Pareto optimal solution.

(2) The Selection and Analysis of the Pareto Solution Set for Optimizing Mix Proportion.

The Pareto solutions for optimizing each concrete mix proportion are derived through a meticulous consideration of the trade-offs inherent in multiple objectives. Put differently, there exists no singular solution capable of concurrently attaining both high concrete durability and low economic costs. Consequently, when confronted with diverse engineering projects, the imperative lies in selecting solutions that are aptly tailored to their specific requirements. In order to explicate this nuanced equilibrium, we designate Point A as the optimal balance obtained through the application of the ideal point method, comprehensively assessing the equilibrium between concrete durability and cost. Point B is delineated as the optimal solution, focusing predominantly on the performance of concrete durability, while Point C embodies the optimal solution, ensuring the minimization of economic costs in concrete. The intricate details of the specific parameters corresponding to Points A, B, and C along the Pareto boundary are meticulously elucidated in Table 6. These parameters serve as a scholarly guide for achieving optimal solutions amid diverse objective trade-offs, facilitating judicious decision-making in the realm of specific engineering projects.

Table 6. Pareto points on selected points and corresponding specific parameters.

Indicator	Item	Units	A	B	C
			Min	Max	Ave
Input indicator	x_1	kg/m^3	135	135	135
	x_2	kg/m^3	385	435	385
	x_3	kg/m^3	96	126	40
	x_4	kg/m^3	690	700	700
	x_5	kg/m^3	1129	1116	1116
	x_6	kg/m^3	4.96	4.16	4.16
	x_7	kg/m^3	48.72	48.3	42.1
	RD	%	89.20	90.91	87.24
	CP	10^{-12} m^2/s	2.18	1.76	2.91
	CO	yuan	851.85	878.78	766.56

(3) Validation of hybrid framework optimization effectiveness

To validate the efficacy of the hybrid algorithm optimization, we prepared concrete specimens in accordance with the aforementioned mixing ratio scheme and subsequently conducted durability tests on them. Table 7 presents a comparative analysis of predicted and experimental values for the concrete durability indicators.

Table 7. Analysis of predicted and experimental values of concrete durability indicators.

Optimization Plan for Mix Proportion	Anticipated Outcomes		Experimental Values		Errors	
	RD	CP	RD	CP	RD	CP
A	89.20	2.18	91.44	2.07	2.45%	5.31%
B	90.91	1.76	92.21	1.69	1.41%	4.14%
C	87.24	2.91	86.22	3.11	1.18%	6.43%

The disparity between the predicted values of RD and CP for concrete and their corresponding experimental values is minimal. In Scheme A, the errors for these two indicators are 2.45% and 5.31%, respectively. In Scheme B, the errors are 1.41% and 4.14%, respectively. Meanwhile, in Scheme C, the errors stand at 1.18% and 6.43% for these two indicators. The aforementioned outcomes substantiate the precision and dependability of the multi-objective optimization model founded on NSGA-III.

5. Discussion

The established multi-objective optimization model based on NSGA-III enables the optimization of three objectives, addressing the multi-objective conflicts encountered in practical engineering scenarios. To substantiate the heightened efficacy of the three-objective optimization relative to both single-objective and two-objective optimization, the optimization process was executed, considering frost resistance, chloride ion impermeability, and economic cost as objectives across varying quantities of objectives. The outcomes of the single-objective, two-objective, and three-objective optimization endeavors are delineated in Table 8.

Table 8. Optimization results of various quantitative indicators.

Optimization Objective		Anticipated Outcomes		
		CP	RD	CO
Single objective	CP	1.68	90.11	878.97
	RD	1.76	91.60	879.33
	CO	2.91	87.24	766.56
Two objectives	CD + RD	1.76	90.91	878.78
	CP + CO	1.72	89.95	871.36
	RD + CO	1.79	91.22	872.49
Three objectives	CP + RD + CO	2.18	89.20	851.85
Actual average		3.52	83.79	873.95

The results summarized in Table 8 indicate the following:

(1) In optimizing the mix proportion of high-performance fiber-reinforced concrete, the application of hybrid algorithms has yielded noteworthy outcomes. Whether pursuing single-objective optimization or multi-objective optimization, the achieved optimization values surpass the average experimental data, signifying the substantial advantages of this optimization method in enhancing concrete durability. Taking three-objective optimization as an illustration, the relative dynamic elastic modulus and chloride ion permeability coefficient are 89.20% and 2.18×10^{-12} m^2/s, respectively. In comparison, the corresponding average experimental values stand at 83.79% and 3.52×10^{-12} m^2/s. This marked improvement underscores the efficacy of multi-objective optimization. Consequently, this research offers robust theoretical underpinnings and practical insights for refining the mix proportion of high-performance fiber-reinforced concrete.

(2) Specificity of single-objective optimization: When using a genetic algorithm for single-objective optimization, the results were most tailored to the respective objective. The target values for chloride ion permeability, relative dynamic elastic modulus, and concrete economic cost obtained through single-objective genetic algorithm optimization were better than those from multi-objective genetic algorithm optimization, with optimized results of 1.68×10^{-12} m^2/s, 91.60%, and 766.56 yuan, respectively.

6. Conclusions

Currently, high-performance fiber-reinforced concrete (HPFRC) finds extensive applications in both domestic and international practical engineering projects. However, as the service environments for concrete structures become increasingly harsh, there is a growing demand for enhanced durability. Rational concrete mix design plays a crucial role in improving the durability of high-performance fiber-reinforced concrete, increasing the service life of concrete components, and reducing the overall life-cycle maintenance costs. Nevertheless, another essential objective in concrete mix design is cost reduction, which can sometimes conflict with the goal of improving durability. Traditional concrete mix design methods suffer from issues such as low efficiency and suboptimal results, making the multi-objective optimization of durability and economic costs for high-performance fiber-reinforced concrete a challenging task. Therefore, this article introduces an intelligent optimization framework based on hybrid algorithms. A Latin hypercube experimental design method is employed, considering factors such as water content, cement content, fly ash content, fine aggregate content, coarse aggregate content, superplasticizer dosage, and fiber dosage. Evaluation criteria include the relative dynamic modulus of elasticity, chloride ion resistance, and economic considerations. Response surface prediction models are established for each evaluation criterion. The NSGA-III algorithm is utilized within the RSM model to autonomously search for the optimal mix design that maximizes overall performance. Based on the optimization results and comparative experiments, the intelligent framework proposed in this article, leveraging hybrid algorithms, effectively optimizes the mix proportion of high-performance fiber-reinforced concrete. It not only meets durability requirements to a certain extent but also ensures cost control.

The hybrid algorithm proposed in this article can achieve multi-objective optimization of high-performance fiber-reinforced concrete within a certain range, but it also has certain limitations. The performance of machine learning models is usually influenced by the amount of training data, reliability, and complexity of the data. Even larger datasets may not necessarily improve the accuracy of the model. In machine learning, this can refer to "Kolmogorov complexity", which is the length of the shortest computer program that produces output. By reducing the complexity of the dataset, we can reduce the computational burden on the model when processing data, making it easier to capture patterns and correlations in the data. In this way, the model can predict and classify more accurately, thereby improving its accuracy. Therefore, when designing and preparing datasets, we should, to some extent, simplify the structure and features of the dataset to improve the performance of deep learning models.

The 36 sets of data used for training the model in this study were all from the same engineering project. Therefore, trained models may perform poorly in predicting specific data for other projects. In future research, collecting more diverse and specific data can better cover the characteristics and changes of different engineering projects and, to some extent, improve the generalization ability of the model. This means that the trained model can more accurately predict the specific data of other projects, rather than being limited to engineering projects with training data sources; it can reduce the bias and variance of the model and improve its reliability. This means that the model is more accurate and reliable in predicting and optimizing engineering materials. At the same time, constructing a hybrid algorithm that considers more parameters and objectives, further improving the effectiveness of multi-objective optimization, and promoting the development of new material design and optimization methods is our next research direction.

Author Contributions: Conceptualization, X.W. and F.C.; methodology, X.W. and F.C.; software, X.W.; validation, X.W., F.C. and D.J.; formal analysis, X.W.; investigation, F.C. and L.C.; resources, X.W.; data curation, F.C.; writing—original draft preparation, X.W.; writing—review and editing, L.C.; visualization, X.W.; supervision, D.J.; project administration, X.W. and F.C.; funding acquisition, X.W. All authors have read and agreed to the published version of the manuscript.

Funding: The research described in this paper was supported by the "Shandong Natural Science Foundation Project" (Project No. 60000101032). The authors greatly acknowledge their financial support.

Institutional Review Board Statement: Not applicable.

Informed Consent Statement: Not applicable.

Data Availability Statement: Data are contained within the article.

Conflicts of Interest: Long Cui was employed by the company Shandong Provincial Academy of Building Research Co., Ltd., Di Jiang was employed by the company Shandong Provincial Academy of Building Research Co., Ltd., the remaining authors declare that the research was conducted in the absence of any commercial or financial relationships that could be construed as a potential conflict of interest.

References

1. Habel, K.; Viviani, M.; Emmanuel, D.; Bruhwiler, E. Development of the mechanical properties of an Ultra-High Performance Fiber Reinforced Concrete (UHPFRC). *Cem. Concr. Res.* **2006**, *36*, 1362–1370. [CrossRef]
2. Habel, K.; Gauvreau, P. Response of ultra-high performance fiber reinforced concrete (UHPFRC) to impact and static loading. *Cem. Concr. Compos.* **2008**, *30*, 938–946. [CrossRef]
3. Abbas, S.; Soliman, A.M.; Nehdi, M.L. Exploring mechanical and durability properties of ultra-high performance concrete incorporating various steel fiber lengths and dosages. *Constr. Build. Mater.* **2015**, *75*, 429–441. [CrossRef]
4. Graybeal, B.A.; Hartmann, J.L. Strength and durability of ultra-high performance concrete. In Proceedings of the Concrete Bridge Conference, Taupo, New Zealand, 3–5 October 2003; p. 20.
5. Athanasopoulou, A.; Parra-Montesinos, G. Experimental Study on the Seismic Behavior of High-Performance Fiber-Reinforced Concrete Low-Rise Walls. *Aci Struct. J.* **2013**, *110*, 767–777.
6. Shehab, H.; Eisa, A.; Wahba, A.M.; Sabol, P.; Katunský, D. Strengthening of Reinforced Concrete Columns Using Ultra-High Performance Fiber-Reinforced Concrete Jacket. *Buildings* **2023**, *13*, 2036. [CrossRef]
7. Sheikh, S.A. Performance of concrete structures retrofitted with fiber reinforced polymers. *Eng. Struct.* **2002**, *24*, 869–879. [CrossRef]
8. Zhang, C.; Nerella, V.N.; Krishna, A.; Wang, S.; Zhang, Y.; Mechtcherine, V.; Banthia, N. Mix design concepts for 3D printable concrete: A review. *Cement Concr. Compos.* **2021**, *122*, 15. [CrossRef]
9. De Maeijer, P.K.; Craeye, B.; Snellings, R.; Kazemi-Kamyab, H.; Loots, M.; Janssens, K.; Nuyts, G. Effect of ultra-fine fly ash on concrete performance and durability. *Construct. Build. Mater.* **2020**, *263*, 13. [CrossRef]
10. Medvedev, V.; Pustovgar, A. A Review of Concrete Carbonation and Approaches to Its Research under Irradiation. *Buildings* **2023**, *13*, 1998. [CrossRef]
11. Niu, Z.; Lu, X.; Luo, Y. The Effects of a Multifunctional Rust Inhibitor on the Rust Resistance Mechanism of Carbon Steel and the Properties of Concrete. *Coatings* **2023**, *13*, 1375. [CrossRef]
12. Li, W.; Hu, S. Fracture Behavior of Concrete under Chlorine Salt Attack Exposed to Freeze–Thaw Cycles Environment. *Materials* **2023**, *16*, 6205. [CrossRef] [PubMed]
13. Dai, J.; Wang, Q.; Zhang, B. Frost resistance and life prediction of equal strength concrete under negative temperature curing. *Constr. Build. Mater.* **2023**, *396*, 132278. [CrossRef]
14. Wang, Y.; Liu, Z.; Fu, K.; Li, Q.; Wang, Y. Experimental studies on the chloride ion permeability of concrete considering the effect of freeze–thaw damage. *Constr. Build. Mater.* **2020**, *236*, 117556. [CrossRef]
15. Liu, D.; Tu, Y.; Shi, P.; Sas, G.; Elfgren, L. Mechanical and durability properties of concrete subjected to early-age freeze–thaw cycles. *Mater. Struct.* **2021**, *54*, 211. [CrossRef]
16. Nosouhian, F.; Mostofinejad, D.; Hasheminejad, H. Influence of biodeposition treatment on concrete durability in a sulphate environment. *Biosyst. Eng.* **2015**, *133*, 141–152. [CrossRef]
17. Wu, Y.; Ren, Q.; Zhang, X. Compressive behavior and freeze-thaw durability of concrete after exposure to high temperature. *Eur. J. Environ. Civ. Eng.* **2022**, *26*, 6830–6844. [CrossRef]
18. Kou, S.C.; Poon, C.S.; Chan, D. Influence of fly ash as cement replacement on the properties of recycled aggregate concrete. *J. Mater. Civ. Eng.* **2007**, *19*, 709–717. [CrossRef]
19. Mehta, P.K. Influence of fly ash characteristics on the strength of portland-fly ash mixtures. *Cem. Concr. Res.* **1985**, *15*, 669–674. [CrossRef]
20. Atiş, C.D.; Karahan, O. Properties of steel fiber reinforced fly ash concrete. *Constr. Build. Mater.* **2009**, *23*, 392–399. [CrossRef]

21. Karahan, O.; Atiş, C.D. The durability properties of polypropylene fiber reinforced fly ash concrete. *Mater. Des.* **2011**, *32*, 1044–1049. [CrossRef]
22. *JGJ 55—2011*; Specification for Mix Proportion Design of Ordinary Concrete. Ministry of Housing and Urban-Rural Construction of the People's Republic of China: Beijing, China, 2011.
23. Chopra, P.; Kumar, R.; Kumar, M. Artificial Neural Networks for the Prediction of Compressive Strength of Concrete. *Int. J. Appl. Sci. Eng.* **2015**, *13*, 187–204.
24. Alsanusi, S.; Bentaher, L. Prediction of Compressive Strength of Concrete from Early Age Test Result Using Design of Experiments (RSM). *Int. J. Civ. Environ. Struct. Constr. Archit. Eng.* **2015**, *9*, 1522–1526.
25. Qurishee, M.A.; Iqbal, I.T.; Islam, M.S.; Islam, M.M. Use of Slag As Coarse Aggregate and Its Effect on Mechanical Properties of Concrete. In Proceedings of the 3rd International Conference on Advances in Civil Engineering, CUET, Chittagong, Bangladesh, 21–23 December 2016.
26. Xu, Y.; Li, M.; Zhao, X.; Lu, F. The response curved surface regression analysis technique the application of a new regression analysis technique in materials research. *Rare Met. Mater. Eng.* **2001**, *30*, 428–432.
27. Li, L.; Zhang, S.; He, Q.; Hu, X.B. Application of response surface methodology in experiment design and optimization. *Res. Explor. Lab.* **2015**, *34*, 41–45.
28. Zhang, L.; Sojobi, A.; Kodur, V.; Liew, K. Effective utilization and recycling of mixed recycled aggregates for a greener environment. *J. Clean. Prod.* **2019**, *236*, 117600. [CrossRef]
29. Alyamac, K.E.; Ghafari, E.; Ince, R. Development of eco-efficient self-compacting concrete with waste marble powder using the response surface method. *J. Clean. Prod.* **2017**, *144*, 192–202. [CrossRef]
30. Tyagi, M.; Rana, A.; Kumari, S.; Jagadevan, S. Adsorptive removal of cyanide from coke oven wastewater onto zero-valent iron:Optimization through response surface methodology, isotherm and kinetic studies. *J. Clean. Prod.* **2018**, *178*, 398–407. [CrossRef]
31. Simşek, B.; Uygunoğlu, T.; Korucu, H.; Kocakerim, M.M. Analysis of the effects of dioctyl terephthalate obtained from polyethylene terephthalate wastes on concrete mortar: A response surface methodology based desirability function approach application. *J. Clean. Prod.* **2018**, *170*, 437–445. [CrossRef]
32. Taherkhani, H.; Noorian, F. Investigating permanent deformation of recycled asphalt concrete containing waste oils as rejuvenator using response surface methodology (RSM). *Iran. J. Sci. Technol. Trans. Civ. Eng.* **2021**, *45*, 1989–2001. [CrossRef]
33. Bheel, N.; Mohammed, B.S.; Liew, M.S.; Zawawi, N.A.W.A. Effect of Graphene Oxide as a Nanomaterial on the Durability Behaviors of Engineered Cementitious Composites by Applying RSM Modelling and Optimization. *Buildings* **2023**, *13*, 2026. [CrossRef]
34. Wang, J.; Wang, W. Response surface based multi−objective optimization of basalt fiber reinforced foamed concrete. *Mater. Rep.* **2019**, *33*, 4092–4097.
35. Zhang, Q.; Feng, X.; Chen, X.; Lu, K. Mix design for recycled aggregate pervious concrete based on response surface methodology. *Constr. Build. Mater.* **2020**, *259*, 119776. [CrossRef]
36. DeRousseau, M.A.; Kasprzyk, J.R.; Srubar, W.V. Multi-objective optimization methods for designing low-carbon concrete mixtures. *Front. Mater.* **2021**, *8*, 13. [CrossRef]
37. Nguyen, T.T.; Thai, H.T.; Ngo, T. Optimised mix design and elastic modulus prediction of ultra-high strength concrete. *Construct. Build. Mater.* **2021**, *302*, 124150. [CrossRef]
38. Sun, H.; Burton, H.V.; Huang, H.L. Machine learning applications for building structural design and performance assessment: State-of-the-art review. *J. Build. Eng.* **2021**, *33*, 14. [CrossRef]
39. Zavala, G.R.; Nebro, A.J.; Luna, F.; Coello Coello, C.A. A survey of multi-objective metaheuristics applied to structural optimization. *Struct. Multidiscip. Optim.* **2014**, *49*, 537–558. [CrossRef]
40. Srinivas, N.; Deb, K. Multiobjective function optimization using nondominated sorting genetic algorithms. *Evol. Comput.* **1995**, *2*, 221–248. [CrossRef]
41. Deb, K.; Pratap, A.; Agarwal, S.; Meyarivan, T. A fast and elitist multiobjective genetic algorithm: NSGA-II. *IEEE Trans. Evol. Comput.* **2002**, *6*, 182–197. [CrossRef]
42. Bolon-Canedo, V.; Remeseiro, B. Feature selection in image analysis: A survey. *Artif. Intell. Rev.* **2020**, *53*, 2905–2931. [CrossRef]
43. Kabir, H.; Garg, N. Machine learning enabled orthogonal camera goniometry for accurate and robust contact angle measurements. *Sci. Rep.* **2023**, *13*, 1497. [CrossRef]
44. Deb, K.; Jain, H. An Evolutionary Many-objective optimization algorithm using reference-point based non-dominated sorting approach part I: Solving problems with box constraints. *IEEE Trans. Evol. Comput.* **2014**, *18*, 577–601. [CrossRef]
45. Zhang, J.; Wang, S.; Tang, Q.; Zhou, Y.; Zeng, T. An improved NSGA-III integrating adaptive elimination strategy to solution of many-objective optimal power flow problems. *Energy* **2019**, *172*, 945–957. [CrossRef]
46. Ruan, F.; Gu, R.; Huang, T.; Xue, S. A big data placement method using NSGA-III in meteorological cloud platform. *EURASIP J. Wirel. Commun. Netw.* **2019**, *2019*, 143. [CrossRef]
47. Yuan, X.; Tian, H.; Yuan, Y.; Huang, Y.; Ikram, R.M. An extended NSGA-III for solution multi-objective hydro-thermal-wind scheduling considering wind power cost. *Energy Convers. Manag.* **2015**, *96*, 568–578. [CrossRef]
48. Liu, Y.; You, K.; Jiang, Y.; Wu, Z.; Liu, Z.; Peng, G.; Zhou, C. Multi-objective optimal scheduling of automated construction equipment using non-dominated sorting genetic algorithm (NSGA-III). *Autom. Constr.* **2022**, *143*, 104587. [CrossRef]

49. Jafari, H.; Nazif, S.; Rajaee, T. A multi-objective optimization method based on NSGA-III for water quality sensor placement with the aim of reducing potential contamination of important nodes. *Water Supply* **2022**, *22*, 928–944. [CrossRef]
50. Zaifang, Z.; Feng, X.; Xiwu, S. Multi-objective Optimization of Hydroforming Process of Rocket Tank Bottom. *J. Mech. Eng.* **2022**, *58*, 78–86.
51. Kumar, U.; Klefsjö, B. Reliability analysis of hydraulic systems of LHD machines using the power law process model. *Reliab. Eng. Syst. Saf.* **1992**, *35*, 217–224. [CrossRef]
52. Sriravindrarajah, R.; Wang, N.D.H.; Ervin, L.J.W. Mix design for pervious recycled aggregate concrete. *Int. J. Concr. Struct. Mater.* **2012**, *6*, 239–246. [CrossRef]
53. Stein, M. Large Sample Properties of Simulations Using Latin Hypercube Sampling. *Technometrics* **1987**, *29*, 143–151. [CrossRef]
54. Owen, A.B. A Central Limit Theory for Latin Hypercube Sampling. *J. R. Stat. Soc. Ser. B* **1992**, *54*, 541–551.
55. Olsson, A.; Sandberg, G.; Dahlblom, O. On Latin hypercube sampling for structural reliability analysis. *Struct. Saf.* **2003**, *25*, 47–68. [CrossRef]
56. GBT 50476-2019; Durability Design Standard for Concrete Structures. Ministry of Housing and Urban-Rural Development of the People's Republic of China: Beijing, China, 2019.

Disclaimer/Publisher's Note: The statements, opinions and data contained in all publications are solely those of the individual author(s) and contributor(s) and not of MDPI and/or the editor(s). MDPI and/or the editor(s) disclaim responsibility for any injury to people or property resulting from any ideas, methods, instructions or products referred to in the content.

Article

The Effect of Different Particle Sizes of SiO$_2$ in Sintering on the Formation of Ternesite

Fengyu Song, Didi Huo, Yanmin Wang, Dunlei Su and Xiaocun Liu *

School of Civil Engineering, Shandong Jiaotong University, 5 Jiaoxiao Road, Jinan 250357, China; 21107004@stu.sdjtu.edu.cn (F.S.); h2139199519@163.com (D.H.); 212013@sdjtu.edu.cn (Y.W.); sudunlei@163.com (D.S.)
* Correspondence: 204127@sdjtu.edu.cn

Abstract: Ternesite is synthesized through sintering a mixture of CaCO$_3$, SiO$_2$, and CaSO$_4$ in a molar ratio of 4:2:1. Ternesite has a hydration rate between ye'elimite and belite in an aluminum-containing environment, and is considered to be a new material that can be used to enhance the performance of calcium sulphoaluminate cements. This experiment investigated the influence of different particle sizes of SiO$_2$ on ternesite formation. Controlled partial pressure sintering was employed within the temperature range from 1100 °C to 1200 °C, with a 72 h incubation period. The highest purity of ternesite in the samples reached 99.47% (500 nm SiO$_2$ sample). The analysis results from scanning electron microscopy and an energy dispersive spectrometer indicated that the particle size of SiO$_2$ exerted a significant influence on the formation of ternesite. In the preparation of ternesite from 10 μm particle size SiO$_2$, traces of calcium silicate were found in the product. The results of a thermal analysis further demonstrated significant distinctions in the thermal stability of ternesite prepared with SiO$_2$ of different particle sizes. Additionally, the crystallinity of ternesite was influenced by the particle size of SiO$_2$, consequently impacting the hydration performance of ternesite–calcium sulphoaluminate cement.

Keywords: ternesite; particle size; controlled partial pressure; crystallinity; thermal stability

---

Citation: Song, F.; Huo, D.; Wang, Y.; Su, D.; Liu, X. The Effect of Different Particle Sizes of SiO$_2$ in Sintering on the Formation of Ternesite. *Coatings* **2023**, *13*, 1826. https://doi.org/10.3390/coatings13111826

Academic Editor: Ionut Ovidiu Toma

Received: 9 October 2023
Revised: 18 October 2023
Accepted: 23 October 2023
Published: 25 October 2023

Copyright: © 2023 by the authors. Licensee MDPI, Basel, Switzerland. This article is an open access article distributed under the terms and conditions of the Creative Commons Attribution (CC BY) license (https://creativecommons.org/licenses/by/4.0/).

1. Introduction

Ternesite was initially discovered within the low-temperature zone of cement rotary kilns [1–3], and it was later confirmed to have a structure similar to silicocarnotite(Ca$_5$(PO$_4$)$_2$SiO$_4$) [4]. Ternesite possesses the chemical formula Ca$_5$(SiO$_4$)$_2$SO$_4$, denoted as C$_5$S$_2$S$^-$, and belongs to the orthorhombic crystal system with the space group *Pnma* [5]. The macroscopic morphology of C$_5$S$_2$S$^-$ presents itself as a light green solid, while its microscopic morphology exhibits a notable diversity, encompassing various forms such as irregular plate-like crystals, columnar crystals, and polyhedral crystals. These morphological variations are influenced by the temperature and duration of the sintering process of C$_5$S$_2$S$^-$ [2,5–7]. Initially, C$_5$S$_2$S$^-$ was regarded as an inert mineral with a negligible technological significance when encountered as an intermediate phase during the calcination of calcium sulfoaluminate (CSA) cement clinker [8–13]. However, recent research has unveiled that C$_5$S$_2$S$^-$ exhibits hydration activity in aluminum-containing environments, positioning it as an intermediary between ye'elimite (C$_4$A$_3$S$^-$) and belite (C$_2$S) [6,14–19].

Henceforth, the incorporation of C$_5$S$_2$S$^-$ into CSA clinker, primarily composed of C$_2$S and C$_4$A$_3$S$^-$, serves to bridge the substantial gap in the hydration activities between C$_2$S and C$_4$A$_3$S$^-$. This strategic integration enables components with distinct hydration propensities to collaborate synergistically, thereby enhancing the overall performance of the cement [7,17,20–22]. Furthermore, due to the considerably lower sintering temperatures required for C$_5$S$_2$S$^-$ in CSA clinker as compared to Portland cement (OPC), ternesite–calcium sulphoaluminate cement (T-CSA) founded on the C$_5$S$_2$S$^-$ within CSA cement emerges as a highly promising low-carbon construction material. Upon establishing the hydration activity of $C_5S_2S^-$ in aluminum-rich environments, numerous scholars have redirected their focus toward the comprehensive exploration of this mineral, instigating a plethora of investigations encompassing its formation and hydration kinetics.

The formation and microscopic morphology of $C_5S_2S^-$ have been subject to extensive scholarly investigations. $C_5S_2S^-$ can be synthesized through the reaction between C_2S and $CaSO_4$, with the formation temperature range spanning from 900 °C to 1200 °C. Notably, decomposition commences when the temperature surpasses 1200 °C, and upon exceeding 1250 °C, $C_5S_2S^-$ undergoes complete decomposition into C_2S and $CaSO_4$ [2,20,23–25]. Jing et al. explored the impact of different crystalline forms of C_2S on the synthesis of $C_5S_2S^-$, revealing that both β-C_2S and γ-C_2S were capable of yielding $C_5S_2S^-$. Notably, β-C_2S exhibited a higher reactivity in this process compared to γ-C_2S, while α'L-C_2S demonstrated no discernible influence on $C_5S_2S^-$ generation [26]. Skalamprinos delved into the effects of elemental doping on $C_5S_2S^-$ formation, elucidating that the introduction of MgO, Na_2O, K_2O, SrO, MnO_2, TiO_2, and ZnO could induce a transition in the microscopic morphology of $C_5S_2S^-$ from polyhedral crystals to nodular crystals [7]. Liu et al. discerned that both sintering temperature and sintering duration exerted a pronounced impact on the microscopic morphology of $C_5S_2S^-$. Elevating the sintering temperature and extending the sintering duration facilitated the transformation of $C_5S_2S^-$ from polyhedral crystals into columnar crystals, and subsequently into irregular plate-like crystals [2]. Zhang et al. reported that $C_5S_2S^-$ grains produced through primary sintering exhibited an irregular morphology with generally larger sizes (~10 μm). In contrast, secondary sintering could reduce the grain size of $C_5S_2S^-$ (~5 μm) and result in a more regular morphology. Rapid cooling with blast air at the conclusion of sintering was observed to reduce the crystallinity of $C_5S_2S^-$ [27]. Furthermore, Zhang et al. investigated the influence of alkali metal doping on the crystal structure of $C_5S_2S^-$. Their findings revealed that the solid solution of alkali metals within the crystal structure of $C_5S_2S^-$ led to unit cell contraction or expansion, contingent upon the radius of the dopant ions, ultimately resulting in a reduction in the overall crystallinity of $C_5S_2S^-$ [28]. Meanwhile, plenty of investigations have been conducted on the purity of $C_5S_2S^-$. Hanein et al. determined that the partial pressures of both SO_2 and O_2 during sintering, in addition to the sintering temperature, played pivotal roles in controlling the synthesis of $C_5S_2S^-$. Through the careful control of partial pressures (1175 °C, 72 h), they successfully synthesized $C_5S_2S^-$ samples with a purity as high as 98% in a single step [5]. Likewise, Liu et al. employed a controlled partial pressure method, utilizing reagent-grade $CaCO_3$, $CaSO_4 \cdot H_2O$, and SiO_2, resulting in $C_5S_2S^-$ samples with up to a 96.3% purity after 12 h of sintering at 1200 °C [2]. Wang et al. opted for a one-step calcination method within their hydrothermal synthesis approach, subjecting raw materials to a hydrothermal reaction at 120 °C for 3 h in a hydrothermal reactor, followed by compression and a subsequent 2 h sintering phase. They determined that the optimal sintering temperature for this method was 1150 °C, achieving a purity level of 94.9% for $C_5S_2S^-$ [29].

In their research [2], Liu et al. observed that the formation of $C_5S_2S^-$ occurs through a reaction in which SiO_2 serves as the core component. At 1100 °C, SiO_2 forms the outer layer, which is subsequently enveloped by calcium silicate(CS) and C_2S. Notably, the formation of CS as a precursor to C_2S precedes the formation of C_2S. As the temperature is raised to 1200 °C, and with the extension of sintering time, all the previously generated C_2S and the outer layer of CS gradually transform into $C_5S_2S^-$, starting from the outermost layer and progressing inward. However, it is important to note that Liu et al.'s study did not account for the influence of the raw material particle size on the reaction, which can significantly affect the reaction activation energy and specific surface area. Both of these factors collectively impact the overall reaction efficiency. Simultaneously, in an aluminum-rich environment, the hydration rate of $C_5S_2S^-$ falls between that of C_2S and $C_4A_3S^-$. This property can be harnessed to bridge the substantial discrepancy in the hydration reaction rates between C_2S and $C_4A_3S^-$, ultimately improving the compressive strength at mid ages in T-CSA cement. Consequently, in this experiment, four different particle sizes of SiO$_2$ were deliberately selected to systematically investigate the impact of varying SiO$_2$ particle sizes, acting as the reaction core, on the formation of C$_5$S$_2$S$^-$. So, in this experiment, four different particle sizes of SiO$_2$ were selected to investigate how SiO$_2$, as the reaction nucleus, affects the purity and crystallinity of C$_5$S$_2$S$^-$. Additionally, the influence of ternesite prepared from SiO$_2$ with different particle sizes on the hydration performance of T-CSA cement was also studied.

2. Experimental Section

2.1. Materials and Samples Preparation

Given the chemical formula of C$_5$S$_2$S as Ca$_5$(SiO$_4$)$_2$(SO$_4$), it is evident that the ratio of Ca, Si, and S is 5:2:1. This allows us to determine the ratio of raw materials as CaCO$_3$:SiO$_2$:CaSO$_4$ = 4:2:1. In this experiment, high-purity reagent-grade CaCO$_3$ and CaSO$_4$ were chosen as raw materials, both of which had purity levels of 99+%. Additionally, four different types of high-purity SiO$_2$ with varying particle sizes were selected as variables. These reagents were sourced from Beijing InnoChem Science & Technology Co. Ltd., Beijing, China. Figure 1 illustrates the particle size distribution curves for the SiO$_2$ with varying particle sizes.

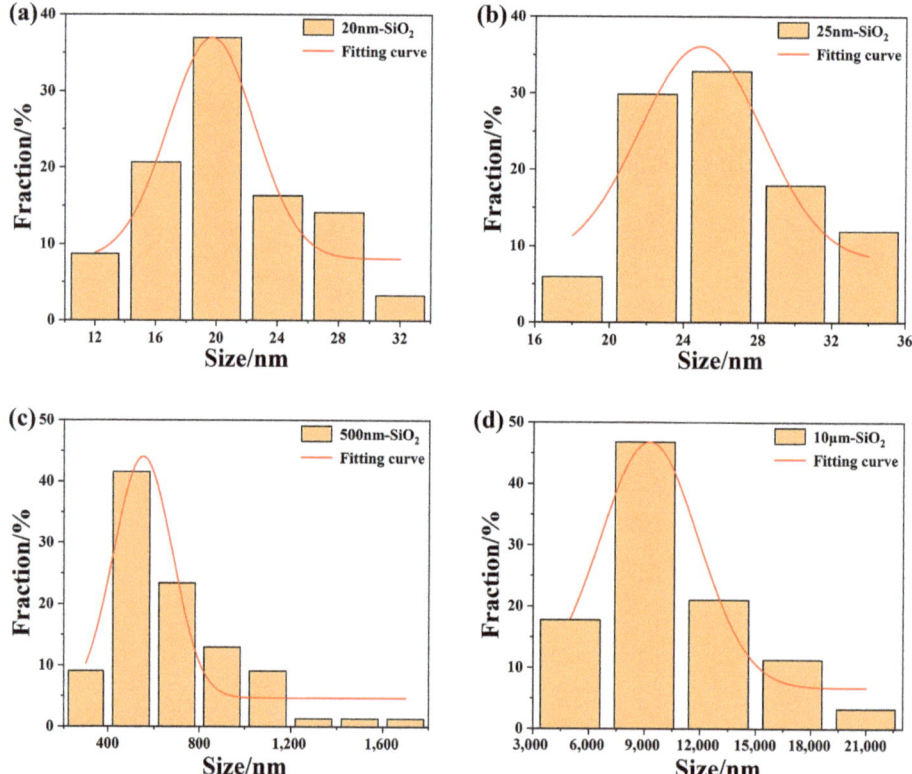

Figure 1. Particle size distribution curves of SiO$_2$ with different particle size grades. (**a**) 20 nm SiO$_2$; (**b**) 25 nm SiO$_2$; (**c**) 500 nm SiO$_2$; and (**d**) 10 μm SiO$_2$.

In accordance with the ratio of CaCO$_3$:SiO$_2$:CaSO$_4$ = 4:2:1, the constituent raw materials were introduced into a ball mill and alcohol was used as a grinding medium for the comminution process. This milling process lasted for a period of 3 h, resulting in a homogeneous blend of the raw materials. Subsequently, the treated raw materials were removed from the ball mill and subjected to a drying process in an oven set at 50 °C for a duration of 24 h. During this experiment, the partial pressure of SO_2 was carefully controlled under one-step sintering conditions. The thoroughly mixed raw material was then placed into a mold and compressed into a tablet with a diameter of 2 cm and a thickness of 3 mm at a pressure of 15 Mpa [30]. Following the compression of the tablet, the SO_2 partial pressure was controlled as depicted in Figure 2. The tablet was then placed into a muffle furnace and maintained at the target temperatures, ranging from 1100 °C to 1200 °C (1100 °C, 1110 °C, 1120 °C... 1190 °C, and 1200 °C). The heating rate employed for this process was 3.3 °C/min, maintained at the target temperature for 72 h with a cooling rate of 6 °C/min.

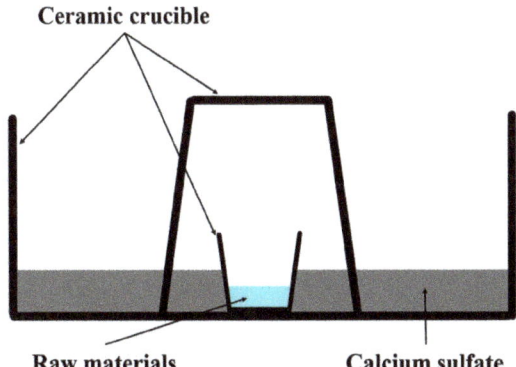

Figure 2. A schematic of the experimental set-up used for the synthesis of $C_5S_2S^-$.

2.2. Characterization

2.2.1. X-ray Diffraction (XRD) Characterization and Analysis

To explore the impact of the SiO_2 with varying particle sizes on the formation of $C_5S_2S^-$, the samples underwent a thorough comminution process followed by an X-ray Diffraction (XRD) analysis. The XRD analysis was conducted using a Bruker D8 Advance XRD diffractometer(Bruker Corporation, Karlsruhe, Germany), and the tests were carried out at room temperature. The XRD tests for $C_5S_2S^-$ involved scanning in the 10°–80° (2θ) range with a step size of 0.02° and a scanning rate of 2°/min. This analytical approach allowed for the assessment of the crystallographic structure and phase composition of the samples, providing valuable insights into the formation of $C_5S_2S^-$ and any potential variations associated with different SiO_2 particle sizes.

The prepared samples were subjected to a quantitative analysis using the Rietveld method with the assistance of the FullProf 64b software. In this analytical process, the data obtained from the XRD experiments were fitted using the Pearson VII function. Additionally, the crystal structure files from Table 1 were refined based on information from the ICSD (Inorganic Crystal Structure Database) database. This methodology allowed for a detailed examination and refinement of the crystallographic properties and structures of the samples, providing a comprehensive understanding of their composition and characteristics.

Table 1. ICSD codes used for Rietveld refinements.

Mineral Name	Phase	ICSD Codes
Ternesite	$C_5S_2S^-$	85123
Belite	C_2S	79552
Anhydrite	$CaSO_4$	40043
Lime	CaO	52783

2.2.2. Scanning Electron Microscope (SEM) and Energy Dispersive Spectrometer (EDS) Analysis

A HITACHI Regulus8100 scanning electron microscope(HITACHI, Tokyo, Janpan) was used to characterize the microscopic morphology of the prepared samples with an accelerating voltage of 5 kV and a working distance of 8.3 mm.

2.2.3. Thermal Analysis

Differential thermal analysis-thermogravimetry (DTA-TG) was tested using a NETZSCH STA2500 Differential Thermal Analyzer(NETZSCH, Free State of Bavaria, Germany) with a ramp rate of 10 °C/min under a nitrogen atmosphere.

2.2.4. Hydration Heat

An eight-channel isothermal calorimeter of TMA AIR was used to test the heat of hydration of the CSA cement blended with 10% of the sample with a water–cement ratio of 0.4.

3. Results and Discussion

3.1. Formation of Ternesite

After 72 h of sintering, a light green colored sample was prepared. To confirm the successful preparation of $C_5S_2S^-$ and assess the purity of $C_5S_2S^-$, the samples were subjected to an XRD analysis. A quantitative analysis was conducted using the Rietveld method with the assistance of the full prof software, as illustrated in Figures 3 and 4. The obtained Rwp values fell within the acceptable range [31]. From Figure 3, it can be observed that the purity of the $C_5S_2S^-$ samples synthesized using the 500 nm SiO_2 particles (referred to as the 500 nm_1170 °C_sample) reached an impressive 99.47% at a sintering temperature of 1170 °C. Furthermore, the highest purities of $C_5S_2S^-$ among the samples prepared using SiO_2 particles of different sizes were as follows: 87.22% for the 20 nm_1160 °C_sample, 94.02% for the 25 nm_1140 °C_sample, and 67.01% for the 10 μm_1170 °C_sample, respectively. Additionally, the presence of CS [32] was solely detected in the sample with the 10 μm particle size SiO_2 as the raw material. The content of CS amounted to a mere 0.15%, and this occurred only after an increase in the sintering temperature to 1120 °C. These findings align with Liu's research [2], indicating that CS acts as a precursor to C_2S and is fully converted into C_2S upon further elevation of the sintering temperature. Conversely, no traces of CS were detected in the $C_5S_2S^-$ samples derived from the other three particle sizes of SiO_2. This discrepancy may be attributed to the significantly larger particle size and reduced specific surface area of the 10 μm SiO_2, resulting in an elevated reaction activation energy and, consequently, a reduced reaction efficiency. In general, the purity of $C_5S_2S^-$ in the 10 μm sample was consistently low across the sintering temperatures ranging from 1100 °C to 1200 °C. This supports the notion of a low reaction efficiency associated with SiO_2 particles of a 10 μm size. On the other hand, when comparing the samples prepared from the SiO_2 particles of different sizes, the 500 nm particle size exhibited the highest purity at all sintering temperatures. Notably, the 500 nm_1170 °C_sample achieved an impressive purity of 99.47%.This intriguing phenomenon leads to speculation: the 500 nm particle size SiO_2 possessed a lower reaction activation energy when compared to the 10 μm particle size SiO_2. Additionally, compared to the SiO_2 particles of 20 nm and 25 nm sizes, the 500 nm particle size SiO_2 offered a larger reaction contact area. Consequently, under the synergistic effect of these two factors, the 500 nm SiO_2 outperformed the other three particle sizes in terms of reaction efficiency, making it the most suitable choice for the preparation of $C_5S_2S^-$. In addition, except for the 500 nm SiO_2 sample, CaO was observed in all other samples. This is a product of the decomposition of $CaCO_3$ at high temperatures. Due to an inadequate reaction activation energy and reaction contact area for the SiO_2 with particle sizes of 20 nm, 25 nm, and 10 μm, the raw materials did not react fully, resulting in the presence of residual CaO. Meanwhile, the 10 μm SiO_2 exhibited the smallest specific surface area and highest reaction activation energy, leading to the lowest purity of $C_5S_2S^-$ in the samples it produced.

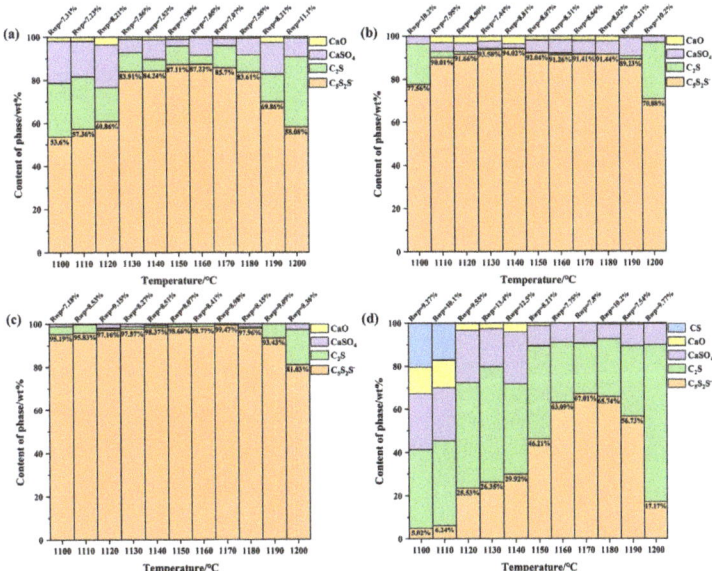

Figure 3. Refinement results for samples prepared with different particle sizes of SiO_2 as raw material. (**a**) 20 nm SiO_2; (**b**) 25 nm SiO_2; (**c**) 500 nm SiO_2; and (**d**) 10 μm SiO_2.

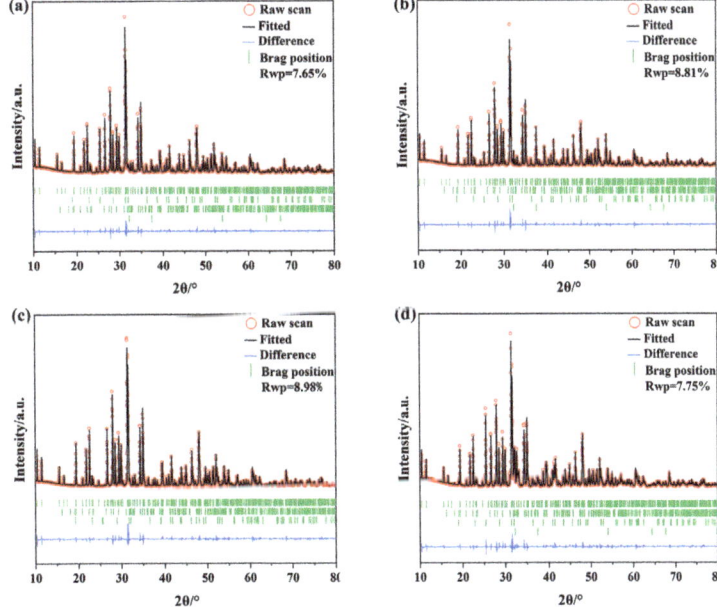

Figure 4. Fitted plots of the highest purity samples of ternesite prepared from SiO_2 of different particle sizes. (**a**) 20 nm_1160 °C_sample; (**b**) 25 nm_1140 °C_sample; (**c**) 500 nm_1170 °C_sample; and (**d**) 10 μm_1170 °C_sample.

3.2. SEM/EDS

From Figure 5a–c, it can be observed that the morphology of the $C_5S_2S^-$ prepared in this experiment bears a resemblance to the samples prepared by Skalamprinos et al. [7], exhibiting an irregular granular appearance. The 10 μm_1170 °C_sample shown in Figure 5d contains numerous spherical and variably sized block-like particles. The EDS analysis in Figure 6d reveals that the central parts of these spherical particles are primarily composed of Ca and Si elements (note that Figures 5 and 6 are not the same location). Considering the sintering conditions and raw materials used, it can be inferred that the main constituent of the central parts of these spherical particles is C_2S. At the edges of the spherical particles, three elements are concurrently present: Ca, Si, and S. This suggests the formation of $C_5S_2S^-$ at the periphery of the spherical particles. This is in accordance with the findings of Liu et al. [2], where the formation of $C_5S_2S^-$ occurred through a reaction initiated at the core of SiO_2. With an increase in the sintering temperature and a prolonged sintering duration, the C_2S enveloping the exterior of SiO_2 reacts with $CaSO_4$ to form $C_5S_2S^-$. The EDS results for the samples prepared with the other three particle sizes of SiO_2 indicate that the distribution of S elements aligns with the distribution of Ca and Si. This is attributed to the higher specific surface area of the smaller SiO_2 particles, which is more conducive to the reaction leading to the formation of $C_5S_2S^-$. Smaller SiO_2 particles facilitate extensive contact and reaction with other raw materials, resulting in an even distribution of S elements in the product. In conclusion, in conjunction with the refinement results in Figure 3, it can be inferred that the formation process of $C_5S_2S^-$ begins with SiO_2 as the nucleus, where CS is formed first on the outer layer, followed by the formation of C_2S, and finally, C_2S reacts with $CaSO_4$ to produce $C_5S_2S^-$. Using SiO_2 as the reaction nucleus imposes requirements on the particle size and specific surface area of SiO_2, making 500 nm SiO_2 suitable for the preparation of $C_5S_2S^-$.

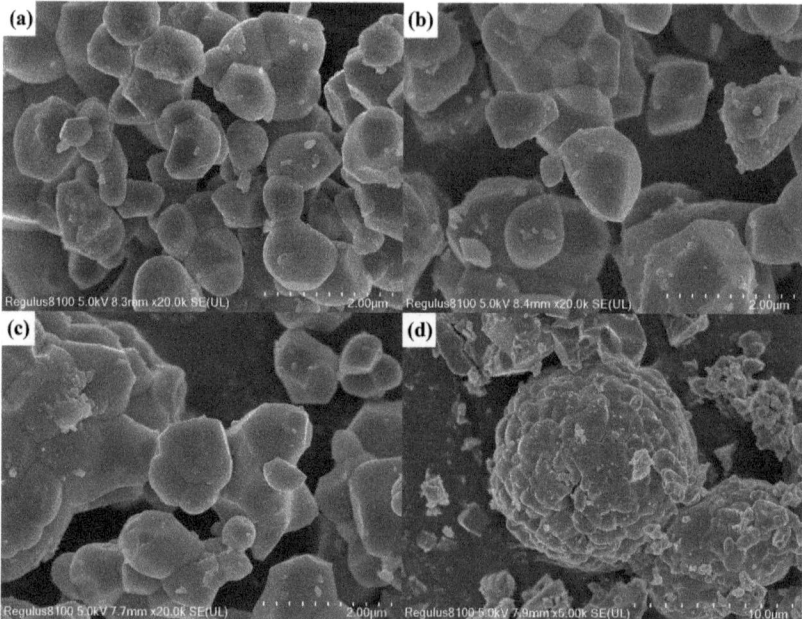

Figure 5. The microscopic morphology of $C_5S_2S^-$ prepared with SiO_2 of different particle sizes. (**a**) 20 nm_1160 °C_sample; (**b**) 25 nm_1140 °C_sample; (**c**) 500 nm_1170 °C_sample; and (**d**) 10 μm_1170 °C_sample.

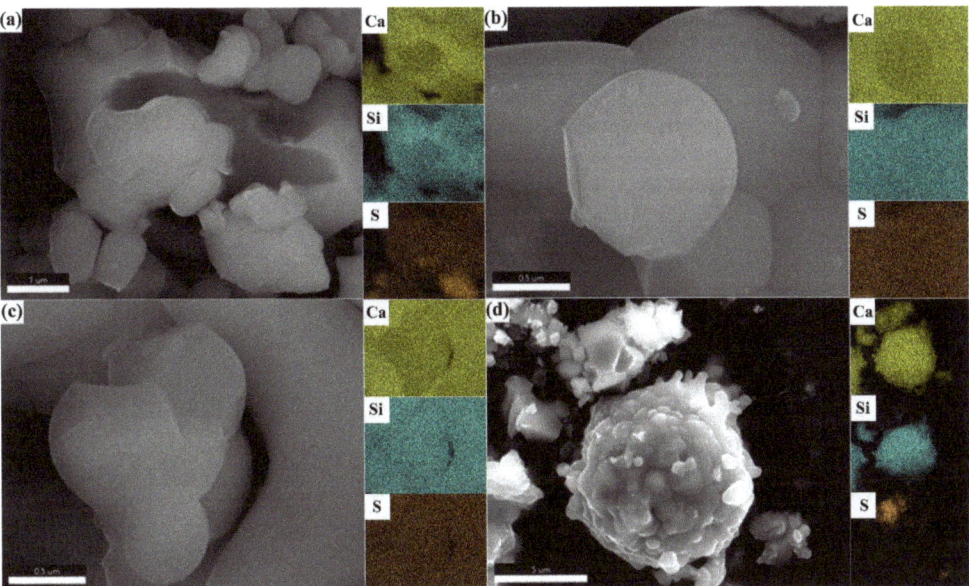

Figure 6. The EDS results of $C_5S_2\bar{S}$ prepared with SiO_2 of different particle sizes. (**a**) 20 nm_1160 °C_sample; (**b**) 25 nm_1140 °C_sample; (**c**) 500 nm_1170 °C_sample; and (**d**) 10 μm_1170 °C_sample.

3.3. Thermal Analysis

Due to the relatively low purity of the $C_5S_2\bar{S}$ (67.01%) in the sample prepared from SiO_2 particles with a size of 10 μm, this particular sample was excluded from a direct comparative weight loss analysis with the high-purity samples. Consequently, only three samples with a high $C_5S_2\bar{S}$ purity, namely the 20 nm_1160 °C_sample, 25 nm_1140 °C_sample, and 500 nm_1170 °C_sample, were selected for a thermogravimetric analysis. Hanein et al. conducted a similar experimental setup, sintering the raw material in a two-stage process at 1175 °C for 3 days and 1 day, respectively [5]. This resulted in the preparation of a $C_5S_2\bar{S}$ sample with a purity of 99% and an upper limit decomposition temperature of approximately 1290 °C. Furthermore, Liu et al. demonstrated that the thermal stability of $C_5S_2\bar{S}$ is directly proportional to the holding time during sintering [33]. Longer sintering durations promote an improved crystal development in $C_5S_2\bar{S}$, thereby enhancing the thermal stability of the samples. In the current experiment, the holding time for all the samples during sintering was 72 h. Therefore, the particle size of the SiO_2 employed in the preparation of $C_5S_2\bar{S}$ emerges as the primary determining factor for the thermal stability of the resulting samples. It appears that, in the thermogravimetry test shown in Figure 7a, there is minimal difference in the weight loss between the 20 nm_sample and 25 nm_sample, while the 500 nm_sample experiences a relatively low weight loss. The refinement results in Figure 3 indicate that the main impurities in the sample are $CaSO_4$ and C_2S. The decomposition temperature of $CaSO_4$ falls within the range of 1350–1400 °C, while the decomposition temperature of C_2S is above 1400 °C. Therefore, a temperature of 1340 °C was chosen for the calculation of the weight loss. At this temperature, only $C_5S_2\bar{S}$ undergoes decomposition. Based on these findings, the 500 nm_1170 °C_sample demonstrates a high purity and superior thermal stability. The sintering conditions of 1170 °C for 72 h are found to be particularly suitable for the 500 nm particle size SiO_2 to undergo the necessary reaction for the formation of $C_5S_2\bar{S}$ and crystal development. This combination results in a higher purity and optimal crystal development, thereby providing the best high-temperature stability. Consequently, the 500 nm_1170 °C_sample is less likely to decompose at high temperatures and exhibit mass loss in the form of SO_2

gas [2] (as indicated by "Equation (1)"). Additionally, the decomposition temperature of the 500 nm_1170 °C_sample, as revealed by the DTA (differential thermal analysis) results shown in Figure 7b, is the highest among the samples, estimated to be around 1280 °C. This further supports the superior thermal stability of the 500 nm_1170 °C_sample. In summary, based on the thermogravimetric and refinement results, along with the DTA analysis, the 500 nm_1170 °C_sample exhibits a high purity, excellent crystal development, and superior thermal stability, making it less prone to decomposition at high temperatures.

$$Ca_5(SiO_4)_2SO_4 \rightarrow 2Ca_2SiO_4 + CaO + SO_2 + 1/2O_2 \qquad (1)$$

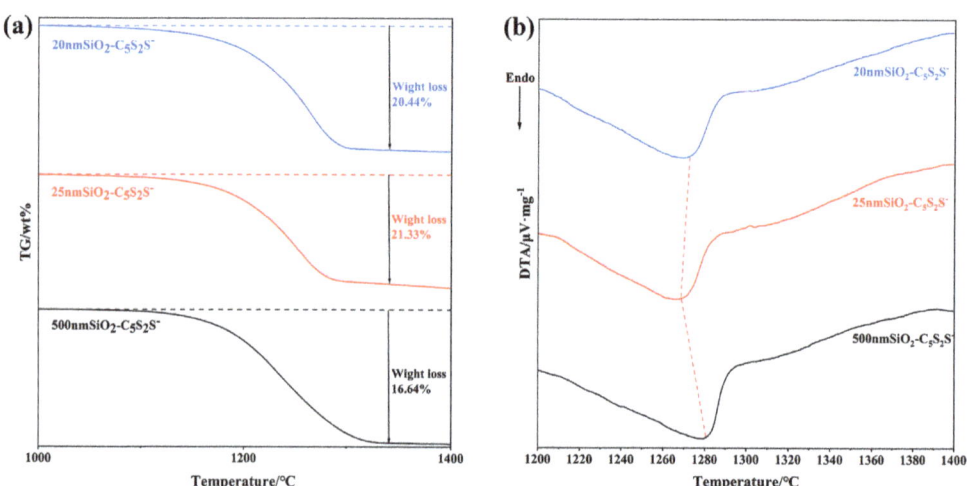

Figure 7. (a) TG curves and (b) DTA curves of 20 nm_1160 °C_sample, 25 nm_1140 °C_sample, and 500 nm_1170 °C_sample.

3.4. Hydration Heat

After Montes et al. demonstrated that $C_4A_3S^-$ can effectively stimulate the hydration activity of $C_5S_2S^-$ [15], Zhang et al. further investigated this phenomenon [27]. Their study revealed that $C_5S_2S^-$ with a higher crystallinity exhibits poorer early hydration activity, while $C_5S_2S^-$ with a lower crystallinity displays higher early hydration activity. In Figure 8a, it is evident that the sample represented by 500 nm SiO_2 particles has the lowest exothermal peak intensity and the latest formation time compared to the other three particle sizes. This is attributed to the highest purity of the samples being prepared from the 500 nm SiO_2 particles, sintered at 1170 °C for 72 h, representing the most suitable sintering conditions for the formation and development of $C_5S_2S^-$ crystals. In other words, the 500 nm_1700 °C_sample exhibits the best crystallinity, indicating lower early hydration activity. The cumulative heat curves in Figure 8b also support this observation, as the 500 nm sample exhibits the lowest total exothermic amount, while the exothermic amounts of the other three groups of samples are relatively similar, further confirming the lower early hydration activity of the 500 nm_1700 °C_sample. Additionally, the three groups of samples prepared with the 20 nm, 25 nm, and 10 μm SiO_2 particles show a shoulder peak before 1 h. This is attributed to the presence of a small amount of CaO impurities in the $C_5S_2S^-$ prepared with these SiO_2 particle sizes (20 nm, 25 nm, and 10 μm). These impurities react exothermically with water to form $Ca(OH)_2$, leading to the appearance of the shoulder peak. The intensity of the shoulder peak is influenced by the CaO content. The 25 nm_1140 °C_sample has the highest CaO content, leading to the highest intensity of its shoulder peak. In summary, due to its superior crystallinity, the 500 nm_1700 °C_sample exhibits the slowest early hydration rate and the lowest hydration reaction intensity, resulting in the lowest cumulative heat. The slower hydration rate and lowest cumulative heat of 500 nmSiO$_2$-T-CSA can mitigate, to some extent, the disadvantage of poor dimensional stability in alumina cement.

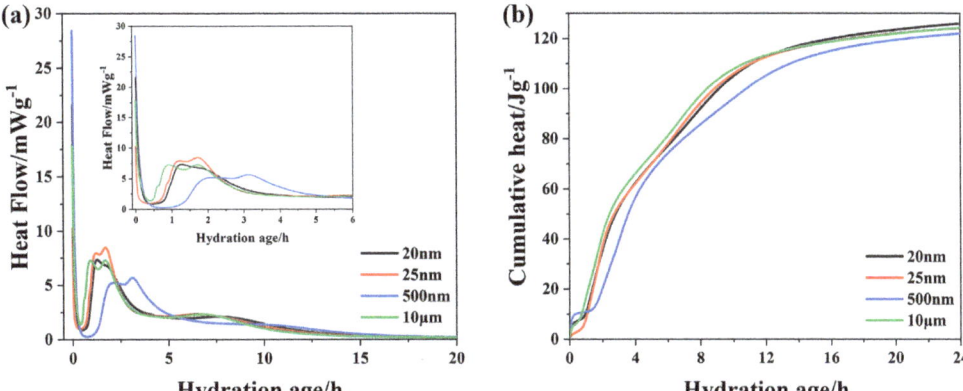

Figure 8. (**a**) Heat of hydration curves and (**b**) cumulative heat curves of C$_5$S$_2$S$^-$ preparation with different particle sizes of SiO$_2$ in CSA cement—20 nm_1160 °C_sample; 25 nm_1140 °C_sample; 500 nm_1170 °C_sample; and 10 μm_1170 °C_sample.

4. Conclusions

In this experiment, we addressed previously overlooked aspects by utilizing SiO$_2$ of varying particle sizes as raw materials to investigate their impact on the formation of C$_5$S$_2$S$^-$, changes in morphology, thermal stability, and the influence on the hydration heat of T-CSA cement. This study yields the following conclusions:

I. Different particle sizes of SiO$_2$ have a significant impact on the formation of C$_5$S$_2$S$^-$. Specifically, C$_5$S$_2$S$^-$ prepared using 500 nm SiO$_2$ as a raw material can achieve a purity as high as 99.47%, whereas C$_5$S$_2$S$^-$ prepared with 10 μm SiO$_2$ as a raw material exhibits a purity of only 67.01%. Additionally, the presence of CS is detected in the samples prepared with 10 μm SiO$_2$.

II. In the samples prepared using SiO$_2$ particles with sizes of 20 nm, 25 nm, and 500 nm, S elements are evenly distributed within each particle, facilitating the relatively smooth formation of C$_5$S$_2$S$^-$. Conversely, in the samples prepared with 10 μm SiO$_2$ particles, C$_5$S$_2$S$^-$ containing S elements coats the edges of the spherical particles dominated by SiO$_2$ and C$_2$S. Due to the influence of the SiO$_2$'s specific surface area, the formation of C$_5$S$_2$S$^-$ is relatively challenging.

III. In the TG and DTA tests, the 500 nm_1700 °C_sample exhibits the best thermal stability, primarily due to its superior crystallinity. In the hydration heat test, also influenced by the better crystallinity of the 500 nm_1700 °C_sample, CSA cement doped with this sample displays the slowest hydration rate and lowest cumulative heat release.

In this experiment, the optimal SiO$_2$ particle size and temperature for ternesite preparation were identified, and the underlying reaction mechanisms were analyzed. Additionally, the hydration performance of ternesite was studied for potential applications, offering valuable insights for future research on low-carbon construction materials and alumina cement.

Author Contributions: Conceptualization, X.L. and Y.W.; methodology, X.L. and F.S.; validation, X.L. and Y.W.; formal analysis, F.S.; investigation, X.L. and F.S.; resources, X.L.; writing—original draft preparation, F.S.; writing—review and editing, F.S. and D.H.; visualization, D.S.; supervision, F.S.; funding acquisition, D.S. All authors have read and agreed to the published version of the manuscript.

Funding: The National Natural Science Foundation of China (No. 52208265).

Institutional Review Board Statement: Not applicable.

Informed Consent Statement: Not applicable.

Data Availability Statement: Not applicable.

Conflicts of Interest: The authors declare no conflict of interest.

References

1. Huang, Y.; Dong, D.; Wang, X.; Zhang, Z.; Zhao, P.; Cui, N.; Lu, L. Ternesite-calcium sulfoaluminate cement: Preparation and hydration. *Constr. Build. Mater.* **2022**, *344*, 128187. [CrossRef]
2. Liu, L.; Zhang, W.; Ren, X.; Ye, J.; Zhang, J.; Qian, J. Formation, structure, and thermal stability evolution of ternesite based on a single-stage sintering process. *Cem. Concr. Res.* **2021**, *147*, 106519. [CrossRef]
3. Sherman, N.; Beretka, J.; Santoro, L.; Valenti, G.L. Long-term behaviour of hydraulic binders based on calcium sulfoaluminate and calcium sulfosilicate. *Cem. Concr. Res.* **1995**, *25*, 113–126. [CrossRef]
4. Dickens, B.; Brown, W.E. The Crystal Structure of $Ca_5(PO_4)_2SiO_4$ (Silieo-Carnotite). *Miner. Petrol.* **1971**, *16*, 1–27. [CrossRef]
5. Hanein, T.; Galan, I.; Glasser, F.P.; Skalamprinos, S.; Elhoweris, A.; Imbabi, M.S.; Bannerman, M.N. Stability of ternesite and the production at scale of ternesite-based clinkers. *Cem. Concr. Res.* **2017**, *98*, 91–100. [CrossRef]
6. Shen, Y.; Wang, P.; Chen, X.; Zhang, W.; Qian, J. Synthesis, characterisation and hydration of ternesite. *Constr. Build. Mater.* **2021**, *270*, 121392. [CrossRef]
7. Skalamprinos, S.; Jen, G.; Galan, I.; Whittaker, M.; Elhoweris, A.; Glasser, F. The synthesis and hydration of ternesite, $Ca_5(SiO_4)_2SO_4$. *Cem. Concr. Res.* **2018**, *113*, 27–40. [CrossRef]
8. Beretka, J.; de Vito, B.; Santoro, L.; Sherman, N.; Valenti, G.L. Utilisation of industrial wastes and by-products for the synthesis of special cements. *Resour. Conserv. Recycl.* **1993**, *9*, 179–190. [CrossRef]
9. Beretka, J.; de Vito, B.; Santoro, L.; Sherman, N.; Valenti, G.L. Hydraulic behaviour of calcium sulfoaluminate-based cements derived from industrial process wastes. *Cem. Concr. Res.* **1993**, *23*, 1205–1214. [CrossRef]
10. Bullerjahn, F.; Zajac, M.; Ben Haha, M. CSA raw mix design: Effect on clinker formation and reactivity. *Mater. Struct.* **2015**, *48*, 3895–3911. [CrossRef]
11. Al Horr, Y.; Elhoweris, A.; Elsarrag, E. The development of a novel process for the production of calcium sulfoaluminate. *Int. J. Sustain. Built Environ.* **2017**, *6*, 734–741. [CrossRef]
12. Adolfsson, D.; Menad, N.; Viggh, E.; Björkman, B. Steelmaking slags as raw material for sulphoaluminate belite cement. *Adv. Cem. Res.* **2007**, *19*, 147–156. [CrossRef]
13. Pliego-Cuervo, Y.B.; Glasser, F.P. The role of sulphates in cement clinkering Subsolidus phase relations in the system CaO-Al_2O_3-SiO_2-SO_3. *Cem. Concr. Res.* **1979**, *9*, 51–55. [CrossRef]
14. Carmona, Q.P.M.; Montes, M.; Pato, E.; Fernández, J.A.; Blanco, V.M.T. Study on the activation of ternesite in $CaO·Al_2O_3$ and $12CaO·7Al_2O_3$ blends with gypsum for the development of low-CO_2 binders. *J. Clean. Prod.* **2020**, *291*, 125726. [CrossRef]
15. Montes, M.; Pato, E.; Carmona-Quiroga, P.M.; Blanco-Varela, M.T. Can calcium aluminates activate ternesite hydration? *Cem. Concr. Res.* **2018**, *103*, 204–215. [CrossRef]
16. Ben, H.M.; Bullerjahn, F.; Zajac, M. On the Reactivity of Ternesite. In Proceedings of the 14th International Congress on the Chemistry of Cement, Beijing, China, 13–16 October 2015; p. 1.
17. Dienemann, W.; Schmitt, D.; Bullerjahn, F.; Haha, M.B. Belite-Calciumsulfoaluminate-Ternesite (BCT)—A new low-carbon clinker Technology. *Cem. Int.* **2013**, *11*, 100–109.
18. Maheswaran, S.; Kalaiselvam, S.; Karthikeyan, S.K.S.S.; Kokila, C.; Palani, G.S. β-Belite cements (β-dicalcium silicate) obtained from calcined lime sludge and silica fume. *Cem. Concr. Compos.* **2016**, *66*, 57–65. [CrossRef]
19. Hargis, C.W.; Telesca, A.; Monteiro, P.J.M. Calcium sulfoaluminate (Ye'elimite) hydration in the presence of gypsum, calcite, and vaterite. *Cem. Concr. Res.* **2014**, *65*, 15–20. [CrossRef]
20. Bullerjahn, F.; Schmitt, D.; Haha, M.B. Effect of raw mix design and of clinkering process on the formation and mineralogical composition of (ternesite) belite calcium sulphoaluminate ferrite clinker. *Cem. Concr. Res.* **2014**, *59*, 87–95. [CrossRef]
21. Shen, Y.; Qian, J.; Huang, Y.; Yang, D. Synthesis of belite sulfoaluminate-ternesite cements with phosphogypsum. *Cem. Concr. Compos.* **2015**, *63*, 67–75. [CrossRef]
22. Shen, Y.; Chen, X.; Zhang, W.; Li, X.; Qian, J. Influence of ternesite on the properties of calcium sulfoaluminate cements blended with fly ash. *Constr. Build. Mater.* **2018**, *193*, 221–229. [CrossRef]
23. Böhme, N.; Hauke, K.; Neuroth, M.; Geisler, T. In Situ Hyperspectral Raman Imaging of Ternesite Formation and Decomposition at High Temperatures. *Minerals* **2020**, *10*, 287. [CrossRef]
24. Dvořák, K.; Gazdič, D.; Fridrichová, M. The influence of firing parameters on the crystallinity of ternesite. *J. Cryst. Growth* **2020**, *542*, 125691. [CrossRef]
25. Hou, P.; Qian, J.; Wang, Z.; Deng, C. Production of quasi-sulfoaluminate cementitious materials with electrolytic manganese residue. *Cem. Concr. Compos.* **2012**, *34*, 248–254. [CrossRef]
26. Jing, G.; Zhang, J.; Lu, X.; Xu, J.; Gao, Y.; Wang, S.; Cheng, X.; Ye, Z. Comprehensive evaluation of formation kinetics in preparation of ternesite from different polymorphs of Ca_2SiO_4. *J. Solid State Chem.* **2020**, *292*, 121725. [CrossRef]

27. Zhang, W.; Liu, L.; Ren, X.; Ye, J.; Zhang, J.; Cao, L.; An, N.; Qian, J. The Impact of Calcination Regimen on the Structure and Hydration Performance of Calcium Silicate Sulfate. *J. Chin. Ceram. Soc.* **2022**, *10*, 2712–2721. [CrossRef]
28. Zhang, W.; Liu, L.; Ren, X.; Zhang, H.; Ye, J.; Zhang, J.; Cao, L.; An, N.; Qian, J. Mechanisms of Alkali Metal Doping Effects on the Hydration Performance of Calcium Silicate Sulfate. *J. Chin. Ceram. Soc.* **2023**, *51*, 290–302. [CrossRef]
29. Wang, X.; Guo, W.; Hu, Y.; Chen, Q.; Qiu, J.; Li, Z.; Chen, J.; Guan, R. Synthesis and Hydration Performance Study of Calcium Silicate Sulfate. *Mater. Rep.* **2020**, *34*, 169–172, 188.
30. Shi, L.; Zhao, Y.; Xue, J.; Zhang, X. Effect of Particle Size and Compaction Pressure on Sintering Behavior of α-Al_2O_3 Powder. *Dev. Appl. Mater.* **2020**, *35*, 1–4.
31. Toby, B.H. R factors in rietveld analysis how good is good enough. *Powder Diffr.* **2012**, *21*, 67–70. [CrossRef]
32. Papynov, E.K.; Shichalin, O.O.; Apanasevich, V.I.; Portnyagin, A.S.; Yu, M.V.; Yu, B.I.; Merkulov, E.B.; Kaidalova, T.A.; Modin, E.B.; Afonin, I.S.; et al. Sol-gel (template) synthesis of osteoplastic $CaSiO_3$/HAp powder biocomposite: "In vitro" and "in vivo" biocompatibility assessment. *Powder Technol.* **2020**, *367*, 762–773. [CrossRef]
33. Liu, L.; Zhang, W.; Ren, X.; Ye, J.; Zhang, J.; Luo, Z.; Qian, J. Sintering behaviour and structure-thermal stability relationships of alkali-doped ternesite. *Cem. Concr. Res.* **2023**, *164*, 107043. [CrossRef]

Disclaimer/Publisher's Note: The statements, opinions and data contained in all publications are solely those of the individual author(s) and contributor(s) and not of MDPI and/or the editor(s). MDPI and/or the editor(s) disclaim responsibility for any injury to people or property resulting from any ideas, methods, instructions or products referred to in the content.

Article

Eco-Innovative Concrete for Infrastructure Obtained with Alternative Aggregates and a Supplementary Cementitious Material (SCM)

Ofelia Corbu [1,2], Attila Puskas [1,*], Mihai-Liviu Dragomir [1,*], Nicolae Har [3] and Ionuț-Ovidiu Toma [4]

[1] Faculty of Civil Engineering, Technical University of Cluj-Napoca, 28 Memorandumului Street, 400114 Cluj-Napoca, Romania; ofelia.corbu@staff.utcluj.ro
[2] Research Institute for Construction Equipment and Technology, ICECON S.A. Bucharest, 266, Pantelimon Road, 2nd District, CP 3-33, 021652 Bucharest, Romania
[3] Faculty of Biology and Geology, Babeș-Bolyai University, 400347 Cluj-Napoca, Romania; nicolae.har@ubbcluj.ro
[4] Faculty of Civil Engineering and Building Services, The "Gheorghe Asachi" Technical University of Iasi, 700050 Iași, Romania; ionut.ovidiu.toma@tuiasi.ro
* Correspondence: attila.puskas@dst.utcluj.ro (A.P.); mihai.dragomir@cfdp.utcluj.ro (M.-L.D.)

Abstract: Concrete is a heterogeneous material, one of the most widely used materials on the planet, and a major consumer of natural resources. Its carbon emissions are largely due to the extensive use of cement in its composition, which contributes to 7% of global CO_2 emissions. Extraction and processing of aggregates is another source of CO_2 emissions. Many countries have succeeded in moving from a linear economy to a circular economy by partially or fully replacing non-renewable natural materials with alternatives from waste recycling. One such alternative consists of partially replacing cement with supplementary cementitious materials (SCMs) in concrete mixes. Thus, this work is based on the experimental investigation of the fresh and hardened properties of road concrete in which crushed river aggregates were replaced with recycled waste aggregates of uncontaminated concrete. At the same time, partial replacement of cement with a SCM material in the form of glass powder improved the durability characteristics of this sustainable concrete. The microstructure and compositional features of the selected optimum mix have also been investigated using polarized light optical microscopy (OM), scanning electron microscopy (SEM), and X-ray diffraction by the Powder method (PXRD) for the qualitative analysis of crystalline constitutive materials.

Keywords: concrete waste; alternative aggregate; supplementary cementitious material; SEM; PXRD

Citation: Corbu, O.; Puskas, A.; Dragomir, M.-L.; Har, N.; Toma, I.-O. Eco-Innovative Concrete for Infrastructure Obtained with Alternative Aggregates and a Supplementary Cementitious Material (SCM). *Coatings* **2023**, *13*, 1710. https://doi.org/10.3390/coatings13101710

Academic Editor: Valeria Vignali

Received: 17 August 2023
Revised: 20 September 2023
Accepted: 26 September 2023
Published: 28 September 2023

Copyright: © 2023 by the authors. Licensee MDPI, Basel, Switzerland. This article is an open access article distributed under the terms and conditions of the Creative Commons Attribution (CC BY) license (https://creativecommons.org/licenses/by/4.0/).

1. Introduction

Transitioning from "Linear Economy" [1] towards the waste management concept of European Directive 2008/98/EC [2] and the Industrial Emissions Directive, known as European Directive 2010/75/EU, in favor of "Circular Economy" [3,4] is an absolute necessity. The new paradigm is based on reducing the consumption of natural resources through the use of recycled waste from various industries and industrial by-products, which might become alternatives to the raw materials used in the construction industry. A direct, long term result will be the creation of more ecological construction products. The uncontrolled consumption of natural resources may have led to the climate and economic changes we are already facing, which are the result of the still-existing linear economies in certain areas of the world [4]. The increasingly large volume of waste in landfills, the lack of storage space, and the stringent reduction of non-renewable natural resources are strong motivations for researchers to mobilize and support recyclers in finding innovative ways to use recycled waste as raw materials in new concrete mixes, which represents one of the most eco-friendly options [5–10].

Conventional concrete is the most widely used artificial material in the construction industry and worldwide, second only in total use to water. Its production involves the depletion of the natural resources and increased use of cement, the production of which requires significant amount of electrical energy and limestone, leading to significant environmental impacts and CO_2 generation in addition the high amount of released emissions [11–17].

Recycled wastes used in the production of concrete, such as glass, plastic, concrete, construction waste, rubber, and ceramics, may facilitate the transition from a linear economy to a circular one [4]. Extensive research has been conducted on concrete mixtures using glass powder as partial replacement for cement in conventional concrete [18]. It has proven to be an innovative method of supporting the recycling companies' in reducing of waste in landfills [19] and their transformation into final recyclers by putting into production the new eco-friendly concrete for various construction elements (paving curbs, hollow blocks, platforms, sidewalks) [20–23]. The recycling of concrete from demolitions or that obtained by crushing prefabs is equally welcomed. If the volume allows, concrete obtained from tests of samples subjected to destructive or non-destructive laboratory determinations, with or without a known track, can be used [24].

Numerous studies have been conducted worldwide with respect to the use of glass powder or recycled concrete aggregates individually. However, studies of the mixture containing both in a single composite are relatively limited in number.

Glass powder resulting from recycling (WGP—waste glass powder), used as a substitute for cement in proportions of 10–25% in concrete/mortar mixes, has been highlighted in recent studies [25–30]. It has also been used in geopolymer concrete in proportions of 20–80% [31–33]. Furthermore, studies have been conducted on the substitution of up to 60% of materials and the monitoring of strength development up to 90 days [34,35] focusing on the efficiency of pozzolanic activity of cementitious materials [32,36–39]. The properties of mortars with cathode ray tube (CRT) glass waste (Pb-containing) added in the amounts of 30%, 40%, and 50% by weight percent as replacement for river sand were also investigated. The experimental results indicated that all the samples containing glass waste achieved higher compressive strength compared to the control mortar. After 14, 28, and 42 days of maintaining the mortar in water, no evidence of Pb was detected in the solutions [40].

All these studies can encourage recyclers to create their own SMEs and to produce and implement such eco-friendly concretes in the production of various precast elements, like paving or curbs, Lego-type hollow blocks, sidewalks, bike lanes, industrial platforms, parking lots, etc., depending on the specific requirements and regulations of each region.

The purpose of this study is to evaluate the possibility of integrating waste materials such as WGP (waste glass powder) and RCA (recycled concrete aggregates) into concrete and quantify their influence on mechanical and durability properties, particularly in terms of abrasion resistance.

As a result, the findings of this study confirm a favorable contribution to the mechanical characteristics and abrasion resistance of road concrete when using WGP waste materials. The current use of WGP-RCA in concrete composition may redirect the construction industry towards a circular and sustainable economy in line with the main directions and requirements of EU.

2. Materials and Methods

2.1. Materials

An in-process flow diagram of substituting cement with WGP and coarse aggregates with PCA in the concrete mix is presented in Figure 1.

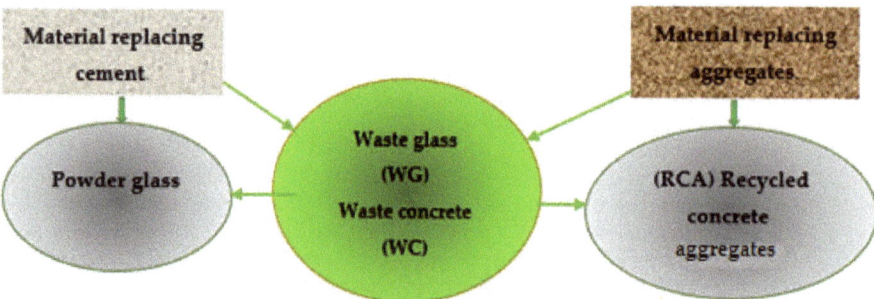

Figure 1. Diagram of raw material substitution.

2.1.1. Aggregates

The aggregates used for the control road concrete and for the eco-friendly road concrete are presented synthetically in Table 1.

Table 1. Types of aggregates used in the design of road concrete.

Size of Aggregate [mm]	Type of Aggregate
0/4	Natural river aggregate (NRA) for all mixes (gravel)
4/8	Crushed river aggregates (CRA) (crushed gravel)
4/8	Recycled concrete aggregates (RCA)
8/16 mm, 16/25	Crushed aggregates/chippings (CAC) for all mixes

In Figure 2, images of the aggregates used in the design for the control/reference concrete (Figure 2a) and the recycled coarse aggregate (4/8 mm) RCA that replaces the natural aggregate 4/8 mm (Figure 2b) are presented.

Figure 2. Aggregates used in the concrete mixes: (a) natural aggregates (NA), (b) recycled concrete aggregates (RCA).

In the present study, concrete mix design was carried out through careful analysis of the characteristics of the used aggregates, especially for the recycled aggregates [41], followed by the adjustment of appropriate proportions in such a way that the designed mixes comply with the requirements of the current standards in Romania for the production of road concrete [42,43].

Special attention was paid to the total aggregate volume, as it represents approximately 60%–80% of the concrete composition. Recycled concrete aggregate exhibits inconsistency in quality due to its various sources often lacking a known history. In this study, RCA (recycled concrete aggregate) comes from a local laboratory that consistently verifies the quality of the designed/implemented materials, ensuring that they have not been contaminated in storage. The RCA were obtained by crushing C25/30-concrete-class cubic specimens [41].

The design of concrete mixes took into consideration the widest possible applicability of alternative aggregates to replace traditional, natural aggregates. The popularity of the new concrete is increasing as it is used as a secondary raw material in construction products. Through the sustainable design of the resulting concrete, a smaller carbon footprint is achieved by using RCAs. They are approximately 10% lighter compared to traditional mineral aggregates due to their porosity and the presence of cement paste. However, high porosity in aggregates might result in higher water absorption, lower mechanical strengths, and reduced resistance to repeated freeze–thaw cycles. Therefore, strict quality control is essential for recycled aggregates compared to traditional, natural aggregates. Some studies indicate methods of hydrophobization or increasing the density of aggregates through various techniques to close the pores and, as a consequence, to reduce water absorption [44–47].

In the present study, the criteria imposed on aggregates (according to SR EN 933-1:2012) [48] for single-layer road wearing courses have been chosen. The criteria are identical to the requirements for the types of aggregates used for double-layer road-wearing courses, including the wearing layer, according to NE 014:2002 [42].

In the following, the determinations performed on aggregates in the laboratory will be presented.

The Granularity of the Aggregates

The granularity was determined in accordance with the SR EN 933-1:2012 specifications [48]. The test consists of separating the material into several sizes with decreasing dimensions based on grain size by using a series of sieves. The masses of particles retained on the different sieves are reported relative to the initial mass of the material. The cumulative percentages of passage through each sieve are presented in numerical form (Tables 2–5). The obtained results are the average of 3 determinations.

Table 2. Passage of 0/4 mm natural river aggregate for all the mixes (NRA).

Aggregate	Passes, in %, through the Size Sieve (mm):								
	0.125	0.250	0.500	1	2	4	8	16	31.5
0/4 mm	4.23	15.18	38.30	64.70	86.30	99.43	100	100	100

Table 3. Passage of 4/8 mm, 8/16 mm crushed river aggregates (CRA).

Aggregate	Passes, in %, through the Size Sieve (mm):								
	0.125	0.250	0.500	1	2	4	8	16	31.5
4/8 mm	0.19	0.22	0.24	0.27	1.33	27.50	96.90	100	100
8/16 mm	0.05	0.06	0.06	0.07	0.07	0.09	1.76	94.62	100

Table 4. Passage of 4/8 mm, recycled concrete aggregates (RCA).

Aggregate	Passes, in %, through the Size Sieve (mm):								
	0.125	0.250	0.500	1	2	4	8	16	31.5
4/8 mm	0.02	0.02	0.03	0.03	0.04	0.12	79.90	100	100

Real Density and Water Absorption Coefficient of RCA

The determination was carried out according to SR EN 1097-6: 2013 [49].

The real density was calculated based on the mass-to-volume ratio. The mass was determined by weighing the saturated test specimen, with dry surface, and again after drying in an oven. The volume was calculated based on the mass of the displaced water, determined by weighing using the pycnometer method (for aggregates of size 0/4 and 4/8) or by reducing the mass using the wire loop method (for aggregates of size 8/16 and 16/25).

Table 5. Passage of 8/16 mm, 16/25 mm crushed aggregates/chippings (CAC).

Aggregate	Passes, in %, through the Size Sieve (mm):									
	0.125	0.250	0.500	1	2	4	8	16	25
8/16 mm	1.00	0.11	0.11	0.11	0.11	0.11	7.27	95.51	100
16/25 mm	0.08	0.09	0.09	0.09	0.10	0.10	0.10	5.02	100

The following parameters have been calculated and are presented in Table 6: absolute volume mass (ρ_a), real volume mass determined via oven drying (ρ_{rd}), real volume mass on the saturated dry surface (ρ_{ssd}), and water absorption coefficient (WA24) (expressed as a percentage of the dry mass after 24 h of immersion in water), according to the equations:

Table 6. Real volumetric mass and coefficient of water absorption of the aggregates.

Symbol Aggreg. Sorts (mm)	Characteristics of Aggregates			
	ρ_a (Mg/m^3)	ρ_{rd} (Mg/m^3)	ρ_{ssd} (Mg/m^3)	WA24 (%)
NRA_0/4	2.70	2.57	2.63	3.00
CRA_4/8	2.68	2.59	2.62	2.40
RCA_4/8	2.70	2.32	2.46	6.00
CAC_8/16	2.65	2.56	2.61	1.40
CAC_16/25	2.67	2.59	2.62	1.20

Absolute volumetric mass:

$$\rho_a = \rho_w \frac{M_4}{M_4 - (M_2 - M_3)} \; Mg/m^3 \tag{1}$$

Actual density determined after drying in oven:

$$\rho_{rd} = \frac{M_4}{M_1 - (M_2 - M_3)} \; Mg/m^3 \tag{2}$$

Actual density on the saturated dry surface:

$$\rho_{ssd} = \frac{M_1}{M_1 - (M_2 - M_3)} \; Mg/m^3 \tag{3}$$

The water absorption coefficient (expressed as a percentage of the dry mass) after 24 h of immersion (WA24) was calculated according to the following equation:

$$WA24 = \frac{100 \times (M_1 - M_4)}{M_4} \% \tag{4}$$

where:

ρ_w—volumetric mass of water at the test temperature (0.973 at T = 24 °C), Mg/m^3;
M_1—mass in air of saturated and superficially dried aggregates, g;
M_2—mass of the pycnometer containing the sample of saturated aggregates, g;
M_3—pycnometer mass filled with water only, g;
M_4—mass in air of the test sample dried in the oven, g.

Resistance to Fragmentation and Abrasion

The determination of resistance to fragmentation of coarse aggregates was performed in accordance with SR EN1097-2:2020 [50] specification. The Los Angeles coefficient (LA) was calculated based on the following equation:

$$LA = \frac{5000 - m}{50} \quad (5)$$

where m—the mass of the material retained on the 1.6 mm sieve, (g).

The obtained results are presented in Table 7:

Table 7. Los Angeles coefficient (LA) for the aggregates.

Symbol of Aggregate	Sorts of Aggregates	LA_{med} (%)	Traffic Class
RCA	4/8 mm	30.9	Reduced
CRA	4/8 mm	31.0	Reduced
CAC	8/16 mm	16.0	Intensive
CAC	16/25 mm	15.0	Intensive

High value means less resistance to crushing.

Abrasion Testing of Coarse Aggregate (MicroDeval Coefficient) for (RCA)

The abrasion testing was performed according to SR EN 1097-1:2011 [51]. The Micro-Deval coefficient (MDE) of the test in the presence of water is calculated according to the following equation:

$$MDE = \frac{500 - m}{5} \quad (6)$$

where m—mass of the retained aggregates on the 1.6 mm sieve. The results, considered as an average of 2 determinations, are presented in Table 8.

Table 8. Micro-Deval (MDE) coefficient in presence of water for (RCA).

Symbol of Aggregate	Sorts of Aggregates	M_{DEmed} (%)	Traffic Class
RCA	4/8 mm	20.8	Medium
CRA	4/8 mm	10.1	Intensive
CAC	8/16, 16/25 mm	14.0	Intensive

Flattening Coefficient of RCA

The flattening coefficient of RCA was determined following the guidelines of SR EN 933-3:2012 [52]. The overall flattening coefficient was calculated as the total mass of particles passing through the grate with slots, expressed as a percentage of the total dry mass of the particles tested.

The overall flattening coefficient (A) was calculated based on the following equation:

$$A = \frac{M_2}{M_1} \times 100 \quad (7)$$

where:

M_1—sum of the aggregate masses of the elements di/Di, (g);
M_2—sum of the masses of the granules passed through the slotted grate corresponding to the opening Di/2, (g).

The obtained results are summarized in Table 9:

Table 9. Overall flattening coefficient (A) for RCA 4/8 mm.

Sorts/Elementary Aggregates	d_i/D_i	The Nominal Opening of the Grill Slots, mm	A_i	M_1	M_2	A
4/8 mm	8/10	5	0	600	76	13
	6.3/8	4	5			
	5/6.3	3.15	29			
	4/5	2.5	26			

2.1.2. Cement

A CEM I 42.5R rapid-hardening cement was used to cast the road concrete considered in this research. The cement properties are presented in Table 10.

Table 10. Cement characteristics for CEM I 42, 5R.

Characteristics CEM I 42, 5R		Value	According to
Composition	Clincher Portland (%)	95–100	SR EN 197-1 [53]
	Minor component (%)	0–5	SR EN 197-1 [53]
Chemical Characteristics	Sulphate content (in the form of SO, %)	≤ 4	SR EN 196-2 [54]
	Chloride content (%)	≤ 0.1	SR EN 196-2 [54]
	Loss of calcination (%)	≤ 5	SR EN 196-2 [54]
	Insoluble residue (%)	≤ 5	SR EN 196-2 [54]
Physicomechanical Characteristics	Setting time (min.)	≥ 60	SR EN 196-3 [55]
	Stability (mm)	≤ 10	SR EN 196-3 [55]
	Compressive strength at 2 days (MPa)	≥ 20	SR EN 196-1 [56]
	Compressive strength at 28 days (MPa)	$\geq 42.5 \leq 62.5$	SR EN 196-1 [56]

The road concrete classes to be designed with this type of cement are: BcR3.5, BcR4, BcR4.5 and BcR5. The quantity of cement in the road concrete mix for class BcR4 is set by the normative NE 014:2002 to a quantity of 330 kg/m^3 [42]. The concrete shrinkage is influenced by the mineralogical nature of Portland cement—its specific surface area and cement ratio.

2.1.3. WGP—Waste Glass Powder

Recycled glass waste, a pozzolanic material, in the form of powder [57] was obtained through grinding in a ball mill. When incorporated into cement, mortar, or concrete compositions, pozzolanic materials have a major role in reducing carbon dioxide (CO_2) emissions and are referred to in the literature as "Supplementary Cementitious Materials (SCMs)" or "additional constituents with cementitious (hydration or pozzolanic) characteristics". Depending on their size, glass particles transition from being inert minerals to becoming reactive materials during cement hydration in the concrete mixture, especially when ground into a fine powder with reduced particle sizes [58–60].

For particle sizes smaller than 0.250 mm [57] alkali–silica reactions (ASR) are cancelled [61,62]. Moreover, this particle size improves the properties of cement-based materials, such as mortar and concrete (strength and wear resistance), due to the pozzolanic reaction which leads to the development of secondary reaction products during and after the hydration process of cement.

Pozzolanicity is the ability of a natural or artificial material to react with $Ca(OH)_2$ in the presence of water. Pozzolan reaction rate depends on the intrinsic characteristics of pozzolan itself, such as specific surface area, chemical composition, and content of the active phase [63]. According to ASTM C618 prescriptions [64], the following condition should be fulfilled for a material to be considered a pozzolan: $SiO_2 + Al_2O_3 + Fe_2O_3 \geq (50–70)\%$.

SEM (scanning electron microscopy) investigations were previously conducted on the plain cement matrix and cement mixed with waste glass powder [57].

Scanning electronic microscopy was performed using SEM Tescan VEGA II LSH to investigate the microstructure and the raw material. The test was carried out using secondary as well as backscattered electron detectors.

The mortar mixes were tested at the age of 28 days during a previous stage of the research showed that when substituting 10% of the cement content with glass waste in the form of powder (CEM—10% WGP1), the compressive strength was slightly higher compared to the conventional control mix. This confirmed the effective pozzolanic activity of the glass powder at the 10% substitution ratio.

The SEM images highlighted that the microstructure showed a complete consumption of fine glass particles in the mixture with 10% WGP, compared to the one with 20% WGP [57], due to the pozzolanic reaction during cement hydration favoring the evolution of compressive strengths [65]. In both cases, the fracture surfaces of the mortar specimens indicated a compact microstructure [66].

The chemical compositions of the considered WGP and CEM I 42.5R cement were determined using X-ray fluorescence (XRF-Qualitax, Italy) at the School of Materials Engineering, University of Malaysia Perlis (UniMAP), Perlis, Malaysia [67] and are presented in Table 11.

Table 11. XRF Analysis result—chemical composition (%) of the cement CEM I 42.5R and WGP.

Oxides	SiO_2	K_2O	Fe_2SiO_3	CaO	Al_2O_3	MgO	Na_2O	Oder
CEM I 42.5R	14.30	1.08	3.70	71.46	2.90	0.86	5.70	-
WGP	77.70	1.01	0.44	13.6	0.06	0.01	5.27	1.92

In Table 12, the fineness of the glass powder is observed. Due to its small particle size, it reacts very effectively within the concrete matrix during cement hydration process, similar to other supplementary cementitious materials (SCMs) [68–70]. The XRD pattern showed the amorphous feature of the glass waste and the presence of C2S, C3S, C3A, C4AF, and gypsum as the main mineralogical phases in the CEM I 42.5 R cement [31].

Table 12. ≤0.125 mm waste glass powder (WGP).

WGP	Passing (in %) through the Sieve (Size in mm)			
	0.63	0.125	0.250	0.500
≤0.125 mm	43.80	100.00	100	100

C_3S (tricalcium silicate) is the most important component, showing optimal characteristics for road cements from all perspectives, and it is recommended to be equal to or greater than 55% (Table 13) [71].

Table 13. Favorable behavior of the mineralogical component C_3S of cement [70,71].

Properties	Mechanical Strength	Shrinkage	Abrasion Resistance	Freeze–Thaw Resistance	Modulus of Elasticity	Hydration Rate
C_3S	Very high	Low	Good	Very good	Very high	Moderate

2.1.4. Water and Additives

The water–cement ratio (w/c) is imposed by the current norms [42,43] for the designed concrete class BcR4 as 0.45 for the content of aggregates with continuous gradation. The water fulfills the requirements set in SR EN 1008:2003 [72].

Two types of additives were used in the composition of the road concrete mixtures. The first one was a superplasticizer for high-performance concrete (MasterGlenium

115 BASF), and the second one was an air-entraining admixture (MICROAir 107-2 BASF), which is imposed by the current standard [43] to provide the concrete with better freeze–thaw resistance. The additives are compatible with Lafarge CEM I 42.5R cement and meet the requirements of the standard SR EN 934-2 + A1:2012 [73]. The necessary amount of the additive was calculated as a percentage of the cement mass.

2.2. Methods

2.2.1. Design and Preparation of BcR4 Class Road Concrete Mix

Three road concrete mixes were designed in Variant I, referred to as Var. I. Mix 1—the control mix BcR-NA with natural aggregates (river aggregates, river crushed aggregates, and crusher aggregates/chippings). The second mix design (BcR-RCA) differs from the control mix by replacing the river crushed aggregate with alternative aggregates obtained from recycled concrete waste (RCA) in a sorted form in accordance with the normative for road concrete NE 014:2002 [42] and the harmonized standards SR EN 12620 + A1:2008 [74] and SR EN 12043:2013 [75]. The third mix (BcR-RCA-WGP) includes the same types of aggregates as BcR-RCA but with the addition of substituting 10% of the cement, by mass, with glass waste in the form of powder (WGP).

The gradations of aggregates for the concrete mixes for road pavements was chosen in accordance with NE 014:2002 [42], as shown in Table 14. Thus, the chosen variant coincides with both methods of laying road concrete either in a single layer or in two layers for the wearing layer.

Table 14. Sorts of aggregates used in the layers of road wearing course in accordance with NE 014:2002 [42].

Pavements Realized	Nature of Aggregates	Sorts of Aggregates	Gradation of Total Aggregates
Single layer	Natural Sand	0/4	
Two layers	Crushed Gravel	4/8	0/25
	Chipping	8/16	
Wearing course	Chipping	16/25	

The natural sand of 0/4 mm fraction was used in each mix to support good compaction and workability of the concrete due to the rounded shape of the particles (all other aggregates in the concrete mix were obtained by crushing). The proportions of the aggregates were the same for all mixes in Variant I: 30% for 0/4 mm natural river sand (NRA); 16% for 4/8 mm crushed gravel (CRA); 24% for 8/16 mm crushed aggregate (CAC); 30% for 16/25 mm crushed aggregate (CAC). These percentages have been calculated to ensure that the total gradation curve falls within the "Favorable Zone", as presented in Table 14.

In a similar manner, the water/cement ratio of 0.45—in accordance with NE 014:2002 [42] and SR EN 206-1:2021 [43]—was maintained for all mixes to observe any changes in workability/consistency of the concrete based on the variations in the mixture.

The percentages of the aggregates have been calculated in such a way to ensure that the total gradation curve falls within the "Favorable Zone", which is delimited by the limits of granulometry curves of the total aggregate for road concretes made with continuous gradation aggregates of 0/25 mm. The representation of the total curve can be found in Figure 3.

In Table 15 the total gradation curves are presented for the control mix and the other two mixes design, where the modification involves replacing the 4/8 mm crushed river aggregate (CRA) with recycled concrete aggregate (RCA). This replacement was carried out as a first phase of the study. The replacement is relevant because the two types of aggregates have very similar Los Angeles abrasion resistance values (31.0% and 30.9%, respectively) and have been obtained through crushing processes.

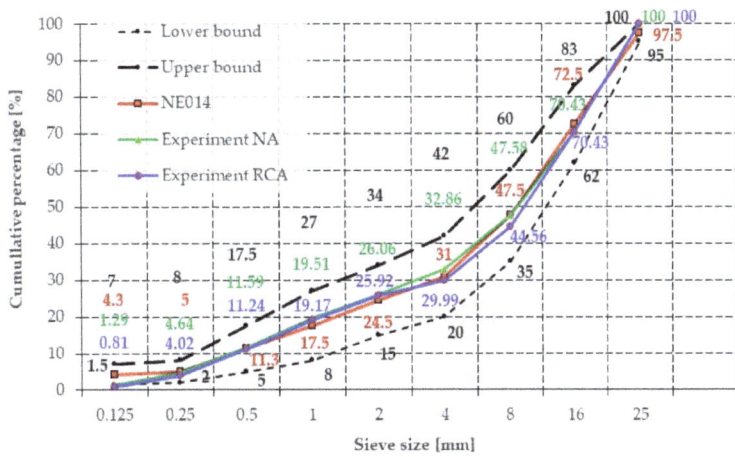

Figure 3. The total gradation curve for the control and RCA mixes of road concrete in Var. I.

Table 15. The total gradation curves for the variant with natural aggregates and recycled aggregates.

Total Gradation Curves for the Concrete Mixes	Passing % through Sieve with Size (mm)								
	0.125	0.250	0.5	1	2	4	8	16	25
Concrete mixture with natural aggregates [38]	1.29	4.64	11.59	19.51	26.06	32.86	47.58	70.43	100
Concrete mixture with recycled aggregates	0.81	4.02	11.24	19.17	25.92	29.99	44.56	70.43	100
Lower limit	1.5	2	5	8	15	20	35	62	100
Upper limit	7	8	17.5	27	34	42	60	83	100

Design parameters according to standards and cement quantities (kg/m^3) for preparing concretes are presented in Table 16.

Table 16. Design parameters for road concrete BcR, imposed by NE 014:2002 [42] norm.

Design Parameters BcR4	Min. Cement Ratio CEM I 42,5	(w/c)	Consistency Class S1 (mm)	Air Void Content (%)	Freeze-Throw Circles	f_c 28 Days MPa	$f_{ct,fl}$ 28 Days MPa
NE 014 [42] and SR EN 206-1: 2021 [43]	330 kg/m^3	max. 0.45	10–40	3.5 ± 0.5	100	min. 35 max. 50	min. 4 max. 5

In Table 17, the compositions of the designed/calculated and realized mixes are presented, where it can be remarked that the difference in the cement amount occurs in the concrete made with glass waste (in the form of powder with particle size < 0.125 mm), where it substitutes 10% of the cement content compared to the control mix.

Table 17. The composition of the road concretes in Variant I (kg/m^3).

Mix Components	BcR-NA-1	BcR-RCA-1	BcR-RCA-WGP10
Water/Cement ratio	0.45	0.45	0.45
Cement I 42.5R	330	330	297
DSP(WGP) < 0.125 mm—10%	-	-	33
NRA—0/4 mm	569	569	569

Table 17. Cont.

Mix Components	BcR-NA-1	BcR-RCA-1	BcR-RCA-WGP10
CRA—4/8 mm	303	-	-
RCA—4/8 mm	-	303	303
CAC—8/16 mm	455	455	455
CAC—16/25 mm	569	569	569
Admixture 1 (Master Glenium 115)—1.80%	5.94	5.94	5.94
Admixture 2 (MICROAir 107-2)—0.25%	0.285	0.285	0.285

Aiming to use of as much recycled material in the road concrete mix as possible and since the total aggregate curve allows entry into the favorable zone, new percentages for aggregates have been adapted, as shown in Figure 4. Additionally, 20% of the cement mass was substituted with WGP in a second mix design variant (Var. II), presented in Table 18.

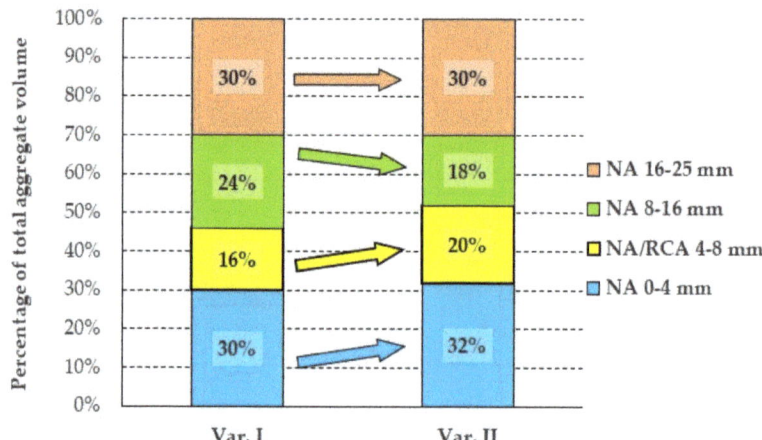

Figure 4. Aggregate percentages in Var. II.

Table 18. The composition of the road concretes in Variant II (kg/m^3).

Mix Components	BcR-NA-2	BcR-RCA-2	BcR-RCA-WGP20
Water/Cement ratio	0.55	0.55	0.55
Cement I 42.5R	330	330	264
DSP(WGP) < 0.125 mm—20%	-	-	66
NRA—0/4 mm	607	569	569
CRA—4/8 mm	379	-	-
RCA—4/8 mm	-	379	379
CAC—8/16 mm	341	341	341
CAC—16/25 mm	569	569	569
Admixture 1 (Master Glenium 115)—2.30%	7.59	7.59	7.59
Admixture 2 (MICROAir 107-2)—0.25%	0.285	0.285	0.285

In Variant II, the first mixtures have been prepared using natural river aggregates (NRA) and crushed aggregates / chipping (CAC), along with recycled concrete aggregates (RCA) with a size of 4/8 mm. In the third mixture, 20% of the cement was substituted with WGP to enhance wear resistance while maintaining the water-to-cement ratio (W/C) and admixture dosage. However, after preparing the mixtures, a significant reduction in workability was observed, as expected, due to the 6% water absorption of RCA, which necessitated an increase in the superplasticizer admixture (Admixture 1) from 1.8% to

2.3% relative to the cement quantity as well as maintenance of the W/C ratio of 0.55. Consequently, a consistency class of S1 (10–40 mm) was achieved in accordance with NE 014:2002 [42].

Several studies have highlighted that crushed concrete aggregates in the size range of 4 to 8 mm exhibit the highest amount of adhered mortar, which implies that aggregate size has a significant effect on water absorption and concrete strength [76,77]. The determined water absorption value for RCA (sort 4/8 mm) was 6.0%, while for NRA it was 3%.

2.2.2. Determinations of Fresh Properties of BcR Concrete

In the study the following determinations have been conducted:
- Concrete temperature (°C), in accordance with SR EN 206-1:2021 [43];
- Consistency using the slump test (mm), in accordance with SR EN 12350-2:2019 [78];
- Density (kg/m^3), in accordance with SR EN 12350-6:2019 [79];
- Air content (%), in accordance with SR EN 12350-7:2019 [80].

The temperature was measured using a specialized thermometer for concrete determinations, at various points on the poured material. The consistency, which indicates workability, was determined using the slump cone method. Three determinations were performed for each mix. The density of fresh concrete mixes was determined using a vessel of known volume, which was weighed before and after filling. The difference in masses compared to volume, provided the density value.

The air content was determined using equipment with a manometer. The vessel was filled in three layers and the material was vibrated until no more air bubbles escaped. The vessel was then covered with a lid equipped with valves, an air release valve, and a manometer indicating the percentage of entrained air.

2.2.3. Mechanical Properties of BcR Concrete

The following determinations were conducted for BcR concrete:
- Flexural strength ($f_{ct,fl}$) in accordance with SR EN 12390-5:2019 [81];
- Compressive strength (f_{cm}) in accordance with SR EN 12390-3:2009 [82];
- Splitting tensile strength ($f_{ct,sp}$) in accordance with SR EN 12390-6:2010 [83];
- Hardened concrete density (ρ_a) in accordance with SR EN 12390-7:2019 [84].

For determining the flexural strength, a minimum of three prismatic specimens measuring 150 mm × 150 mm × 600 mm were considered for each BcR concrete mix. The results from this test are used as a reference for establishing the class of BcR concrete compared to conventional concretes, where the compressive strength on cylinders or cubes is relevant. The hardened specimens were stored in water at a controlled temperature of 20 ± 2 °C immediately after demolding until the reference age of 28 days and then tested using the four-point loading test, Figure 5a, in a universal testing machine. The loading rate was set at 0.04–0.06 MPa/s, as per SR EN 12390-5:2019 [81].

(a)

(b)

(c)

Figure 5. Schematic representation of force positioning and action: (**a**) 4-point bending test, (**b**) compression test, (**c**) splitting test [61].

The flexural strength was determined using the equation:

$$f_{ct,fl} = \frac{F \times l}{d_1 \times d_2^2} \qquad (8)$$

where:

$f_{ct,fl}$ is the tensile strength in MPa/s (N/mm²·s);
F is the maximum load in N;
l the span between the supports in mm;
d_1 and d_2 are the cross-sectional dimensions of the specimen in mm (as shown in Figure 6b).

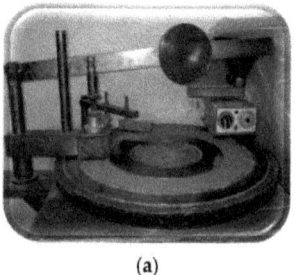
(a)

(b)

(c)

Figure 6. The Böhme equipment (**a**), the track sprinkled with standard abrasive material (**b**), and the positioning slot for the cubic specimen (**c**).

The flexural strength was expressed rounded to the nearest 0.1 MPa (N/mm²).

For determining the compressive strength (Figure 5b), in addition to the series of cubes taken with a side length of 150 mm, determinations also on the prism fragments have been conducted using non-deformable metal plates for the testing surface.

The loading rate should remain constant in the range of 0.6 ± 0.2 MPa/s (N/mm².s). After applying the initial load, which should not exceed approximately 30% of the ultimate load, the load is applied without shock to the specimen, and it is continuously increased at the chosen constant rate $\pm 10\%$ until the specimen can no longer withstand a higher load.

The compressive strength was determined by the equation:

$$f_c = \frac{P}{A_c} \qquad (9)$$

where f_c is the compressive strength in MPa (N/mm²), P is maximum load at yield in N, and A_c is the cross-section of the specimen perpendicular to the load direction in mm².

The obtained result was rounded to 0.1 N/mm². The loading rate was 0.5 MPa/s according to SR EN 12390-3:2009 [82].

The determination of splitting tensile strength was carried out on cubes with sides of 150 mm and on prism fragments (Figure 5c). The specimen was centrally positioned, and the direction of applied force was perpendicular to the direction of concrete casting. The prismatic specimens were subjected to a compressive force applied over a narrow region along their length.

Loading strips made of hard wood with density > 900 kg/m³ and dimensions of width: (10 ± 1) mm, thickness: (4 ± 1) mm—with the length greater than the length of the contact line of the tested specimen—were used to transfer the applied load from the testing machine to the specimen. These strips were used only once. The loading strips were positioned along the upper and lower parts of the specimen loading plane. The loading rate was 0.04 MPa/s according to SR EN 12390-6:2010 [81,83].

The splitting tensile strength was determined using the formula:

$$f_{ct,sp} = \frac{F \times l}{d_1 \times d_2^2} \tag{10}$$

where:

$f_{ct,sp}$—tensile splitting strength in MPa;
F—maximum load in N;
l—length of the contact line of the specimen in mm;
d_1 and d_2—size of the cross-section in mm.

The obtained value was rounded to the nearest 0.05 MPa.

All mechanical property tests were performed using the Advantest 9 digital press with 2000 kN capacity, provided by the CONTROLS company.

The hardened concrete density was determined by weighing and dividing the mass of each specimen by the volume of the cube with a side length of 150 mm.

2.2.4. Durability Properties of BcR Concrete

The following determinations were conducted for BcR concrete:
- Loss of strength after 100 freeze–thaw cycles in accordance with SR 3518:2009 [85];
- Abrasion resistance in accordance with SR EN 1338:2004 [86], SR EN 1339:2004 [87], SR EN 1340:2004 [88];
- Carbonation depth determination according to SR CR 12793:2002 [89].

The loss of strength due to freeze–thaw cycles was determined on concrete specimens kept for 7 days in water at a temperature of $T_{water} = 20 \pm 2$ °C and for 21 days in air at $T_{air} = 20 \pm 2$ °C and relative a humidity of URA = 65 ± 5%.

Specimens aged at a minimum of 28 days were placed in a water bath at a temperature of 20 ± 5 °C for saturation 4 days before the start of the test. Water was poured into the bath up to 1/4 of the height of the specimens; after 24 h, it was added up to 1/2 of the height, and after another 24 h, up to 3/4 of the height. Three days after introducing them into the bath, the water level should be at least 20 mm above the height of the specimens, and this level should be maintained for 24 h. Afterward, the specimens are considered saturated.

Specimens intended for freeze–thaw cycles were placed in the thermostat cabinet. Control specimens were continuously kept in water.

Saturated specimens were introduced into the refrigeration chamber at −17 ± 2 °C and kept for 4 h for the freezing cycle. The thawing cycle was carried out at a temperature of 20 ± 5 °C and humidity of 95% for 4 h. The specimens must be arranged so that they are completely surrounded by air. The distance between specimens and between specimens and the walls of the installations must be at least 20 mm. The support on which the specimens are placed must be constructed in such a way that the contact surface with the base of the specimens is minimal.

After subjecting the specimens to 100 freeze–thaw cycles, the loss of compressive strength was determined by subjecting them to a compressive test. The number of specimens tested was the same as the number of control specimens. The test was stopped after reaching the number of 100 cycles or if the loss of compressive strength exceeded 25% compared to the control specimens of the same age.

The loss of compressive strength (η) was determined by the relationship:

$$\eta = \frac{Rm - Ri}{Rm} \times 100 \tag{11}$$

where:

Rm—Arithmetic mean value of the compressive strengths of the control specimens, in N/mm².

Ri—Arithmetic mean of the compressive strength values of the freeze-thaw specimens, in N/mm² or MPa.

For the determination of the Böhme abrasion resistance expressed as volume loss relative to 5000 mm², cubes with a side length of 71.0 ± 1.5 mm were placed on the abrasive disk of the Böhme apparatus (Figure 6), on the testing track where a standard abrasive was spread. The disk was rotated, and the specimens were subjected to an abrasive load of 29 ± 3 N for a specified number of cycles. The abrasive wear on the specimen leads to volume loss. The contact face and the opposite face of the specimen must be parallel to each other and flat. Before the test, the density of the specimen, ρ_R, was determined by measuring to the nearest millimeter and by weighing to the nearest 0.1 g. The standard abrasive used should be fused alumina (artificial corundum) designed to produce an abrasion of 1.10 mm to 1.3 mm when specimens are tested.

Before the abrasion test and after every four cycles, the specimen was weighed with an accuracy of 0.1 g. Additionally, 20 g of standard abrasive was placed on the testing track.

The specimen was tested for 16 cycles, each consisting of 22 rotations. After each cycle, the disk and the contact face were cleaned, the specimen was progressively rotated by 90°, and a new abrasive was placed on the testing track.

The abrasion was calculated after 16 cycles as the average of the lost volume ΔV, using the equation:

$$\Delta V = \frac{\Delta m}{\rho_R} [mm^3] \qquad (12)$$

where:

ΔV is the volume loss after 16 cycle, in mm³;
Δm is the loss of mass after 16 cycles in g;
ρ_R is the density of the specimen in g/mm³.

The abrasion was expressed by the volume loss of the specimen after it was subjected to wear.

The depth of the carbonation layer in hardened concrete was determined using the phenolphthalein method in accordance with SR CR 12793: 2002 [89].

The carbonation of the concrete specimens was assessed in freshly exposed sections through the reaction with pH indicator substances, such as phenolphthalein, which turns the concrete red purple at a pH value of approximately 9. This measurement can be carried out at various ages (days).

Carbonation was determined at an age of 8 years (2920 days). The specimens were kept in a laboratory room at a temperature of 20 ± 3 °C throughout this period. Before the test, they were placed in water for 28 days, then taken out. Excess water was removed, and then the specimens were split open. A 1% solution of phenolphthalein in 70% ethanol was sprayed onto the fresh concrete section. The depth of carbonation is represented by the distance dk (measured in mm) from the outer surface of the concrete to the edge of the red-purple-colored region. Both the average depth, dk average, and the maximum depth, dkmax, were measured.

2.2.5. Microstructural Determinations

Optical microscopy and the X-ray powder diffraction method (PXRD) for the qualitative analysis of crystalline constitutive materials were used to assess the microstructure of the BcR concrete.

Polarized light optical microscopy was used in order to investigate the concrete/mortars samples. A thin section was prepared from all samples of investigated concrete and mortars, and a Nikon Optiphot T2—Pol was used for optical studies (texture and composition at crossed and parallel pollars, respectively) as well as for taking photos.

X-ray diffraction (XRD) was performed on the concrete/mortars using a Bruker D8 Advance diffractometer with Cu Kα radiation (λ = 1.541874 Å), a 0.01 mm Fe filter, and a LynxEye one-dimensional detector at the Department of Geology, Babeș-Bolyai University (Cluj-Napoca, Romania). The determinations were performed on samples aged for 2920 days, as in the case of carbonation determination.

3. Results

3.1. Characteristics of Fresh State for Road Pavement Concrete

It has to be noted that the designed mixes in Variant II required an increase in the amount of admixture and the water-to-cementitious-materials ratio (w/c) due to the increased proportions of natural river aggregates (NRA) of 0/4 mm by 2% and recycled concrete aggregates (RCA) of 4/8 mm by 4% compared to the mixes in Variant I, as shown in Figure 4. The increase in water content leads to a reduction in the values of mechanical properties.

The results obtained from the determinations on the fresh concrete are presented in Table 19.

Table 19. Fresh BcR composite properties.

Fresh Property	UM	Performance Level	Mix Design					
			BcR-NA		BcR-RCA		BcR-RCA-WGP	
			Var. I	Var. II	Var. I	Var. II	10%	20%
Temperature (T)	°C	5–30	23	22	22	21	23	22
Consistency (S)	mm	10–40	35	40	35	37	27	31
Apparent Density (ρ)	kg/m^3	2400	2374	2370	2364	2352	2358	2347
Entrained Air for Aggreg. dmax-25 mm	%	3.5–4.5 ($\pm$0.5)	4.0	4.2	3.7	3.9	3.8	4.2

3.2. Hardened BcR Composite Properties

The results obtained from mechanical tests on cubic and prismatic concrete specimens for the two mix variants are presented in Table 20.

Table 20. Hardened BcR composite properties.

Hard Property	UM	Performance Level	Mix Design					
			BcR-NA		BcR-RCA		BcR-RCA-WGP	
			Var. I	Var. II	Var. I	Var. II	10%	20%
Flexural strength (fct,fl)	MPa	4.0–5.0	6.7	5.4	5.6	5.5	5.4	4.3
Compressive strength (fc)	MPa	35–45	84.2	69.2	83.1	69.4	80	62.0
Splitting strength (fct,sp)	MPa	-	4.5	3.7	4.4	3.7	4.5	3.5
Density (ρa)	kg/m^3	2400 $\pm$ 40	2430	2417	2425	2410	2420	2406.6
Loss of strength (η)	%	$\leq$25	14.11	16.8	14.60	17.6	16.0	20.2
Volume loss due to abrasion (η)	ΔV/5000 mm^2	ΔV $\leq$ 18,000 mm^3	11,301	9371	11,500	9494	11,220	9436
Depth of carbonation (dk)	mm		-	0.5	-	0.5	0.2	0.5

In Figure 7, it can be observed that in all specimens aged for 2920 days, a slight and non-uniform carbonation outline is present, reaching a maximum depth of 0.5 mm. However, the color intensity remains pronounced after 1 h of spraying. This phenomenon, occurring at a testing age of 28 days under testing/holding conditions according to the standard, would not have existed. It can be hypothesized that the diffusion of carbon dioxide did not take place within the cement matrix, a phenomenon hindered by the compactness of the cementitious stone in the concrete [90,91].

3.3. Microstructural Determinations

3.3.1. Optical Microscopy Using Polarized Light

Taking into account the data presented in Tables 19 and 20 and the main contribution of this research work (e.g., the use of waste glass powder as a substitute for cement and recycled concrete aggregates), it can be concluded that BcR-NA-1 and BcR-RCA-WGP10 mixes exhibited similar values. These results are encouraging from a sustainability perspective for road concrete. Modifying the percentages of aggregates from the entire aggregate volume in order to accommodate more RCA, Var. II in Figure 4, resulted in decreased values of all mixes compared to their counterparts in Var. I.

Figure 7. The appearance of the specimens after spraying with phenolphthalein (**a**) BcR-NA; (**b**) BcR-RCA; (**c**) BcR-RCA-WGP10; (**d**) BcR-RCA-WGP20.

Therefore, the prime candidates for microstructural analyses were BcR-NA-1 and BcR-RCA-WGP10 mixes.

Hence, two samples from each of the above-mentioned concrete mixes were randomly chosen for microstructural analysis. From Figures 8 and 9, it can be observed that both types of samples have porphyroclastic textures defined by the presence of large fragments of aggregates embedded into the microcrystalline groundmass (matrix of the concrete). In the matrix there are pores, spherical in shape, with diameters up to 0.2 mm (Figures 8 and 9). The aggregates consisted of fragments (clasts) of minerals (crystalloclasts), rocks (lithoclasts), and concrete (concreteclasts). The crystalloclasts originated from river sand or resulted from the mechanical crushing of the rock (aggregates) extracted from the quarry.

Figure 8. Microscopic image of the sample BcR-NA1 at crossed pollars with porphyroclastic textures and rounded pores into the matrix. The diameter of the pore is 0.184 mm. Qzt—quartzite, CSH—hydrated calcium silicates, CAH—hydrated aluminum silicates.

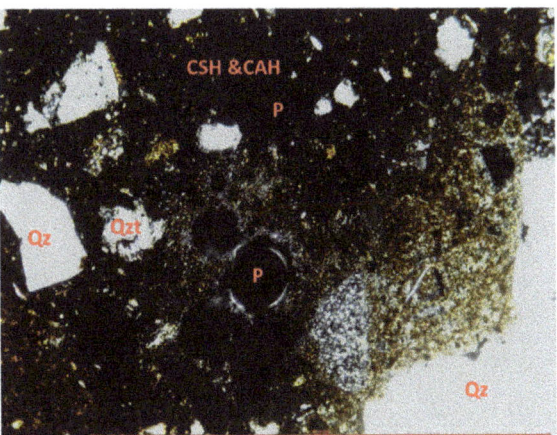

Figure 9. Microscopic image of the sample BcR-RCA-WGP10 at crossed pollars with porphyroclastic textures and pores (P) into the matrix. Qzt—quartzite, CSH—hydrated calcium silicates, CAH—hydrated aluminum silicates. Quartz (Qz). The scale bar is 1 mm.

In the case of BcR-NA-1, the aggregates consist of fragments of minerals and rocks (Figures 10 and 11). The minerals identified into the sample BcR-NA1 are represented by quartz, muscovite, biotite (sometime chloritized), pyroxene, plagioclase, feldspars, etc. The fragments of rocks are represented mainly by dacite and subordinated by quartzite and crystalline schists. The matrix is very fine crystallized (Figure 12) and consists of hydrated calcium and aluminum silicates, calcite, and portlandite. The local brown color of the matrix indicates the presence of iron hydroxides formed on the brownmillerite from the cement. Frequently, the newly formed minerals resulted from hydration processes developed as a rim surrounding the aggregate fragments (Figure 13).

Figure 10. Microscopic image of the sample BcR-NA-1 at crossed pollars showing porphyritic textures with aggregates embedded into a microcrystalline matrix (M). Aggregates consist of dacite, quartzite (Qzt), crystalline schists (Sch), pyroxene (Pyr), quartz (Qz), and chlorite (Chl). The scale bar is 1 mm.

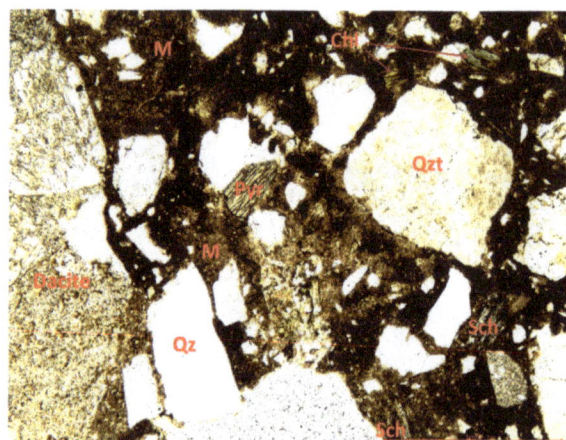

Figure 11. Microscopic image of the sample BcR-NA-1 at parallel pollars showing porphyritic texture aggregates embedded into a microcrystalline matrix (M). Aggregates consist of dacite, quartzite (Qzt), crystalline schists (Sch), pyroxene (Pyr), quartz (Qz), and chlorite (Chl). The scale bar is 1 mm.

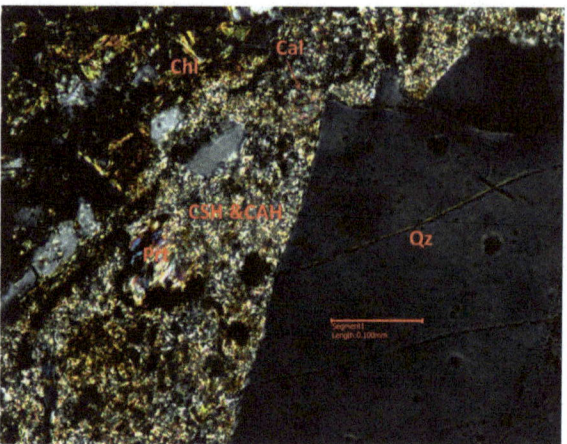

Figure 12. Detailed microscopic image of the sample BcR-NA-1 at crossed pollars with the matrix of concrete/mortar consisting of very fine crystallized mixture of hydrated calcium silicates (CSH), hydrated aluminum silicates (CAH) calcite (Cal), portlandite (Prt). Chl—chlorite, Qz—quartz. The scale bar is 0.100 mm.

In the case of BcR-RCA-WGP10 concrete, the aggregates consisted of fragments of minerals, rocks (Figure 14), and recycled concrete (Figure 15). The fragments of minerals consisted of quartz, muscovite (Figures 16 and 17), pyroxene, and feldspars and originated in river sand or resulted from the mechanical crushing of the rock (aggregates) extracted from the quarry. The lithoclasts consisted of dacite (quarry-crushed aggregates), quartzite, and crystalline schists, as shown in Figures 16 and 17. Recycled concrete aggregates (RCA) are also present in the sample of BcR-RCA-WGP10 mix. Under the microscope, it was well visible that the matrix of RCA was well crystalized compared to the matrix of BcR-RCA-WGP10 sample (Figure 14). BcR-RA-WGP10 matrix was predominantly isotropic (black in color at crossed pollars), indicating the presence of glass powder. Small crystals of portlandite were also visible in the matrix.

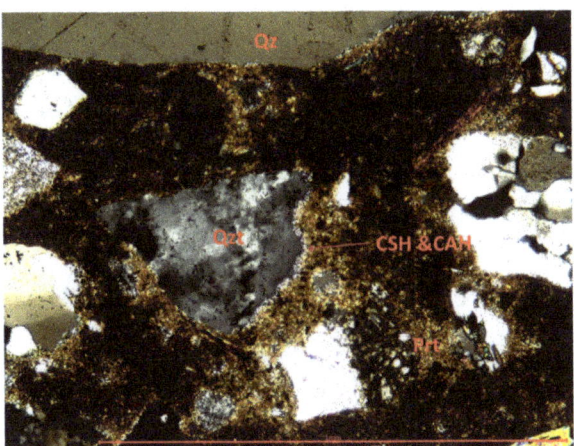

Figure 13. Microscopic image of the sample BcR-NA-1 at crossed pollars with newly formed hydrated calcium silicates (CSH) and hydrated aluminum silicates (CAH) developed as coronas on the quartzite aggregate which represents the crystallization support. The scale bar is 1 mm.

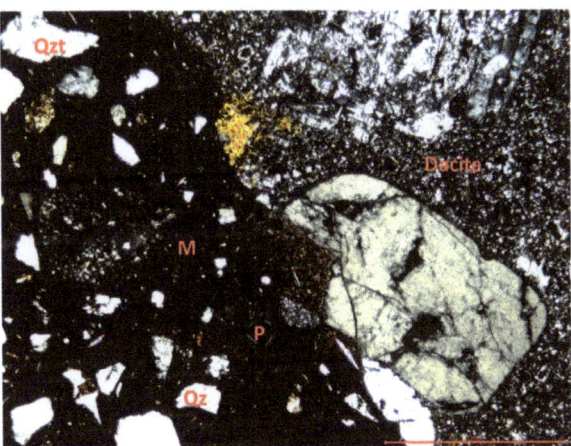

Figure 14. Microscopic image of the sample BcR-RCA-WGP10 at crossed pollars with porphyroclastic texture. The clasts consist of dacite, quartzite (Qzt), fragments of quartz, etc., embedded into the matrix, black in color. The scale bar is 1 mm.

3.3.2. X-ray Diffraction Using the Powder Method (PXRD) for Qualitative Analysis

Samples from both mixes, BcR-NA-1, and BcR-RCA-WGP10, were investigated using X-ray diffraction. The X-ray spectra obtained on the whole sample powder were dominated by the presence of the minerals forming the aggregates, especially that of quartz. In order to investigate the newly formed mineral phases of the matrix the elimination of the aggregates was crucial. Such a separation of the matrix and aggregates is almost impossible as long as some fragments of minerals are microscopic in size. The X-ray investigation performed on the sample from the BcR-NA-1 mix indicated the presence of mineral phases originated from the aggregates (quartz, muscovite, albite, clinochlore, and orthoclase) as well as newly formed minerals, such as portlandite (Ca $(OH)_2$) and gypsum ($CaSO_4·2H_2O$). Calcite ($CaCO_3$) was also present as the result of carbonation of concrete (Figure 18).

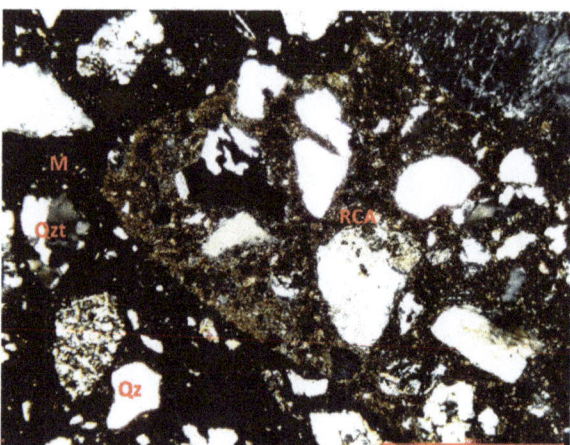

Figure 15. Microscopic image of the sample BcR-RCA-WGP10 at crossed pollars with porphyroclastic texture. The clasts consist of recycled concrete (RCA), quartzite (Qzt), fragments of quartz, etc. embedded into the matrix, black in color. The scale bar is 1 mm.

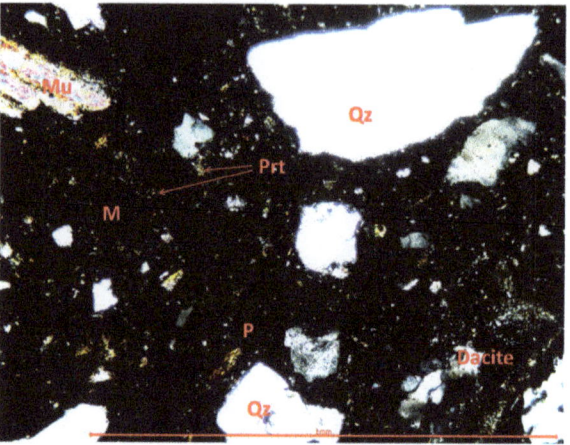

Figure 16. Detailed microscopic image of the sample BcR-RCA-WGP10 at crossed pollars with aggregates of dacite and quartz and fragments of muscovite embedded into the black matrix with newly formed portlandite. The scale bar is 1 mm.

For a better highlighting of portlandite as well as to eliminate the strong signal of quartz, a limited X-ray spectrum between 3.8 and 20 degrees 2θ was collected (Figure 19). Besides portlandite, muscovite, clinochlore, and tourmaline (var. Schol), some typical lines for calcium silicate hydrate (CSH) and aluminum silicate hydrate (ASH) were easily visible.

The X-ray investigation performed on the sample from the BcR-RCA-WGP10 mix indicated the presence of mineral phases originated from the aggregates (quartz, muscovite, plagioclase feldspars, clinochlore, and orthoclase) as well as newly formed minerals as portlandite and gypsum (Figure 20). Calcite was also present as the result of carbonation of concrete. The shape of the spectrum indicates a high degree of matrix crystallization, suggesting the pozzolanic reaction of the glass powder (WGP).

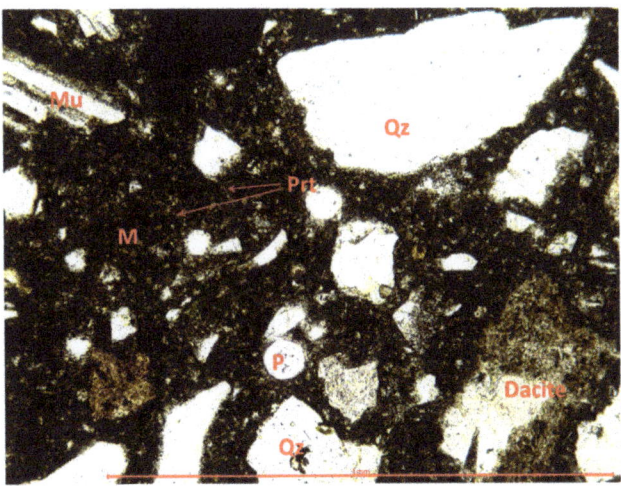

Figure 17. Detailed microscopic image of the sample BcR-RCA-WGP10 at parallel pollars with aggregates of dacite and quartz and fragments of muscovite embedded into the black matrix with newly formed portlandite. The scale bar is 1 mm.

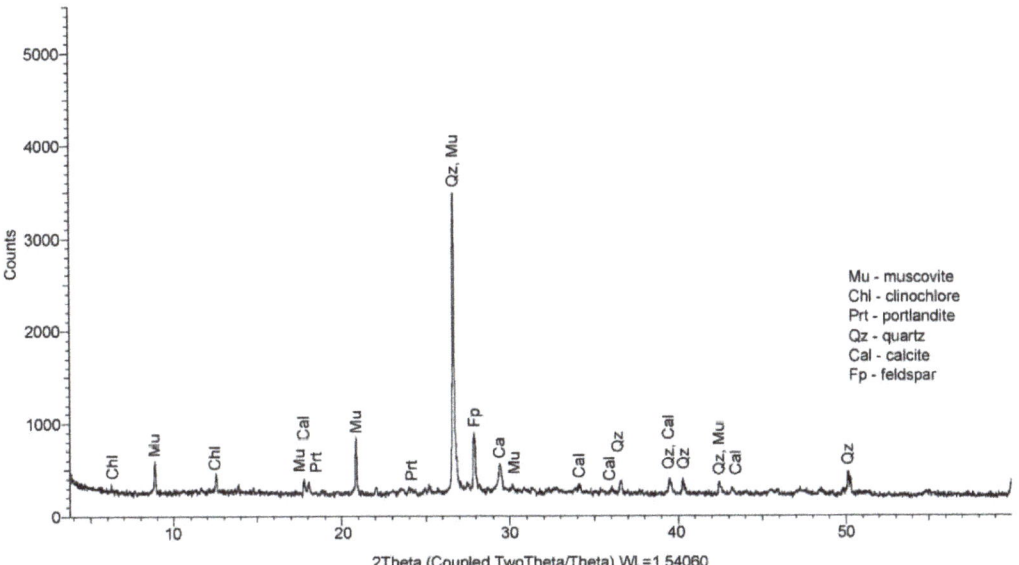

Figure 18. The X-ray spectra of the sample BcR-NA-1 with the typical line for quartz, muscovite, calcite, albite, clinochlore, gypsum, portlandite, and orthoclase.

The X-ray spectrum collected between 3.8 and 20 degrees 2θ for the sample BcR-RCA-WPG10 shows a typical line for muscovite, clinochlore, gypsum, schorl, and portlandite as well as for calcium silicate hydrate (CSH) and aluminum silicate hydrate (ASH) (Figure 21).

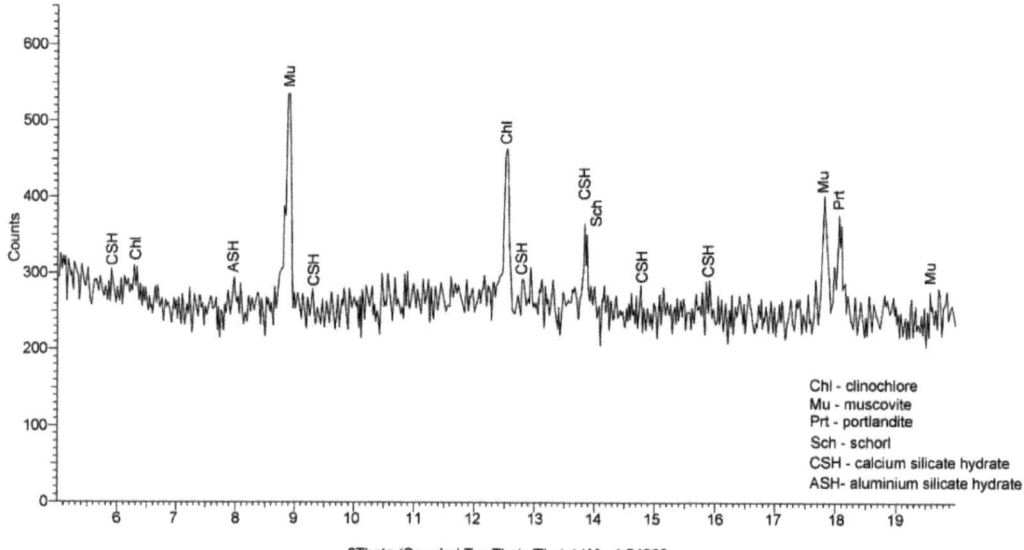

Figure 19. The X-ray spectra of the sample BcR-NA1 collected between 3.8 and 20-degrees 2θ with the typical line for muscovite, clinochlore, portlandite, schorl, CSH, and ASH.

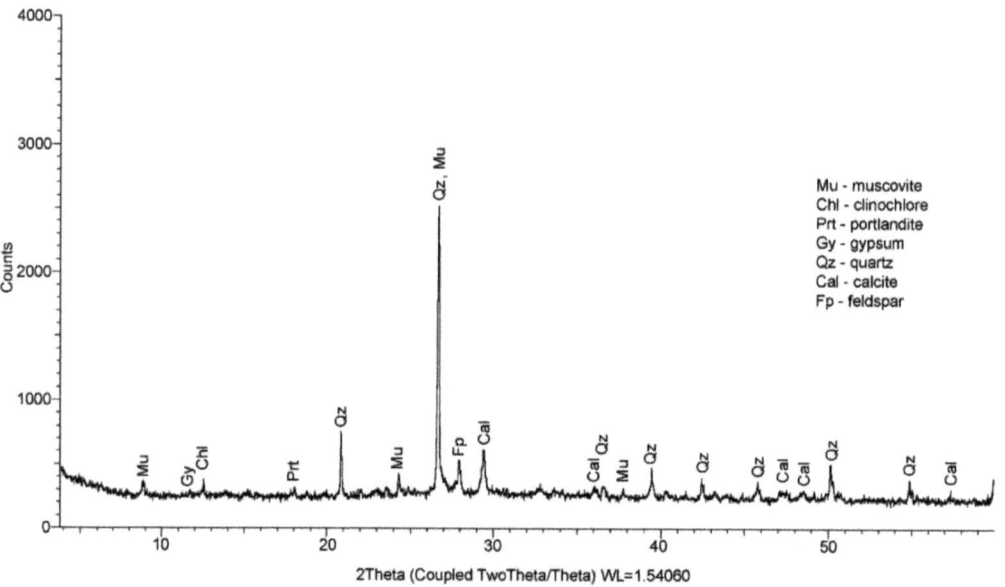

Figure 20. The X-ray spectra of the sample BcR-RCA-WGP1 with the typical line for quartz, muscovite, calcite, plagioclase feldspar (anorthite), clinochlore, gypsum, and portlandite.

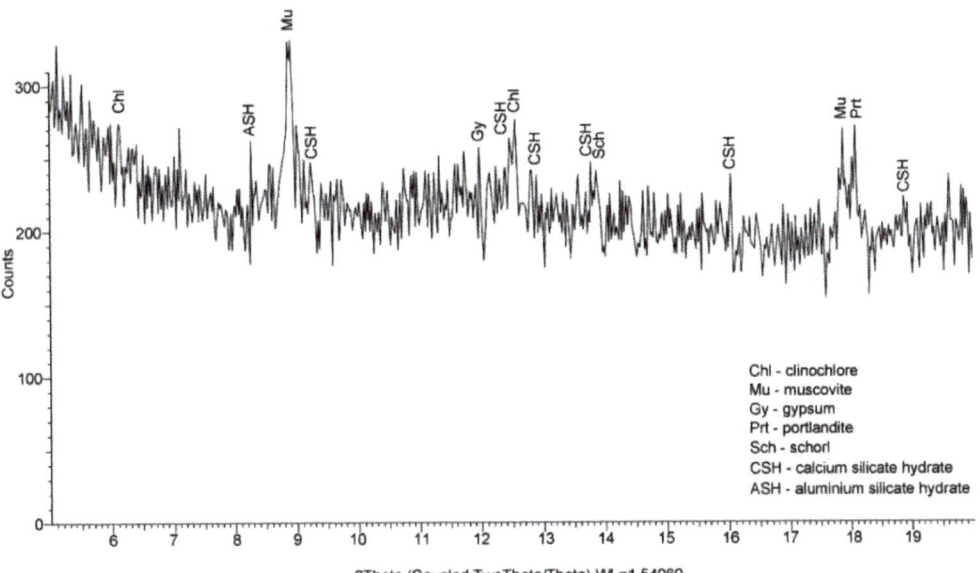

Figure 21. The X-ray spectra of the sample BcR-RCA-WGP10 collected between 3.8 and 20 degrees 2θ, with a typical line for muscovite, clinochlore, portlandite, and gypsum.

4. Discussion

4.1. Performance of the BcR Composites Fresh Properties

Figure 22 shows the fresh properties of the investigated mixtures. As is already known in the scientific literature, both WG particles and quarry aggregates have a detrimental effect on the workability of the material when both natural aggregates and high-water-absorption recycled aggregates (RCA) are used [7,58,92,93].

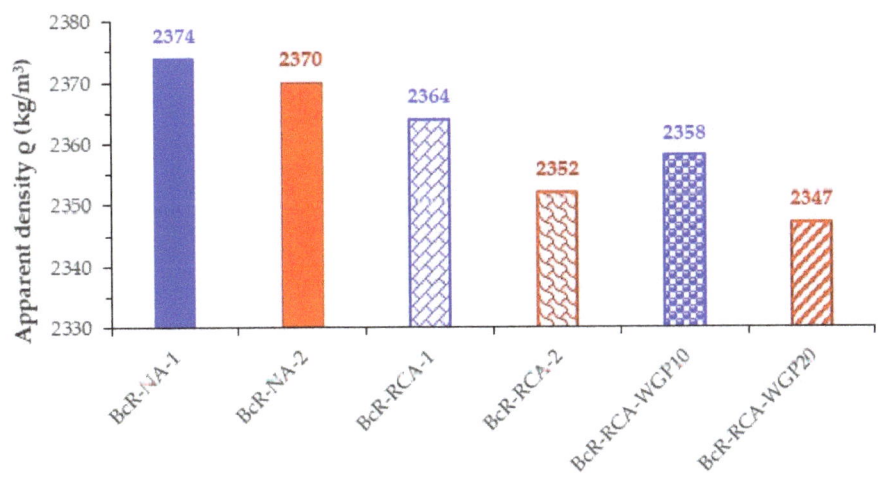

(a) Apparent density

Figure 22. Cont.

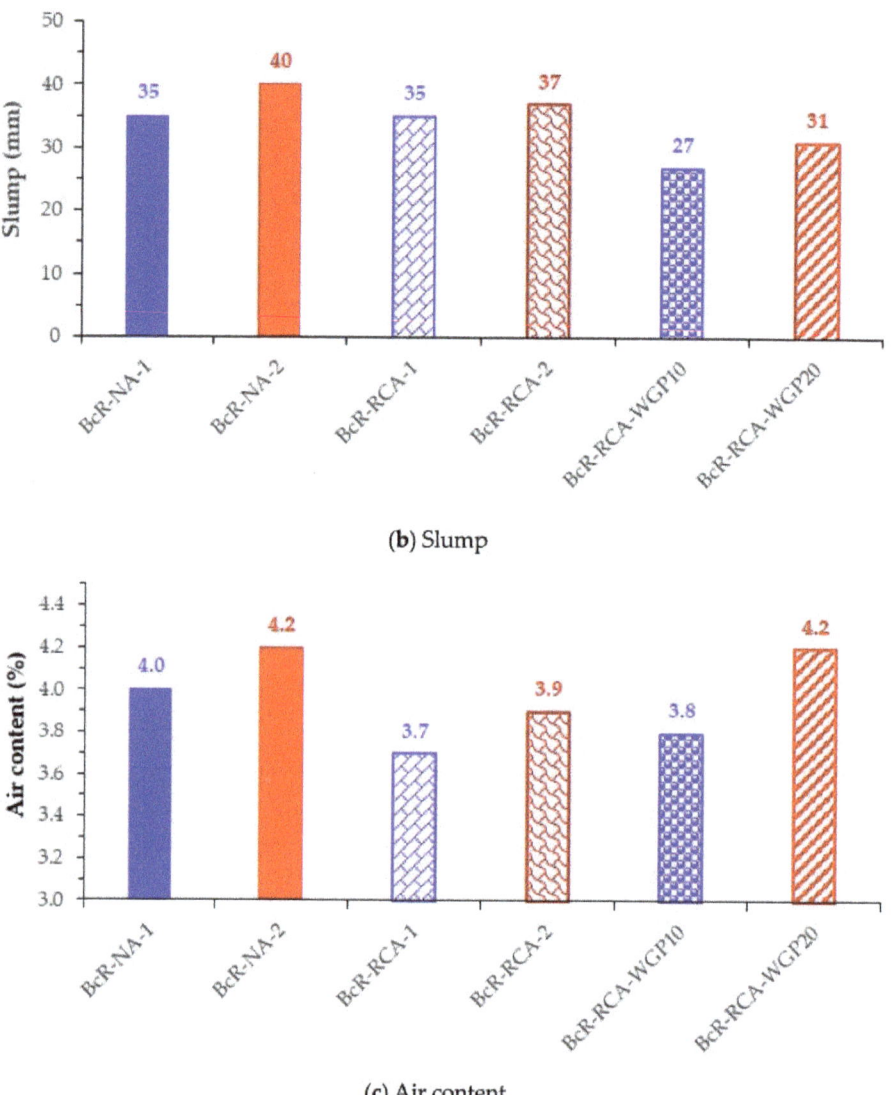

(b) Slump

(c) Air content

Figure 22. Fresh BcR composites' properties.

From Figure 22a, it can be observed that the densities of Var. I mixtures were higher than those of Var. II, which is natural due to the percentage composition of the coarse quarry aggregate (8/16 mm) being 6% higher in Var. I than in Var. II, as shown in Figure 4.

The values of the slump were higher for the Var. II mixes, as seen in Figure 22b, where both the w/c ratios increased from 0.45 to 0.55, and there was an additional superplasticizer input from 1.8% to 2.3%. As previously mentioned, these adjustments were necessary due to the 2% increase in the NRA volume coupled with a 4% increase in RCA volume.

The entrained air content followed the same trend as the slump of the concrete. Although it led to a decrease in mechanical strength values, it may favor a reduction in loss of strength during repeated freeze–thaw cycles (Tables 19 and 20).

4.2. Performance of the BcR Composites' Hard Properties

In Figure 23, the graphical representations of the most relevant values of road composite properties ($f_{c,fl}$, f_c, η, and ΔV) are shown. Additionally, the increasing or decreasing trends of these properties are illustrated with their corresponding equations along with the limit value imposed by the NE 014 standard [38], highlighted by the green line.

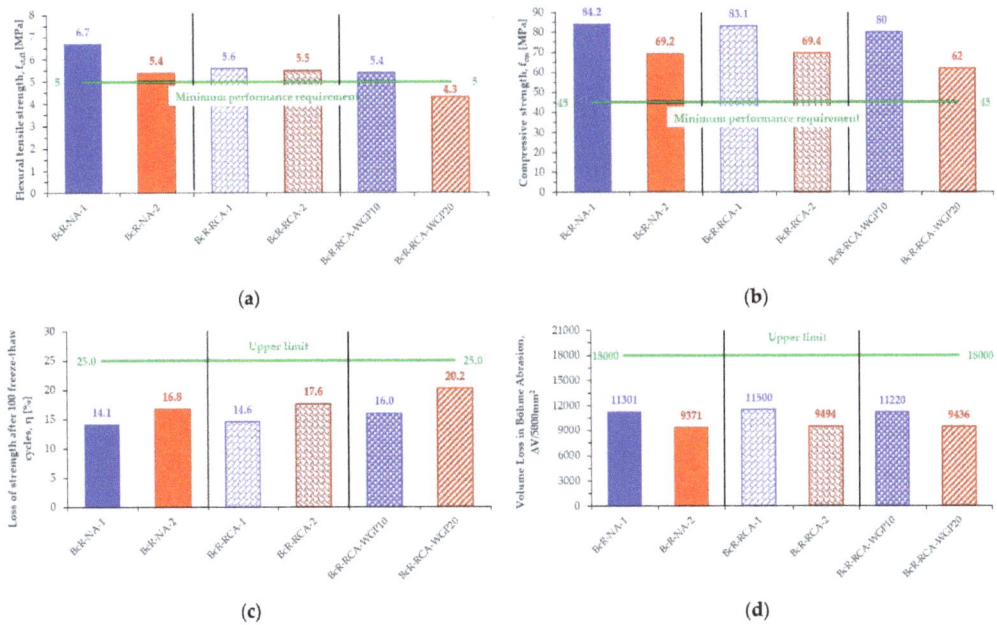

Figure 23. Hardened BcR composites' properties (both variants); (**a**) flexural strength, Var. I and Var. II; (**b**) compressive strength; (**c**) loss of strength after 100 freeze–thaw cycles; (**d**) volume loss in Böhme abrasion.

For road concrete, the flexural strength ($f_{ct,fl}$) classifies the concrete into BcR strength classes, which are presented in Table 21 for the composites considered in this study.

Table 21. BcR Classification Classes according to NE 014:2002 [42].

Hard Property	UM	Performance Level	Mix Design					
			BcR-NA		BcR-RCA		BcR-RCA-WGP	
			Var. I	Var. II	Var. I	Var. II	10%	20%
Flexural Strength ($f_{ct,fl}$)	MPa	4.0–5.0	6.7	5.4	5.6	5.5	5.4	4.3
Compressive Strength (f_c)	MPa	35–45	84.2	69.2	83.1	69.4	80.0	62.0
Loss of Strength (η)	%	25	14.11	16.8	14.60	17.6	16.0	20.2
Achieved Strength Class	BcR	BcR4–BcR5	BcR 5	BcR 5	BcR 5	BcR 5	BcR 5	BcR 4

The design of road concrete mixes meets the requirements of both national NE 012-1:2022 [92,94] and European fib Bulletin 42 [93] standards, as shown in Table 22. For both standards, the acceptable value of the mean compressive strength (f_{cm}) is obtained by adding to the characteristic strength, f_{ck}, a value Δf ranging from 6 to 12 units (MPa). Additionally, it is known that the flexural tensile strength ($f_{ct,fl}$) has values in the range of (1/10–1/20) of f_{cm}, and the values from this study also meet this criterion.

Table 22. Mathematical relation—comparison of European norms and national norms [20].

Source	Mathematical Relation (Cylinders with H/ Φ- 300/150 or Cube with l = 150 mm)
fib Bulletin 42 [95]	(a) $f_{cm} = f_{ck} + \Delta f$, $\Delta f = 8$ MPa
NE 012-1: 2022 [94]	(b) $f_{cm} = f_{ck} + (6\text{–}12)$ MPa

The reported values of the properties of the control composite for each of the variants (Var. I and II), the alternatives with RCA, and the ones with both RCA and WGP—as well as the ratio of properties between the control composites, those with RCA, and those with WGP from the two variants—are expressed by the coefficient of variation (Cv1). In terms of the classification criteria ($f_{ct,fl} > 5.0$ (4.0) MPa; $f_c > 45$ (35) MPa, loss of strength (η) after 100 freeze–thaw cycles < 25%, and the value of volume loss (ΔV) < 18,000 mm^3/5000 mm^2 for classification in the best wear resistance class), they are expressed by the coefficient of variation (Cv2) and presented in Table 23.

Table 23. Coefficients of variation of the properties of interest of road concrete composites.

Mix	$f_{ct,fl}$	Cv1	Cv2	f_c	Cv1	Cv2	η	Cv1	Cv2	ΔV	Cv1	Cv2
Var. I	MPa			MPa			%			/5000 mm^2		
BcR-NA-1	6.7		0.75	84.2		0.65	14.11		1.77	11,301		1.59
BcR-RCA-1	5.6	1.20	0.89	83.1	1.01	0.66	14.60	0.97	1.71	11,500	0.98	1.57
BcR-RCA-WGP10	5.4	1.24	0.93	80.0	1.05	0.69	16.00	0.88	1.56	11,220	1.01	1.60
Var. II												
BcR-NA-2	5.4		0.93	69.2		0.79	16.80		1.49	9371		1.92
BcR-RCA-2	5.5	0.98	0.91	69.4	0.997	0.79	17.60	0.95	1.42	9494	0.98	1.90
BcR-RCA-WGP20	4.3	1.26	0.93	62.0	1.12	0.89	20.20	0.83	1.24	9436	0.99	1.91
Var. I & II												
BcR-NA	6.7			84.2			14.11			11,301		
	5.4	1.24		69.2	1.22		16.80	0.84		9371	1.21	
BcR-RCA	5.6			83.1			14.60			11,500		
	5.5	1.02		69.4	1.20		17.60	0.83		9494	1.21	
BcR-RCA-WGP	5.4			80.0			16.00			11,220		
	4.3	1.26		62.0	1.29		20.20	0.79		9436	1.19	

The value of $f_{ct,fl}$ strength for Var. I obtained for BcR-NA-1 was higher than that of BcR-RCA-1 by 16.4% and higher than that of BcR-RCA-WGP10 by 19.4%. Considering the minimum required value of 5 MPa for classification in BcR5 Class, the value of BcR-NA-1 exceeded the limit by 34%, BcR-RCA-1 by 12%, and BcR-RCA-WGP10 by 8%.

The value of $f_{ct,fl}$ strength for Var. II obtained for BcR-NA-2 was lower than that of BcR-RCA-2 by 1.9% and compared to that of BcR-RCA-WGP20 was higher by 20.4%. Considering the minimum required value of 5 MPa for classification in BcR5 Class, the value of BcR-NA-2 exceeded the limit by 8% and BcR-RCA-2 did so by 10%, whereas BcR-RCA-WGP20 fell into the BcR4 class.

The comparative values of $f_{ct,fl}$ strength between Var. I and Var. II obtained for BcR-NA were higher by 19.4% for BCR-NA-1. At the same time, the value obtained for BcR-RCA-1 was only 1.79% higher than the values obtained for BcR-RCA-2. The BcR-RCA-WGP mixes exhibited similar trends to the values of $f_{ct,fl}$, as BcR-NA mixes, with an increase of 20.37% in the value obtained for BcR-RCA-WGP10 mix.

The values of the flexural tensile strength at the reference age of 28 days of all concrete mixes, except BcR-RCA-WGP20, recorded values greater than >5.0 MPa, corresponding to the requirements for the very heavy traffic class BcR 5.0 [42], the same traffic class as the reference concrete. The increase in the ratio of WGP led to a decrease in the value of the flexural tensile strength, which can be attributed to the higher absorption and porosity compared to natural sand and, consequently, an increased water demand to maintain the desired workability [94].

Apart from the BcR-RCA-WGP20 (BcR4), all other mixes fit into the BcR5 class. The values of f_c strength for Var. I obtained for BcR-NA-1 were higher than the that obtained for the BcR-RCA-1 by 1.3% and compared to that of BcR-RCA-WGP10 by 5.0%. Considering the minimum required value of 45 MPa for classification in BcR5 class, the value of BcR-NA-1 exceeded the limit by 87.11%, that of BcR-RCA-1 did so by 84.67%, and that of BcR-RCA-WGP10 did so by 77.78%.

The value of f_c strength for Var. II obtained for BcR-NA-2 was lower than that of BcR-RCA-2 by 0.3%, and compared to that of BcR-RCA-WGP20, it was 10.4% higher. Considering the minimum required value of 45 MPa for classification in BcR5 class, the value of BcR-NA-2 exceeded the limit by 53.78%, that of BcR-RCA-2 did so by 54.22% and that of BcR-RCA-WGP20 did so by 37.78%.

The comparative value of f_c strength between Var. I and Var. II lead to the conclusion that all mixes belonging to Var. I exhibited consistently higher values of the compressive strength than mixes belonging to Var. II. The differences ranged from 16.5% to 22.5%.

The values of $f_{ct,sp}$ strength associated with Var. I compared to all the $f_{c,sp}$ values associated with composites in Var. II were 17.77% higher for BcR-NA-1, 15.90% higher for BcR-RCA-1, and 22.22% higher for those containing WGP, as seen in Table 20.

The loss of strength (η) due to 100 repeated freeze–thaw cycles of the composites in Variant I was lower than that of those in Var. II, due to the presence of entrained air and the increased w/c ratio for Var. II mixes. The lower η was, the higher the resistance of the composites to freeze–thaw cycles.

The value of η for Var. I obtained for BcR-NA-1 was lower than that of BcR-RCA-1 by 3.47% and than that of BcR-RCA-WGP10 by 13.39%. Considering the minimum required value of 25% for classification in BcR5 class, the value of BcR-NA-1 is lower than the limit by 43.56%, that of BcR-RCA-1 is lower by 41.60%, and the value of BcR-RCA-WGP10 is lower by 36.00%.

The value of η for Var. II obtained for BcR-NA-2 was lower than that of BcR-RCA-2 by 4.76%, and compared to that of BcR-RCA-WGP20, it was lower by 20.24%. Considering the minimum required value of 25% for classification in BcR5 class, the value of BcR-NA-2 was lower than the limit by 32.80%, that of BcR-RCA-2 was lower by 29.60%, and the value of BcR-RCA-WGP20 was lower by 19.20%.

For the comparative value of η between Var. I and Var. II, it can be observed that all values obtained for Var. I mixes were consistently lower than those obtained for Var. II mixes. The difference ranged from 19.06% up to 26.25%.

The results obtained for the mechanical properties indicate that, in general, they decreased as natural aggregate was replaced, cement was substituted with WGP, and the w/c ratio increased. However, all of them highlight better behavior than the limits imposed for the classification into road concrete classes.

The volume loss (ΔV) of the composites in Var. I was greater than that of Var. II due to the increased water content in the designed mixes, but it was also influenced by the strength of the aggregate in the cement matrix and the content of WGP. Composites containing glass powder perform the best in terms of wear, especially the BcR-RCA-WGP20 mix, thus confirming the benefic influence of WGP in the composite and fulfilling the research purpose. Abrasion behavior was also attributed to the characteristics of the constituent aggregates, which form the mineral skeleton incorporated by the cement matrix.

The lower the volume loss, the more resistant the sample was to wear. The greater the amount of FGS (fine glass powder) in the concrete composition, the more resistant the sample subjected to Böhme wear due to the dense composition and the hardness of WGP. All obtained values classify the composites in the best performance class according to the criteria in SR EN 1338:2004 [86], SR EN 1339:2004 [87], and SR EN 1340:2004 [88], specifically in Class 4—Mark I, which requires a volume loss < 18,000 mm^3/5000 mm^2. Regarding the cement replacement variants with 10% and 20% WGP, all mechanical characteristics are higher in the 10% variant except for the volume loss in the wear test (Table 20).

The value of ΔV for Var. I obtained for BcR-NA-1 was lower than that of BcR-RCA-1 by 1.76%, and compared to that of BcR-RCA-WGP10, it was higher by 2.43%. Considering the minimum required value of 18,000 mm^3 for classification in Class 4—Mark I, all mixes from Var. I fulfill the requirement by more than 36%.

The value of ΔV for Var. II obtained for BcR-NA-2 was lower than that of BcR-RCA-2 by 1.31%, and compared to that of BcR-RCA-WGP20, it was higher by 0.69%. Considering the minimum required value of 18,000 mm^3 for classification in Class 4—Mark I, all mixes showed consistently lower values than the upper limit by at least 47%.

The comparative value of ΔV between Var. I and Var. II show that the mixes in Var. II were more resistant to wear. The best wear behavior among all composites was that of BcR-RCA-WGP20.

4.3. Performance of the BcR by PXRD Test

Microscopic studies have highlighted the fact that, from a matrix structure perspective, it was finely crystallized, supporting a compact composite with high mechanical strengths both in the control mix and especially for the one with 10% WGP. Microcrystalline mineral phases were visible, represented by portlandite, calcite, hydrated calcium silicates, and hydrated calcium aluminates. X-ray diffraction indicated a high degree of crystallization in the concrete matrix containing 10% WGP, suggesting a pozzolanic reaction of WGP. Additionally, X-ray diffraction has revealed the presence of calcite, portlandite and gypsum, as well as the presence of calcium silicate hydrate and aluminum silicate hydrate. Thus, the effectiveness of WGP content in BcR composites was confirmed due to its efficient pozzolanic activity and hardness, contributing to durability characteristics [95].

A continuation of this study is planned, involving the assessment of long-term characteristics of the new road concrete with RCA and WGP.

5. Conclusions

This paper focuses on the development of road concrete that aims to use recycled materials from concrete and glass at the end of their life cycle. These materials are incorporated into the cement matrix alongside natural aggregates, with WGP partially substituting Portland cement. The intention is to create a new, sustainable composite that has not been produced yet. Microscopic determinations confirm the pozzolanic efficiency of WGP in concrete through its complete consumption. Additionally, the presence of CSH contributes to the enhancement of mechanical characteristics.

This new composite offers several advantages in terms of reducing existing and potential volumes of mineral waste both economically and visually, reducing pollution, and minimizing the consumption of natural resources and energy (the production of RCA consumes less electrical energy than obtaining natural crushed aggregates, which are rapidly depleted in urban areas undergoing rapid development). Moreover, the potential for practical application is ensured due to meeting durability requirements. Typically, companies incur high fees for the removal, transportation, and disposal of this type of waste at landfills as inert solid waste, reasonable reuse of these waste being of high interest.

This study confirms the appropriate use of RCA and its integration into road concrete. The characteristics of this type of aggregate need more frequent testing compared to natural crushed river aggregate, and they must have values close to those they replace. The fact that RCA contains a quantity of mortar signals the need for water in the mix (effective water) to ensure it is sufficient for a prescribed consistency, considering the increased absorption of RCA. The high absorption of RCA is not detrimental because it can promote the hydration of cement particles that remained un-hydrated for a longer period, contributing to increased mechanical strength even at more advanced ages. This justifies the high characteristic values for composites with RCA content. Another confirmation of the study is the role of WGP in the composites. WGP significantly contributes to the favorable behavior of composites, particularly in terms of wear resistance, which is a key objective of this study. Consequently, these compositions can be used for roads with heavy traffic.

Based on the obtained results, the following conclusion can be drawn:

1. The apparent density of the concrete mix decreases with the substitution of natural river aggregates (4/8 mm) with recycled concrete aggregates and by substituting cement by waste glass powder.
2. High quality recycled concrete aggregates have little influence on the fresh properties of concrete, specifically on the slump values. However, the addition of glass powder leads to a decrease in the slump value due to its higher specific surface area and increased friction force between the particles.
3. Increasing the volume of recycled concrete aggregates and cement replacement percentage by waste glass powder results in decreased values of the slump and significant decrease in workability. In order to counteract this drawback, a higher water/binder ratio is needed coupled with a higher dosage of superplasticizer. This leads to lower values for apparent density but increased air content.
4. Substituting cement with waste glass powder and river aggregates by recycled aggregates results in a decrease in the value of flexural strength. On the other hand, the same substitution has little effect in terms of values of the compressive strength. Increasing the volume of 4/8 mm aggregates, from the total aggregate volume leads to a sharp decrease in the value of the compressive strength as well. Substituting 4/8 mm river aggregates with recycled concrete aggregates and cement with waste glass powder (20% by mass of cement) results in further decrease in mechanical properties.
5. Volume loss in Böhme abrasion and loss of strength after 100 freeze–thaw cycles are significantly lower than the upper limit imposed by existing regulations for all investigated concrete mixes.

Even though the mechanical characteristic values experienced slight decreases for composites with alternative mixtures, all of them fall within the BcR 4 and mostly BcR 5 classes of road concrete, which is the highest road concrete class currently existing in the Romanian standard.

Based on the obtained results in this study, the best concrete mix in terms of mechanical and durability properties and answering to the call for sustainability, is BcR-RCA-WGP10 mix.

6. Patents

Eco-innovative road concrete based on cement, glass powder, and aggregates from recycled concrete waste for applications in the field of constructions "BcR-G":

RO137345A0 • 30 March 2023 • UNIV TEHNICA DIN CLUJ NAPOCA [RO], earliest priority: 29 September 2022 • earliest publication: 30 March 2023. Inventors: Corbu Ofelia Cornelia [RO]; Puskas Attila [RO].

Author Contributions: Conceptualization, O.C. and A.P.; methodology, O.C., A.P. and N.H.; validation, A.P., M.-L.D., N.H. and I.-O.T.; formal analysis, O.C.; investigation, O.C. and N.H.; resources, A.P. and O.C.; writing—original draft preparation, O.C.; writing—review and editing, A.P., M.-L.D., N.H. and I.-O.T.; visualization, I.-O.T.; supervision, A.P.; project administration, O.C. All authors have read and agreed to the published version of the manuscript.

Funding: This research was funded by Project. This paper was supported by the Post-Doctoral Programme POSDRU/159/1.5/S/137516, project cofunded from European Social Fund through the Human Resources Sectorial Operational Program 2007–2013.

Institutional Review Board Statement: Not applicable.

Informed Consent Statement: Not applicable.

Data Availability Statement: All the required data that support the findings are presented in the manuscript.

Conflicts of Interest: The authors declare no conflict of interest.

References

1. Hângănuț, R. Economia de la Liniar la Circular. Available online: https://www.revistasinteza.ro/economia-de-la-liniar-la-circular (accessed on 24 May 2023).
2. Romanian Government ORDONANȚĂ DE URGENȚĂ, nr. 92 din 19 August 2021 Privind Regimul Deșeurilor. Available online: https://legislatie.just.ro/Public/DetaliiDocumentAfis/253009 (accessed on 24 May 2023).
3. Romanian Government Strategia Națională Privind Economia Circulară. Proiect. Available online: https://sgg.gov.ro/1/wp-content/uploads/2022/08/Strategia-economie-circulara_18.08.2022.pdf (accessed on 24 May 2023).
4. Ionescu, B.A.; Barbu, A.-M.; Lăzărescu, A.-V.; Rada, S.; Gabor, T.; Florean, C. The Influence of Substitution of Fly Ash with Marble Dust or Blast Furnace Slag on the Properties of the Alkali-Activated Geopolymer Paste. *Coatings* **2023**, *13*, 403. [CrossRef]
5. Szilagyi, H.; Corbu, O.; Baera, C.; Puskas, A.; Pastrav, M. Opportunities for building materials waste recycling. In Proceedings of the 14th International Multidisciplinary Scientific Geoconference and EXPO, SGEM 2014, Albena, Bulgaria, 17–26 June 2014; Volume 2, pp. 251–258.
6. Toma, I.-O.; Covatariu, D.; Toma, A.-M.; Taranu, G.; Budescu, M. Strength and elastic properties of mortars with various percentages of environmentally sustainable mineral binder. *Constr. Build. Mater.* **2013**, *43*, 348–361. [CrossRef]
7. Burlacu, A.; Racanel, C. Reducing Cost of Infrastructure Works Using New Technologies. In Proceedings of the 3rd International Conference on Road and Rail Infrastructure—CETRA 2014: Road and Rail Infrastructure III, Split, Croatia, 28–30 April 2014; pp. 189–194.
8. Popescu, D.; Burlacu, A. Considerations on the Benefits of Using Recyclable Materials for Road Construction. *Rom. J. Transp. Infrastruct.* **2017**, *6*, 43–53. [CrossRef]
9. Dimulescu, C.; Burlacu, A. Industrial Waste Materials as Alternative Fillers in Asphalt Mixtures. *Sustainability* **2021**, *13*, 8068. [CrossRef]
10. Cadar, R.D.; Boitor, R.M.; Dragomir, M.L. An Analysis of Reclaimed Asphalt Pavement from a Single Source—Case Study: A Secondary Road in Romania. *Sustainability* **2022**, *14*, 7057. [CrossRef]
11. Malhotra, V.M. Making Concrete "Greener" with Fly Ash. *Concr. Int.* **1999**, *21*, 61–66.
12. Corbu, O.C.; Măgureanu, C.; Oneț, T.; Szilágyi, H. Economia de Energie la Realizarea Betoanelor Performante. In *Proceedings of the Știința Modernă și Energia: Producerea, Transportul și Utilizarea Energiei" SME 2010, Ediția XXIX-a, 20–21 May 2010*; Risoprint: Cluj-Napoca, Romania, 2010; pp. 343–354.
13. Althoey, F.; Zaid, O.; Majdi, A.; Alsharari, F.; Alsulamy, S.; Arbili, M.M. Effect of fly ash and waste glass powder as a fractional substitute on the performance of natural fibers reinforced concrete. *Ain Shams Eng. J.* **2023**, 102247. [CrossRef]
14. The European Cement Association (CEMBUREAU) Reaching Climate Neutrality along the Cement and Concrete Value Chain by 2050. Available online: https://www.cembureau.eu/media/kuxd32gi/cembureau-2050-roadmap_final-version_web.pdf (accessed on 10 March 2023).
15. United States Geological Survey (USGS) Mineral Commodity Summaries 2012. Available online: https://d9-wret.s3.us-west-2.amazonaws.com/assets/palladium/production/mineral-pubs/mcs/mcs2012.pdf (accessed on 28 February 2023).
16. Aïtcin, P.-C. Cements of yesterday and today. *Cem. Concr. Res.* **2000**, *30*, 1349–1359. [CrossRef]
17. Sandu, A.V. Obtaining and Characterization of New Materials. *Materials* **2021**, *14*, 6606. [CrossRef]
18. Corbu, O.; Chira, N.; Szilágyi, H. Ecological Concrete by Use of Waste Glass. In Proceedings of the 13th SGEM GeoConference on Nano, Bio and Green—Technologies for a Sustainable Future, Albena, Bulgaria, 16–22 June 2013.
19. The European Commission. *EU Construction & Demolition Waste Management Protocol*; The European Commission: Luxembourg, 2016.
20. Corbu, O.; Puskás, A.; Szilágyi, H.; Baeră, C. C16/20 Concrete Strength Class Design With Recycled Aggregates. *J. Appl. Eng. Sci.* **2014**, *4*, 13–19.
21. Puskás, A.; Corbu, O.; Szilágyi, H.; Moga, L.M. Construction waste disposal practices: The recycling and recovery of waste. In Proceedings of the Sustainable City IX: Urban Regeneration and Sustainability, Siena, Italy, 23–25 September 2014; pp. 1313–1321.
22. Corbu, O.; Puskás, A.; Moga, L.M.; Szilágyi, H. Opportunities for Increasing the Recycling Rate of Mineral Waste in Construction Industry. In Proceedings of the 15th International Multidisciplinary Scientific Geo-Conference Surveying Geology and Mining Ecology Management, Albena, Bulgaria, 18–24 June 2015; pp. 203–210.
23. Frondistou-Yannas, S. Waste Concrete as Aggregate for New Concrete. *ACI J. Proc.* **1977**, *74*, 373–376. [CrossRef]
24. Corbu, O.; Puskás, A.; Sandu, A.V.; Ioani, A.M.; Hussin, K.; Sandu, I.G. New Concrete with Recycled Aggregates from Leftover Concrete. *Appl. Mech. Mater.* **2015**, *754–755*, 389–394. [CrossRef]
25. TrustGod, J.A.; Stevyn, A.I.; Manfred, A.D. Use of Calcined Waste Glass Powder As a Pozzolanic Material. *Eur. J. Eng. Technol. Res.* **2019**, *4*, 53–56. [CrossRef]
26. Raza, A. Mechanical Performance of Lean Mortar by Using Waste Glass Powder as a Replacement of Cement. *Int. J. Res. Appl. Sci. Eng. Technol.* **2022**, *10*, 1447–1453. [CrossRef]
27. Yassen, M.M.; Hama, S.M.; Mahmoud, A.S. Shear behavior of reinforced concrete beams incorporating waste glass powder as partial replacement of cement. *Eur. J. Environ. Civ. Eng.* **2023**, *27*, 2194–2209. [CrossRef]

28. Majeed, W.Z.; Aboud, R.K.; Naji, N.B.; Mohammed, S.D. Investigation of the Impact of Glass Waste in Reactive Powder Concrete on Attenuation Properties for Bremsstrahlung Ray. *East Eur. J. Phys.* **2023**, 102–108. [CrossRef]
29. Mahajan, L.S.; Bhagat, S.R. Utilization of Pozzolanic Material and Waste Glass Powder in Concrete. In *Recent Trends in Construction Technology and Management*; Springer: Singapore, 2023; pp. 201–206.
30. Baikerikar, A.; Mudalgi, S.; Ram, V.V. Utilization of waste glass powder and waste glass sand in the production of Eco-Friendly concrete. *Constr. Build. Mater.* **2023**, *377*, 131078. [CrossRef]
31. Bompa, D.V.; Xu, B.; Corbu, O. Evaluation of One-Part Slag–Fly-Ash Alkali-Activated Mortars Incorporating Waste Glass Powder. *J. Mater. Civ. Eng.* **2022**, *34*, 05022001. [CrossRef]
32. Pashtoon, M.I.; Miakhil, S.; Behsoodi, M.M. Waste Glass Powder "An Alternative of Cement in Concrete": A Review. *Int. J. Curr. Sci. Res. Rev.* **2022**, *05*, 2541–2549. [CrossRef]
33. Manikandan, P.; Vasugi, V. The potential use of waste glass powder in slag-based geopolymer concrete—An environmental friendly material. *Int. J. Environ. Waste Manag.* **2023**, *31*, 291–307. [CrossRef]
34. Shalan, A.H.; El-Gohary, M.M. Long-Term Sulfate Resistance of Blended Cement Concrete with Waste Glass Powder. *Pract. Period. Struct. Des. Constr.* **2022**, *27*, 04022047. [CrossRef]
35. Shankar, H.S.; Gurubasav, S.H.; Shashikanth, A.K.; Akash, A.C. The effect of partial replacement of cement by waste green glass powder with waste plastic as coarse aggregate. *I-Manager's J. Struct. Eng.* **2022**, *11*, 1. [CrossRef]
36. Shi, C.; Zheng, K. A review on the use of waste glasses in the production of cement and concrete. *Resour. Conserv. Recycl.* **2007**, *52*, 234–247. [CrossRef]
37. Federico, L.M.; Chidiac, S.E. Waste glass as a supplementary cementitious material in concrete—Critical review of treatment methods. *Cem. Concr. Compos.* **2009**, *31*, 606–610. [CrossRef]
38. Jani, Y.; Hogland, W. Waste glass in the production of cement and concrete—A review. *J. Environ. Chem. Eng.* **2014**, *2*, 1767–1775. [CrossRef]
39. Abendeh, R.M.; AbuSalem, Z.T.; Bani Baker, M.I.; Khedaywi, T.S. Concrete containing recycled waste glass: Strength and resistance to freeze–thaw action. *Proc. Inst. Civ. Eng.—Constr. Mater.* **2021**, *174*, 75–87. [CrossRef]
40. Hornea, L.; Gorea, M.; Har, N. Study of (Pb, Ba)—CRT glass waste behavior as a partial aggregate replacement in cement mortars. *Stud. Univ. Babeș-Bolyai Chem.* **2017**, *62*, 343–356. [CrossRef]
41. Topçu, İ.B.; Şengel, S. Properties of concretes produced with waste concrete aggregate. *Cem. Concr. Res.* **2004**, *34*, 1307–1312. [CrossRef]
42. NE 014-2002; Normativ Pentru Executarea Îmbrăcăminților Rutiere din Beton de Ciment în Sistem de Cofraje Fixe și Glisante. Ministry of Development, Public Works and Administration: Bucharest, Romania, 2002.
43. SR EN 206-1:2021; Concrete—Specification, Performance, Production and Conformity. ASRO (Romanian Standards Association): Bucharest, Romania, 2021.
44. Tobo, H.; Miyamoto, Y.; Watanabe, K.; Kuwayama, M.; Ozawa, T.; Tanaka, T. Solidification Conditions to Reduce Porosity of Air-cooled Blast Furnace Slag for Coarse Aggregate. *J. Iron Steel Inst. Jpn.* **2013**, *99*, 532–541. [CrossRef]
45. Verian, K.P.; Panchmatia, P.; Olek, J.; Nantung, T. Pavement Concrete with Air-Cooled Blast Furnace Slag and Dolomite as Coarse Aggregates. *Transp. Res. Rec. J. Transp. Res. Board* **2015**, *2508*, 55–64. [CrossRef]
46. Ta, Y.; Tobo, H.; Watanabe, K. Development of Manufacturing Process for Blast Furnace Slag Coarse Aggregate with Low Water Absorption. *JFE Tech. Rep.* **2018**, *23*, 97–101.
47. Roy, S.; Ahmad, S.I.; Rahman, M.S.; Salauddin, M. Experimental investigation on the influence of induction furnace slag on the fundamental and durability properties of virgin and recycled brick aggregate concrete. *Results Eng.* **2023**, *17*, 100832. [CrossRef]
48. SR EN 933-1:2012; Tests for Geometrical Properties of Aggregates. Part 1: Determination of Particle Size Distribution—Sieving Method. ASRO (Romanian Standards Association): Bucharest, Romania, 2012.
49. SR EN 1097-6:2022; Tests for Mechanical and Physical Properties of Aggregates—Part 6: Determination of Particle Density and Water Absorption. ASRO (Romanian Standards Association): Bucharest, Romania, 2022.
50. SR EN 1097-2:2020; Tests for Mechanical and Physical Properties of Aggregates—Part 2: Methods for the Determination of Resistance to Fragmentation. ASRO (Romanian Standards Association): Bucharest, Romania, 2020.
51. SR EN 1097-1:2011; Tests for Mechanical and Physical Properties of Aggregates—Part 1: Determination of the Resistance to Wear (Micro-Deval). ASRO (Romanian Standards Association): Bucharest, Romania, 2011.
52. SR EN 933-3:2012; Tests for Geometrical Properties of Aggregates. Part 3: Determination of Particle Shape—Flatness Index. BASRO (Romanian Standards Association): Bucharest, Romania, 2012.
53. SR EN 197-1; Cement. Part I: Composition, Specifications and Conformity Criteria for Normal Use Cements. ASRO (Romanian Standards Association): Bucharest, Romania, 2011.
54. SR EN 196-2:2013; Method of Testing Cement—Part 2: Chemical Analysis of Cement. ASRO (Romanian Standards Association): Bucharest, Romania, 2013.
55. SR EN 196-3:2017; Methods of Testing Cement. Determination of Setting Times and Soundness. ASRO (Romanian Standards Association): Bucharest, Romania, 2017.
56. SR EN 196-1:2016; Methods of Testing Cement—Part 1: Determination of Strength. ASRO (Romanian Standards Association): Bucharest, Romania, 2016.

57. Corbu, O.; Ioani, A.M.; Al Bakri Abdullah, M.M.; Meiță, V.; Szilagyi, H.; Sandu, A.V. The Pozzolanic Activity Level of Powder Waste Glass in Comparisons with other Powders. *Key Eng. Mater.* **2015**, *660*, 237–243. [CrossRef]
58. Taha, B.; Nounu, G. Properties of concrete contains mixed colour waste recycled glass as sand and cement replacement. *Constr. Build. Mater.* **2008**, *22*, 713–720. [CrossRef]
59. Vafaei, M.; Allahverdi, A. High strength geopolymer binder based on waste-glass powder. *Adv. Powder Technol.* **2017**, *28*, 215–222. [CrossRef]
60. Meena, M.K.; Gupta, J.; Nagar, B. Performance of Concrete by Using Glass Powder—An Experimental Study. *Int. Res. J. Eng. Technol.* **2018**, *5*, 840–844.
61. Corbu, O.; Bompa, D.V.; Szilagyi, H. Eco-efficient cementitious composites with large amounts of waste glass and plastic. *Proc. Inst. Civ. Eng.—Eng. Sustain.* **2022**, *175*, 64–74. [CrossRef]
62. Mansour, M.A.; Ismail, M.H.B.; Imran Latif, Q.B.a.; Alshalif, A.F.; Milad, A.; Bargi, W.A.A. A Systematic Review of the Concrete Durability Incorporating Recycled Glass. *Sustainability* **2023**, *15*, 3568. [CrossRef]
63. Wikipedia Pozzolanic Activity. Available online: https://en.wikipedia.org/wiki/Pozzolanic_activity (accessed on 20 April 2023).
64. *ASTM C618-22*; Standard Specification for Coal Fly Ash and Raw or Calcined Natural Pozzolan for Use in Concrete. ASTM International: West Conshohocken, PA, USA, 2022.
65. Raki, L.; Beaudoin, J.; Alizadeh, R.; Makar, J.; Sato, T. Cement and Concrete Nanoscience and Nanotechnology. *Materials* **2010**, *3*, 918–942. [CrossRef]
66. Shayan, A.; Xu, A. Value-added utilisation of waste glass in concrete. *Cem. Concr. Res.* **2004**, *34*, 81–89. [CrossRef]
67. Khairul Nizar, I.; Al Bakri, A.M.M.; Rafiza, A.R.; Kamarudin, H.; Abdullah, A.; Yahya, Z. Study on Physical and Chemical Properties of Fly Ash from Different Area in Malaysia. *Key Eng. Mater.* **2013**, *594–595*, 985–989. [CrossRef]
68. Liu, S.; Li, L. Influence of fineness on the cementitious properties of steel slag. *J. Therm. Anal. Calorim.* **2014**, *117*, 629–634. [CrossRef]
69. Matos, A.M.; Sousa-Coutinho, J. Waste glass powder in cement: Macro and micro scale study. *Adv. Cem. Res.* **2016**, *28*, 423–432. [CrossRef]
70. Dai, J.; Wang, Q.; Xie, C.; Xue, Y.; Duan, Y.; Cui, X. The Effect of Fineness on the Hydration Activity Index of Ground Granulated Blast Furnace Slag. *Materials* **2019**, *12*, 2984. [CrossRef]
71. Lucaci, G. Imbracaminti rutiere rigide. Available online: https://www.ct.upt.ro/studenti/cursuri/lucaci/Imbr_rutiere_rigide.pdf (accessed on 15 May 2023).
72. *SR EN 1008:2003*; Mixing Water for Concrete. ASRO (Romanian Standards Association): Bucharest, Romania, 2003.
73. *SR EN 934-2+A1:2012*; Concrete Admixtures. ASRO (Romanian Standards Association): Bucharest, Romania, 2012.
74. *SR EN 12620+A1:2008*; Aggregates for Concrete. ASRO (Romanian Standards Association): Bucharest, Romania, 2008.
75. *SR EN 12043:2013*; Natural Aggregates and Processed Stone for Roads. ASRO (Romanian Standards Association): Bucharest, Romania, 2013.
76. Hansen, T.C.; Narud, H. Strength of Recycled Concrete Made from Crushed Concrete Coarse Aggregate. *Concr. Int.* **1983**, *5*, 79–83.
77. Babafemi, A.J.; Sirba, N.; Paul, S.C.; Miah, M.J. Mechanical and Durability Assessment of Recycled Waste Plastic (Resin8 & PET) Eco-Aggregate Concrete. *Sustainability* **2022**, *14*, 5725. [CrossRef]
78. *SR EN 12350-2:2019*; Testing Fresh Concrete—Part 2: Slump-Test. ASRO (Romanian Standards Association): Bucharest, Romania, 2019.
79. *SR EN 12350-6:2019*; Testing Fresh Concrete—Part 6: Density. ASRO (Romanian Standards Association): Bucharest, Romania, 2019.
80. *SR EN 12350-7:2019*; Testing Fresh Concrete—Part 7: Air Content—Pressure Methods. ASRO (Romanian Standards Association): Bucharest, Romania, 2019.
81. *SR EN 12390-5:2019*; Test on Hardened Concrete—Part 5: Bending Tensile Strength of Specimens. ASRO (Romanian Standards Association): Bucharest, Romania, 2019.
82. *SR EN 12390-3/2009*; Testing hardened concrete. Part 3: Compressive Strength of Test Specimens. ASRO (Romanian Standards Association): Bucharest, Romania, 2009.
83. *SR EN 12390-6/2010*; Testing Hardened Concrete. Part 6: Tensile Splitting strength of Test Specimens. ASRO (Romanian Standards Association): Bucharest, Romania, 2010.
84. *SR EN 12390-7:2019*; Testing Hardened Concrete—Part 7: Density of Hardened Concrete. ASRO (Romanian Standards Association): Bucharest, Romania, 2019.
85. *SR 3518:2009*; Tests on Concrete. Determining the Freeze-Thaw Resistance by Measuring the Variation of the Compressive Strength and/or the Relative Dynamic Modulus of Elasticity. ASRO (Romanian Standards Association): Bucharest, Romania, 2009.
86. *SR EN 1338:2004*; Concrete Paving Blocks—Requirements and Test Methods. ASRO (Romanian Standards Association): Bucharest, Romania, 2004.
87. *SR EN 1339:2004*; Concrete Paving Flags—Requirements and Test Methods. ASRO (Romanian Standards Association): Bucharest, Romania, 2004.
88. *SR EN 1340:2004*; Concrete Kerb Units—Requirements and Test Methods. ASRO (Romanian Standards Association): Bucharest, Romania, 2004.
89. *SR CR 12793:2002*; Determination of the Depth of the Carbonation Layer of Hardened Concrete. ASRO (Romanian Standards Association): Bucharest, Romania, 2002.

90. Hosseini, P.; Abolhasani, M.; Mirzaei, F.; Kouhi Anbaran, M.R.; Khaksari, Y.; Famili, H. Influence of Two Types of Nanosilica Hydrosols on Short-Term Properties of Sustainable White Portland Cement Mortar. *J. Mater. Civ. Eng.* **2018**, *30*, 04017289. [CrossRef]
91. Ahmad, J.; Tufail, R.F.; Aslam, F.; Mosavi, A.; Alyousef, R.; Faisal Javed, M.; Zaid, O.; Khan Niazi, M.S. A Step towards Sustainable Self-Compacting Concrete by Using Partial Substitution of Wheat Straw Ash and Bentonite Clay Instead of Cement. *Sustainability* **2021**, *13*, 824. [CrossRef]
92. NE 012/1-2022; Normativ Privind Producerea și Executarea Lucrărilor Din Beton, Beton Armat și Beton Precomprimat—Partea 1: Producerea Betonului. Ministry of Development, Public Works and Administration: Bucharest, Romania, 2022.
93. Fédération Internationale du Béton (FIB). *Constitutive Modelling of High Strength/High Performance Concrete—State of the Art Report*; FIB: Lausanne, Switzerland, 2008.
94. Jain, S.; Santhanam, M.; Rakesh, S.; Kumar, A.; Gupta, A.K.; Kumar, R.; Sen, S.; Ramna, R.V. Utilizationof air-cooled blast furnace slag as a 100% replacement of river sand in mortar and concrete. *Indian Concr. J.* **2022**, *96*, 6–21.
95. Pimienta, P.; Albert, B.; Huet, B.; Dierkens, M.; Fransisco, P.; Rougeau, P. Durability performance assessment of non-standard cementitious materials for buildings: A general method applied to the French context. *RILEM Tech. Lett.* **2016**, *1*, 102–108. [CrossRef]

Disclaimer/Publisher's Note: The statements, opinions and data contained in all publications are solely those of the individual author(s) and contributor(s) and not of MDPI and/or the editor(s). MDPI and/or the editor(s) disclaim responsibility for any injury to people or property resulting from any ideas, methods, instructions or products referred to in the content.

Article

Study of the Design and Mechanical Properties of the Mix Proportion for Desulfurization Gypsum–Fly Ash Flowable Lightweight Soil

Xianglong Zuo [1], Shen Zuo [1], Jin Li [1,*], Ning Hou [2,*], Haoyu Zuo [1] and Tiancheng Zhou [1]

1 School of Civil Engineering, Shandong Jiaotong University, 5 Jiaoxiao Road, Jinan 250357, China; 21107007@stu.sdjtu.edu.cn (X.Z.); zuoshen2006@163.com (S.Z.); 21107008@stu.sdjtu.edu.cn (H.Z.); 21107042@stu.sdjtu.edu.cn (T.Z.)
2 School of Civil Engineering, China University of Geosciences Beijing, 29 Xueyuan Road, Beijing 100083, China
* Correspondence: sdzblijin@163.com (J.L.); sdhouning@163.com (N.H.)

Abstract: In order to solve the global problem of bridge head jumping caused by the insufficient compaction of the roadbed in the transition section of highways and bridges, a desulfurization gypsum–fly ash flowable lightweight soil without vibration, capable of self-compaction, low bulk density, and economic and environmental protection, has been developed. This study selected low-grade cement, industrial waste (fly ash and desulfurization gypsum), and Yellow River silt as the raw materials for the design of the mix ratio of a desulfurization gypsum–fly ash flow-state lightweight soil mix. Through multiple indoor experiments, the influence of cement content, silt content, and the fly ash/desulfurization gypsum quality ratio on its fluidity and mechanical properties was systematically studied. The stress–strain relationship under uniaxial compression was analyzed and the strength formation mechanism was revealed through scanning electron microscopy (SEM). The results show that the mechanical properties of the prepared desulfurization gypsum–fly ash flowable lightweight soil meet the engineering requirements. Increasing both the cement and fly ash content results in the decreased fluidity of the desulfurization gypsum and fluidized fly ash. However, as the mass ratio of fly ash to desulfurization gypsum increases, the fluidity reaches its maximum when the mass ratio of fly ash to desulfurization gypsum is 2:1. Based on the stress–strain relationship test results, a uniaxial compressive constitutive model of the desulfurization gypsum–fly ash flowable lightweight soil was proposed. The model was fitted and analyzed with the test results, and the correlation was greater than 0.96. The high degree of agreement showed that desulfurization gypsum can promote the disintegration of fly ash, thereby increasing the specific surface area. This provides more contact points, promotes the hardening process, and enhances the interlocking force between particles and the formation of cementitious substances, further enhancing strength.

Keywords: road engineering; mix design; mechanical properties; new materials; constitutive model; SEM

Citation: Zuo, X.; Zuo, S.; Li, J.; Hou, N.; Zuo, H.; Zhou, T. Study of the Design and Mechanical Properties of the Mix Proportion for Desulfurization Gypsum–Fly Ash Flowable Lightweight Soil. *Coatings* **2023**, *13*, 1591. https://doi.org/10.3390/coatings13091591

Academic Editors: Ionut Ovidiu Toma and Ofelia-Cornelia Corbu

Received: 16 August 2023
Revised: 4 September 2023
Accepted: 6 September 2023
Published: 12 September 2023

Copyright: © 2023 by the authors. Licensee MDPI, Basel, Switzerland. This article is an open access article distributed under the terms and conditions of the Creative Commons Attribution (CC BY) license (https://creativecommons.org/licenses/by/4.0/).

1. Introduction

As an important component of the highway structure, the subgrade usually needs to be connected to bridges, culverts, retaining walls, etc. The part of the subgrade that connects with the sides of bridges and culverts, as well as the back of retaining walls, is known as the "Three Structures' Fill". Due to significant differences in material strength and stiffness between the subgrade and the structures, as well as limited construction space at the "Three Structures' Fill", insufficient compaction often occurs due to the inability to use heavy compaction machinery on the subgrade. This easily leads to differential settlement issues between the subgrade and the structures, significantly impacting the comfort and safety of highway operation. Therefore, controlling the post-construction and differential settlement of the subgrade and improving the convenience of subgrade construction are essential prerequisites and foundations for ensuring the construction and safe operation of highways.

In recent years, numerous scholars have conducted significant research addressing the problem of subgrade soil treatment in the "Three Structures' Fill". The prevailing approach utilized at present is the application of lightweight subgrade fill materials. Flowable fly ash serves as a characteristic Controlled Low Strength Material (CLSM) [1]. It is predominantly composed of cement, fly ash, and water mixed in specific proportions to create a highly flowable mixture. Following the appropriate curing process, it solidifies into a blended material with a certain level of strength [2]. Yuan Xiaoya [3] found that fly ash has the characteristic of dispersing the cement particle aggregation structure and promoting particle dispersion, thereby improving the internal microstructure of cement-based materials. This improves its mechanical properties. Jia Yan [4] demonstrated in their research that a cement–fly ash slurry exhibited high fluidity and good strength after consolidation, making it suitable for backfilling in subgrade construction. Huang Zhiqin and Yu Yunyan [5] investigated the use of fly ash as an additive in red mudstone and found that it effectively promotes particle aggregation, enhances the unconfined compressive strength, and significantly improves the deformation resistance of the soil. Considering the favorable crack resistance of fly ash concrete, many researchers have focused on increasing the proportion of fly ash in concrete in recent years [6–9]. Ma Chengchang et al. [10] showed that the tensile creep of concrete increases with the increase in fly ash content under the same loading age. Liu Jun et al. [11] studied the mechanical properties of lightweight aggregates under different curing conditions using SEM, X-ray analysis, and other microscopic analysis techniques, considering factors such as aggregate physical and chemical properties. Janardhanatn R et al. [12] found in their study on the mix proportion of flowable fly ash that a fly ash content as high as 90% still met the usage requirements after 14 freeze–thaw cycles, demonstrating its feasibility. Wang Yanzhang [13–15] created a mix proportion design for flowable fly ash. They analyzed the mechanisms of strength development and systematically studied the effects of various raw material proportions on consistency, strength, and shrinkage. Additionally, they performed post-construction settlement calculations, stability analysis, and field tests. Lin et al. [16] conducted laboratory tests to analyze the influence of water content, cement content, and curing environment on the flowability and mechanical properties of flowable fly ash and proposed evaluation indicators. Yang Chunfeng [17–20] carried out a mix proportion design for flowable fly ash, summarized and analyzed the optimal water content for liquid fly ash, and conducted field tests in a reconstruction and expansion project. Sun Jishu et al. [21] designed foam fly ash by incorporating a foaming agent and provided recommended mix proportions. Zhen Wukui et al. [22–24] redesigned the mix proportion of flowable fly ash by adding lime and investigated its engineering properties and freeze–thaw resistance. Although flowable fly ash has shown good performance in subgrade backfilling, it has not been widely promoted in China. The main reason is the rising cost of fly ash in recent years. To better promote the use of flowable fly ash and reduce its production cost, inexpensive coarse aggregates can be added to the mixture. Coarse aggregates can serve as a reinforcing framework and reduce volume changes [25]. Gabr et al. [26] added different amounts of mine tailings pond sludge (AMD) to flowable fly ash and found that it increased the compressive strength by 10–70%. Do et al. [27] discovered that substituting some of the cement in flowable fly ash with reddish clay reduced its fluidity and shortened the setting time. When the substitution rate reached 15%, the 28-day compressive strength increased by 6.3%. In their study, Wu et al. [28] discovered that the inclusion of paper mill sludge at 5% and 10% reduced the strength of flowable fly ash by approximately 20% and 40%, respectively. Additionally, this inclusion led to a decrease in fluidity.

Due to considerations regarding material costs and environmental issues, the use of locally sourced aggregates for the preparation of flowable fly ash has gained widespread attention. Researchers such as Do et al. [29] have studied the use of ponded ash (PA) and excavated soil (ES) as aggregates, and experimental results show that the compressive strength and fluidity of flowable fly ash decrease while the initial setting time lengthens with a decrease in the PA/ES ratio. Tuerkel [30] prepared flowable fly ash by adding limestone sand with particle sizes of 0–5 mm, and the results showed that as the limestone sand content increased, the water absorption of the flowable fly ash increased. Chittoori [31–34] prepared flowable fly ash by adding two types of clay, high-plasticity clay and a 1:1 mixture of high- and low-plasticity clay, and a comparative analysis revealed that flowable fly ash with only high-plasticity clay exhibited better overall performance. It can be seen that adding soil as an aggregate to flowable fly ash is feasible, but the type and nature of the soil have a significant impact on its properties.

Desulfurized gypsum and fly ash are two types of industrial waste that have a significant impact on the environment due to their large output. However, the current utilization rate of these waste materials is very low. Therefore, it is crucial to find rational ways to utilize them in order to protect the environment effectively. In this study, a new type of flowable lightweight soil, called desulfurized gypsum–fly ash flowable lightweight soil, was formulated using fly ash, industrial waste desulfurized gypsum, and powdered soil as a roadbed improvement filling material. Performance tests on different mix proportions were conducted and the effects of factors such as the dosage of each raw material and curing time on the flowability and mechanical properties of the soil were analyzed. This systematic understanding provides valuable insights. Furthermore, by performing uniaxial compressive strength tests, the stress–strain response of the material was plotted and analyzed, and a constitutive model was established. These findings have significant implications for the engineering application and promotion of desulfurized gypsum–fly ash flowable lightweight soil.

2. Test Materials and Test Plans

2.1. Material

2.1.1. Fly Ash

This experiment focused on the secondary fly ash obtained from Shandong Huaneng Jinan Huangtai Power Generation Co., Ltd. (Jinan, China). The chemical composition and performance indicators of this fly ash are presented in Tables 1 and 2 (provided by the supplier). Figure 1 illustrates the macro and micro diagrams of the fly ash, revealing that it primarily consists of spherical particles with a minor presence of impurities. The particle size of spherical particles under SEM is between 0.92 microns and 31 microns. A ZEISS sigma 500 SEM (Shandong Wusheng Information Technology Co., Ltd. Jinan, China) was chosen to observe the microscopic morphology of the fly ash and desulfurization gypsum with an acceleration voltage of 4 kV and a working distance of 6.3 mm. Figure 2 shows the XRD pattern of fly ash. A BRUKER D8 Advance XRD (Jinan Mengmai International Trade Co., Ltd. Jinan, China) was selected for ex situ testing of the fly ash and desulfurization gypsum at room temperature with a 2θ range of 10–80° in steps of 0.02° at 10°/min. Its main crystalline phases are SiO_2 and $Al_6Si_2O_{13}$.

Table 1. Chemical composition of fly ash.

Chemical Composition	Fe_2O_3	CaO	SO_3	Al_2O_3	SiO_2
Mass fraction/%	7.50	4.63	1.50	23.67	56.96

Table 2. Performance indicators of fly ash.

Density/(kg·m^{-3})	Specific Surface Area/(m^2·kg^{-1})	Fineness/%	Loss on Ignition/%
2389	467	16.8	7.5

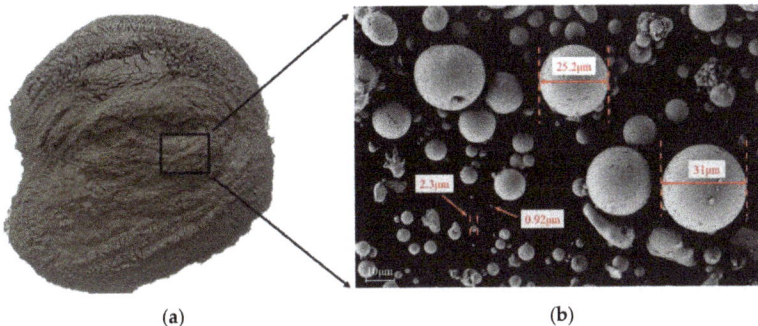

Figure 1. Macro and micro diagrams of fly ash: (**a**) fly ash; (**b**) SEM image.

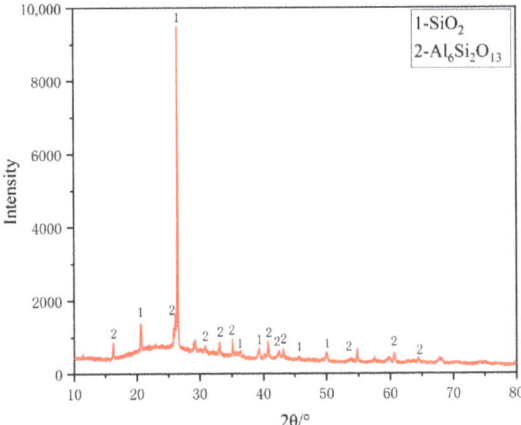

Figure 2. XRD pattern of fly ash.

2.1.2. Desulfurization Gypsum

This study focused on the waste generated from flue gas desulfurization at Shandong Huaneng Jinan Huangtai Power Generation Co., Ltd. The results are presented in Figures 3 and 4, which illustrate the macro and micro characteristics of the desulfurization gypsum, as well as the XRD pattern. It is evident from the figures that the desulfurization gypsum used in this study has a brownish-yellow powder appearance and is primarily composed of elongated blocks. The crystal phase of the gypsum is identified as $Ca_2SO_4 \cdot 2H_2O$.

Figure 3. Macro and micro diagrams of desulfurization gypsum: (**a**) desulfurization gypsum; (**b**) SEM image.

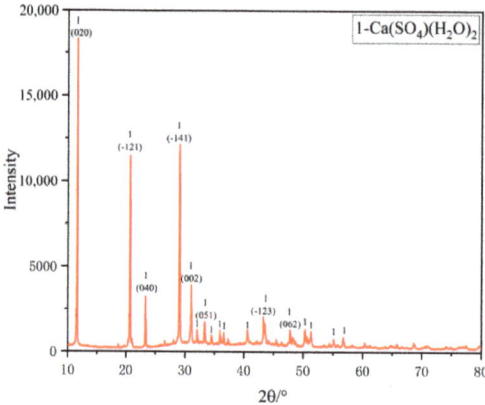

Figure 4. XRD pattern of desulfurization gypsum.

2.1.3. Cement

This experiment used P.O.42.5 ordinary Portland cement. Its performance indicators are shown in Table 3 (provided by the supplier).

Table 3. Cement performance indicators.

Water Requirement of Normal Consistency/%	Setting Time/min		Flexural Strength/MPa			Compressive Strength/MPa		
	Initial	Final	3 d	7 d	28 d	3 d	7 d	28 d
26.9	205	265	5.6	7.5	9.6	30.2	36.7	48.4

2.1.4. Silt

This study focused on the selection of silt from the experimental section of the Beijing Taiwan Expressway renovation and expansion project. Silt is not commonly used as a roadbed filling material due to its unique properties, including low plasticity, poor viscosity, low strength, difficulty in compaction, and susceptibility to liquefaction and erosion. The analysis of Table 4 and Figure 5 reveals that the distribution range of silt particle groups in the experimental section of this project is relatively wide; however, the particle matching is inadequate.

Table 4. The grading criteria of clay soil.

Control Grain Size d_{60}/mm	Effective Grain Diameter d_{10}/mm	Median Diameter d_{30}/mm	Coefficient of Nonuniformity C_u	Coefficient of Curvature C_c
0.092	0.013	0.087	7.1	6.3

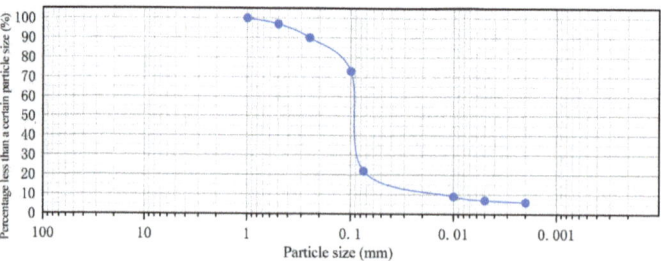

Figure 5. The particle size distribution chart of the clay soil.

2.2. Material Preparation and Mix Ratio Plan

2.2.1. Material Preparation

Due to the small particle sizes of various raw materials, agglomeration is likely to occur during the preparation process. This makes it difficult for water to mix into the solid material, resulting in uneven mixing of the mixture. In this study, the following steps were used for blending: Firstly, weigh the raw materials according to the design mix comparison. Pour the weighed fly ash, desulfurization gypsum, cement, and powder soil into a mixer and mix for about 1 min. Then, add 50% of the required water and stir for 30 s. After that, add all the remaining water and stir for about 3 min. This process forms a desulfurization gypsum–fly ash flowable lightweight soil slurry. Finally, pour the slurry into the trial mold for pouring. The reason for using the method of adding water in stages is that it solves the problem of easy splashing when mixing with water once, which is caused by the difficulty of stirring due to the solid materials and water layers. The specific preparation process is shown in Figure 6.

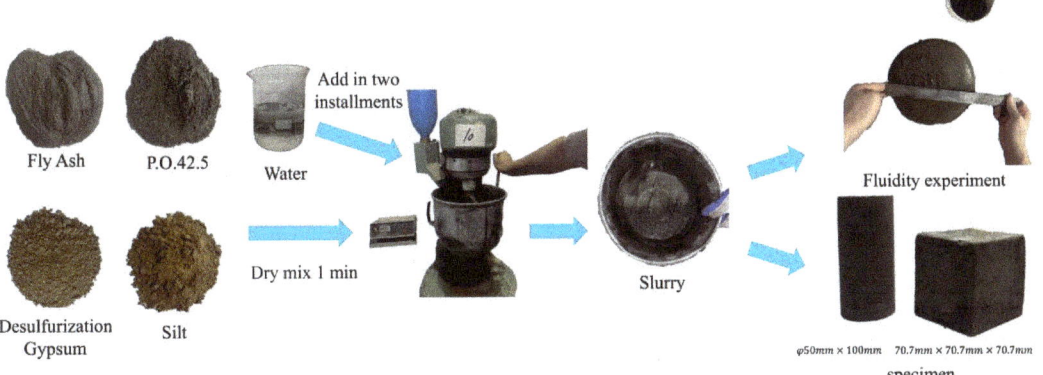

Figure 6. Preparation process.

In this study, two types of specimens were made using desulfurization gypsum and fluidized fly ash. Unconfined compressive strength testing was conducted on 70.7 mm × 70.7 mm × 70.7 mm cube specimens, while uniaxial compression tests were performed on φ 50 mm × 100 mm cylindrical specimens. No vibration was required after pouring the desulfurization gypsum and fluidized fly ash. Once the surface moisture dried slightly, any excess slurry above the mold was flattened using a scraper and the surface was covered with a damp cloth. Next, the specimens were placed in an environment with a room temperature of 20 ± 5 °C and a relative humidity greater than 50%. They were left to stand for 1–2 days before demolding and labeling. After demolding, the specimens were subjected to standard curing conditions. Relevant tests were conducted after 7 or 28 days of curing.

2.2.2. Mix Proportion Design Plan

A mix proportion design plan was adopted to investigate the effects of the mass ratio of fly ash to desulfurization gypsum, cement content, and silt content on the flowability and unconfined strength of the desulfurization gypsum–fluidized fly ash. On the basis of a large number of preliminary trial tests, a fixed water solid ratio of 0.34 was selected, and mass ratios of fly ash to desulfurization gypsum of 1:1, 2:1, and 3:1 were selected. Cement contents of 6%, 8%, and 10% were selected, and silt contents of 0, 20%, 40%, and 60% were selected. The external mixing method was adopted for the silt, and the specific mix design is shown in Table 5.

Table 5. Mix ratio of desulfurization gypsum–fly ash flowable lightweight soil.

Sample	Mass Ratio of Fly Ash to Desulfurization Gypsum	Cement Content	Silt Content	Composition of Fluidized Fly Ash/m^3				
				Cement/kg	Desulfurization Gypsum/kg	Fly Ash/kg	Silt/kg	Water/kg
1		6%	0%	77.02	603.34	603.34		436.46
2			20%	64.94	508.70	508.70	216.47	441.60
3			40%	56.14	439.73	439.73	374.24	445.34
4			60%	49.43	387.22	387.22	494.33	448.19
5	1:1	8%	0%	103.02	592.39	592.39		437.85
6			20%	86.82	449.22	449.22	217.05	442.78
7			40%	75.02	431.37	431.37	375.11	446.38
8			60%	66.04	379.76	379.76	495.34	449.11
9		10%	0%	129.19	581.36	581.36		439.25
10			20%	108.82	489.68	489.68	217.64	443.98
11			40%	93.99	422.98	422.98	375.98	447.42
12			60%	82.73	372.27	372.27	496.35	450.03
13		6%	0%	77.96	814.20	407.10		441.75
14			20%	65.60	685.18	342.59	218.68	446.10
15			40%	56.63	591.46	295.73	377.53	449.26
16			60%	49.82	520.30	260.15	498.16	451.66
17	2:1	8%	0%	104.25	799.24	399.62		443.06
18			20%	87.69	672.28	336.14	219.22	447.21
19			40%	75.67	580.13	290.06	378.34	450.23
20			60%	66.55	510.19	255.10	499.10	452.52
21		10%	0%	130.70	784.20	392.10		444.38
22			20%	109.89	659.31	329.66	219.77	448.33
23			40%	94.79	568.74	284.37	379.16	451.2
24			60%	83.34	500.05	250.02	500.05	453.38
25		6%	0%	78.43	921.56	307.19		444.44
26			20%	65.94	774.78	258.26	219.80	448.39
27			40%	56.88	668.34	222.78	379.20	451.25
28			60%	50.01	587.61	195.87	500.10	453.42
29	3:1	8%	0%	104.87	904.52	301.51		445.71
30			20%	88.13	760.12	253.37	220.32	449.46
31			40%	76.00	655.47	218.49	379.98	452.18
32			60%	66.80	576.15	192.05	501.00	454.24
33		10%	0%	131.47	887.40	295.80		446.99
34			20%	110.42	745.39	248.46	220.85	450.54
35			40%	95.19	642.55	214.18	380.77	453.12
36			60%	83.65	564.65	188.22	501.92	455.07

2.3. Experimental Plan

2.3.1. Flowability Test Method and Evaluation Index

Flowability Test Method

This test was conducted in accordance with ASTM D6103-2004 [35] "Standard Test Method for Flow Continuity of Controllable Low Strength Materials". The fluidity of the slurry is characterized by its free diffusion expansion under self-weight, as shown in Figure 7. Considering that in practical engineering applications, there may be situations where tank truck transportation and other reasons may result in delayed pouring, this study also conducted tests on its flowability over time: that is, testing the flowability after mixing for 0 and 30 min of standing time. After each test of flowability, the desulfurization gypsum–fly ash flowable lightweight soil slurry was loaded back into the container and covered with a damp cloth. It was necessary to stir it again for 1 min before conducting a flowability test.

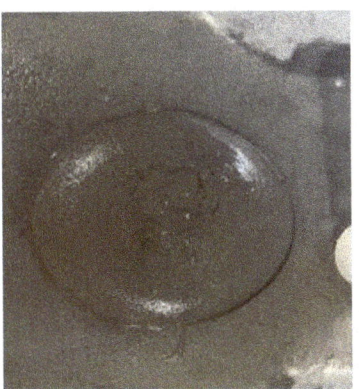

Figure 7. Fluidity experiment.

Liquidity Evaluation Indicators

The research on mobility in this study adopts the internationally recognized CLSM mobility testing procedure ASTM D6103-2017 [36]. This experimental procedure was specifically developed by the American Society for Testing and Materials for CLSM. Based on the results of the tested flowability test, this regulation divides liquidity into three levels of evaluation, as shown in Table 6.

Table 6. CLSM flowability evaluation level.

Level	Flowability/mm	Applicability
Low flowability	<150	Reclamation work for large-scale pipe trenches, roadbeds, etc.
Average flowability	150–200	General backfilling project
High flowability	>200	Backfilling projects with narrow operating space or dead corners

2.3.2. Unconfined Compressive Strength Test

This experiment is based on the compressive strength testing method for building mortar cubes in JCJ/T70-2009 [37] "Standard for Basic Performance Testing of Building Mortar". Unconfined compressive strength testing was conducted on desulfurization gypsum–fly ash flowable lightweight soil. The preparation and curing of the test piece refer to the method described in Figure 6. After the specimen was cured to the specified age (7 days or 28 days), a YAW-300 pressure testing machine(Jinan Zhongluchang Testing Machine Manufacturing Co., Ltd., Jinan, China) was used for continuous and uniform loading, and the loading rate was maintained at 1 mm/min, as shown in Figure 8. The formula for calculating unconfined compressive strength is shown in Equation (1).

$$f_{cu} = \frac{N_u}{A} \quad (1)$$

where f_{cu} is compressive strength, N_u is failure load, and A is the pressure bearing area.

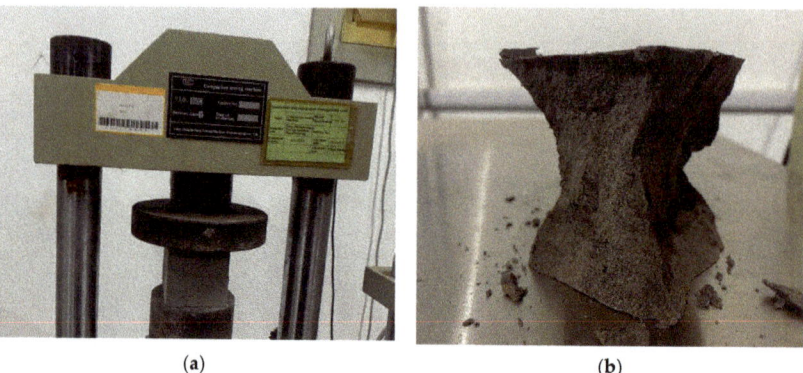

Figure 8. Unconfined compressive strength test: (**a**) specimen loading process; (**b**) typical failure mode of the specimen.

2.3.3. Uniaxial and Biaxial Compressive Test

To evaluate the stiffness characteristics of the desulfurization gypsum–fly ash flowable lightweight soil, we explored the influence of various raw material dosages on their strength and deformation characteristics. The stress–strain test plan was developed, as shown in Table 7. The MTS810 electro-hydraulic servo testing machine(Jiangsu Donghua Testing Technology Co., Ltd., Jiangsu, China) was used in this experiment, and the loading rate was controlled at 1 mm/min. The selection of test pieces, φ 50 mm $\times$ 100 mm cylindrical specimens with 3 formed specimens in each group, were tested after curing for 28 days.

Table 7. Uniaxial compression test plan.

Sample	Mass Ratio of Fly Ash to Desulfurization Gypsum	Cement Content	Silt Content
Y1		6%	
Y2	1:1	8%	
Y3		10%	20%
Y4	2:1		
Y5	3:1		
Y6		8%	0%
Y7	1:1		40%
Y8			60%

3. Results and Discussion

3.1. Flowability Research

3.1.1. The Influence of Cement Content and Silt Content on Fluidity

In this study, flowability tests were conducted on the design mix proportion in Section 2.3.1 to investigate the influence of cement content and silt content in the cementitious material on the flowability of the desulfurization gypsum–fly ash flowable lightweight soil. The relationship between cement content, silt content, and fluidity was analyzed, as shown in Figure 9.

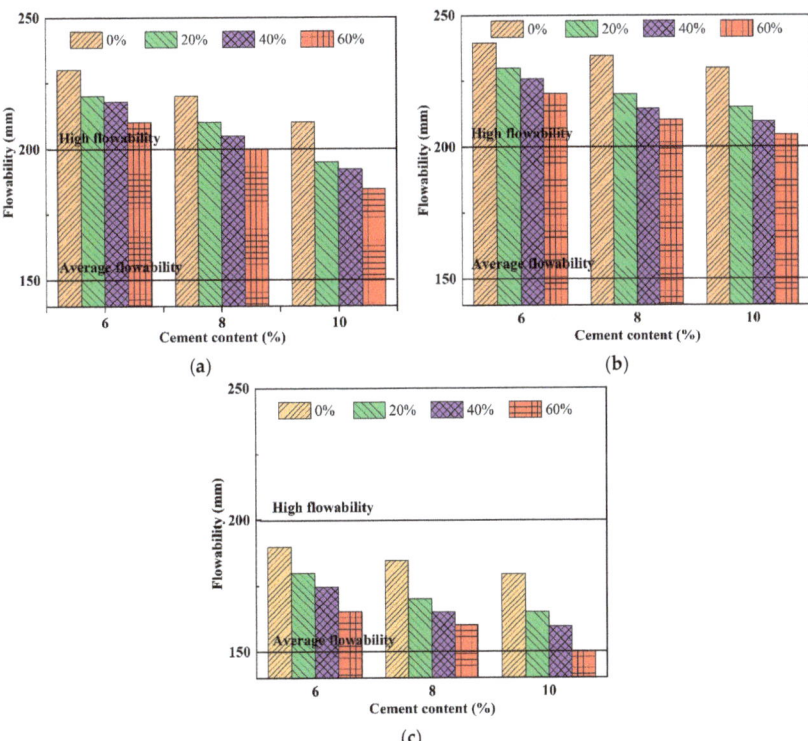

Figure 9. The relationship between cement content, silt content, and fluidity: (**a**) 1:1; (**b**) 2:1; (**c**) 3:1.

From Figure 9, it can be seen that when the content of other raw materials is constant, the flowability of the desulfurization gypsum–fly ash flowable lightweight soil gradually decreases with the increase in cement content. There is a negative correlation between the two. In Figure 9a, the silt content is 40%. When the cement content increases from 6% to 10%, the fluidity decreases from 218 mm to 192 mm. At this point, the relative decrease in mobility is the largest. But the decline rate is only about 6.3%. Although the flowability of the desulfurization gypsum–fly ash flowable lightweight soil decreases with the increase in cement content, the decrease is relatively small. The main reasons for this phenomenon are the following:

(1) Cement has high activity. The hydration reaction can take place in a relatively short time, and the more hydration products generated, the more unfavorable the flowability of the desulfurization gypsum–fly ash flowable lightweight soil.
(2) When the water solid ratio is fixed, the specific surface area of cement is relatively large. An increase in cement content will lead to a decrease in the fluidity of the slurry.
(3) The cement content selected in this study is relatively small, and the increasing gradient of the content is only 2%. Therefore, the influence of cement content on the fluidity of the slurry is also relatively small.

In Figure 9, the condition of a fixed mass ratio of fly ash to desulfurization gypsum and constant cement content can be seen. The flowability of the desulfurization gypsum–fly ash flowable lightweight soil also shows a decreasing trend with the increase in fly ash content. When the content of silt increases from 0% to 20%, the maximum decrease in flowability is 15 mm, with an average decrease rate of 7%. When the content of silt increases from 20% to 60%, the relative decrease in fluidity is relatively small, with an average decrease rate of 3%. The increase in silt content leads to a significant decrease in fluidity. The fine soil particles cause a decrease in the fluidity of the desulfurization gypsum–fly ash flowable lightweight soil, mainly due to their small particle size and large specific surface area. Therefore, the soil mixture has strong water absorption ability.

3.1.2. The Influence of the Mass Ratio of Fly Ash to Desulfurization Gypsum on Fluidity

We investigated the influence of the mass ratio of fly ash to desulfurization gypsum on the flowability of the desulfurization gypsum–fly ash flowable lightweight soil. Based on the test results in Section 3.1.1, the relationship between the mass ratio of fly ash to desulfurization gypsum and flowability was analyzed, as shown in Figure 10.

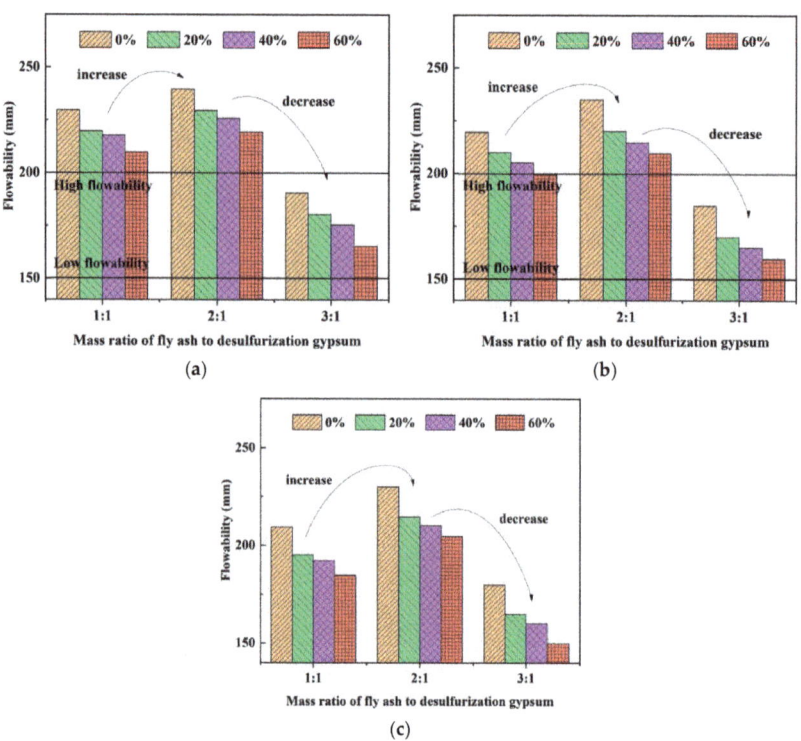

Figure 10. The relationship between the mass ratio of fly ash to desulfurization gypsum and its flowability: (**a**) cement content 6%; (**b**) cement content 8%; (**c**) cement content 10%.

As shown in Figure 10, ceteris paribus, the fluidity of the desulfurization gypsum–fly ash flowable lightweight soil increases first and then decreases with the increase in the mass ratio of fly ash to desulfurization gypsum. When the mass ratio of fly ash to desulfurization gypsum increases from 1:1 to 2:1, its fluidity increase rate is about 5%. But when the cement content is 10%, the increase rate is about 10%. When the mass ratio of fly ash to desulfurization gypsum increases from 2:1 to 3:1, its fluidity decreases by about 23%, with a significant decrease. Especially when the cement content is 10% and the silt content is 60%, the fluidity decreases to 150 mm, with a decrease rate of 27%. A reasonable explanation for this is that the water in the mixture exists in two forms: filling water and free water. Filling water does not affect its fluidity. The fluidity of the slurry mainly depends on free water. Due to the morphological effect and micro-aggregate effect of fly ash, adding an appropriate amount of fly ash can change the filling water between particle voids into free water. This has led to an increase in the proportion of free water. In addition, an appropriate amount of fly ash also acts as a "rolling ball" in the slurry.

This reduces the friction between solid particles. As a result, the fluidity of the slurry is improved. However, due to the large specific surface area of fly ash, excessive fly ash requires a large amount of free water consumption. Under the condition of a constant water to solid ratio, the phenomenon of reduced fluidity occurs when the amount of coal powder added is too high.

3.2. Research on Unconfined Compressive Strength

The Influence of Age and Mass Ratio of Fly Ash to Desulfurization Gypsum on Compressive Strength

The relationship between the compressive strength and the mass ratio of fly ash to desulfurization gypsum of specimens with a curing age of 7 and 28 days was analyzed. The effects of age and the mass ratio of fly ash to desulfurization gypsum on the unconfined compressive strength of the desulfurization gypsum–fly ash flowable lightweight soil were explored, as shown in Figure 11. By analyzing the relationship between the age of the specimen and its compressive strength, the following conclusions can be drawn.

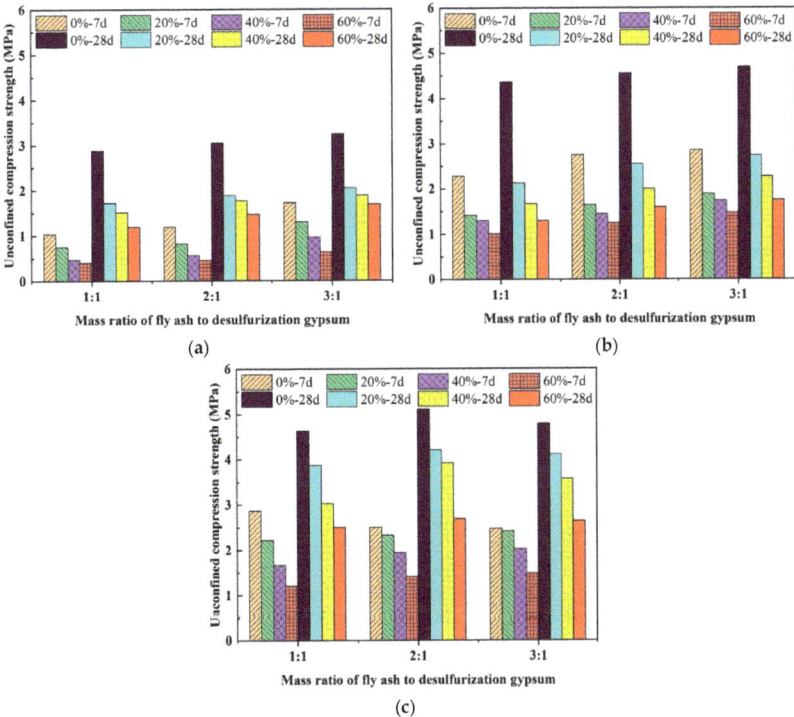

Figure 11. Relationship between age and mass ratio of fly ash to desulfurization gypsum and compressive strength: (a) cement content 6%; (b) cement content 8%; (c) cement content 10%.

The compressive strength range of specimens with a 7 d curing period is 0.38 MPa to 2.89 MPa. The compressive strength range of specimens with a curing age of 28 d is 1.2~5.13 MPa, which meets the requirements (7 d $\geq$ 0.3 MPa, 28 d $\geq$ 0.6 MPa).

The compressive strength of the desulfurization gypsum–fly ash flowable lightweight soil increases with age. Compared to the compressive strength of specimens with a curing age of 7 days, when the cement content is 6%, the compressive strength of the specimens with a curing period of 28 days increases the fastest, with a growth rate of 58~136%. The growth rate ranges from 20% to 107% when the cement content is 8% and 10%. This is mainly because the fly ash volcanic ash reaction needs cement hydration products to provide an alkaline reaction environment, and its reaction rate is slow. When the cement content is low, the early strength of the specimen is mainly contributed by the cement hydration reaction. Its internal alkalinity is weak. The volcanic ash reaction time of fly ash is delayed, resulting in a high growth rate of compressive strength.

Fly ash and desulfurization gypsum are used as mineral admixtures. The activation effect of sulfate activity is one of the main sources of compressive strength of the desulfurization gypsum–fly ash flowable lightweight soil. From the compressive strength of the specimens with a curing age of 28 days in Figure 11, the following can be seen:

(1) In Figure 11a,b, when the cement contents are 6% and 8%, the compressive strength of the specimen increases with the increase in the mass ratio of fly ash to desulfurization gypsum. It has a good positive proportional relationship. In Figure 11c, when the cement content is 10%, the compressive strength of the specimen first increases and then decreases with the increase in the mass ratio of fly ash to desulfurization gypsum. When the mass ratio is 2:1, the compressive strength is the highest.

(2) When the mass ratio of fly ash to desulfurization gypsum increases from 1:1 to 2:1, the compressive strength of the specimen increases significantly. Its growth rate ranges from 5% to 29%. When the mass ratios are 2:1 and 3:1, the compressive strength of the specimen is relatively close. Its amplitude of change is very small.

Based on the above phenomenon, a conclusion can be drawn. As the mass ratio of fly ash to desulfurization gypsum increases, the content of fly ash in the fluidized lightweight soil matrix of desulfurization gypsum and fly ash for the volcanic ash reaction is relatively higher. Most of the unhydrated fly ash particles can effectively fill the gaps between their internal solid particles, gradually densifying the matrix structure. This improves the compressive strength of the specimen. When the mass ratio of fly ash to desulfurization gypsum is 2:1, the proportion of the two mineral admixtures is appropriate. Desulfurization gypsum has the highest degree of volcanic ash activity stimulation on fly ash. At this point, the strength increase provided by the volcanic ash reaction of fly ash is greater than the strength increase provided by its filling effect. This makes the compressive strength close to or even higher than the compressive strength at a mass ratio of 2:1. From this, it can be seen that when considering production costs and higher strength requirements, the optimal mass ratio of fly ash to desulfurization gypsum is 2:1.

3.3. Study on Stress–Strain Relationship under Uniaxial Compression

3.3.1. Complete Stress–Strain Curve

The stress–strain curve is the most basic indicator for describing the mechanical properties of materials. It can well reflect the strength and deformation performance of the desulfurization gypsum–fly ash flowable lightweight soil. Therefore, we explored the stress–strain relationship and failure mode of the desulfurization gypsum–fly ash flowable lightweight soil during the loading process. This is of great significance for fully understanding the mechanical properties of the desulfurization gypsum–fly ash flowable lightweight soil.

Figure 12a–c show the stress–strain curves of the desulfurization gypsum–fly ash flowable lightweight soil under different cement contents, mass ratios of fly ash to desulfurization gypsum, and fly soil contents, respectively. From the figure, it can be seen that the stress–strain development process of the desulfurization gypsum–fly ash flow-state lightweight soil can be roughly divided into four stages. The segmentation position is shown on the Y3 curve. To make the expression clear, only the Y3 curve was segmented and labeled. The following is the specific stage division:

(1) The OA segment of the curve is in the linear elastic stage. The stress–strain curve of the desulfurization gypsum–fly ash flowable lightweight soil changes approximately in a straight line. At this stage, the desulfurization gypsum–fly ash flowable lightweight soil gradually hardens. The stress increases rapidly with strain. The stress at point A is the proportional limit.

(2) The AB segment of the curve represents the plastic yield stage. The slope of the stress–strain curve gradually decreases. The strain growth rate has significantly accelerated. The increase in stress is relatively small. At this stage, cracks or micro defects begin to appear in the internal structure of the desulfurization gypsum–fly ash flowable lightweight soil, and its stress level is close to the material threshold. The stress at point B is the peak stress.

(3) The BC segment of the curve is in the failure stage. The stress–strain curve shows a downward trend. The stress decays continuously with the increase in strain. At this stage, the desulfurization gypsum–fly ash flow-state lightweight soil specimen exhibits cracks and reaches failure.

(4) The CD segment of the curve represents the residual strength maintenance stage. The stress–strain curve can be approximately represented as a horizontal straight line. During this stage, the stress of the desulfurization gypsum–fly ash flow-state lightweight soil is a relatively small and fixed value. The strain keeps on rising. The stress depicted in point D of the figure corresponds to residual strength.

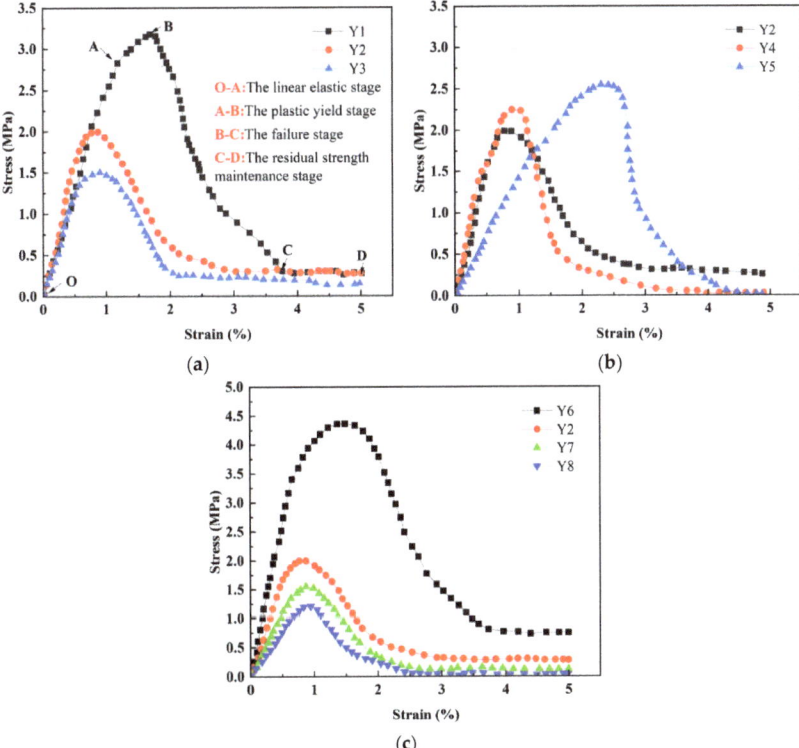

Figure 12. Complete stress–strain curve of desulfurization gypsum–fly ash flowable lightweight soil: (a) different cement content; (b) mass ratio of different fly ash to desulfurization gypsum; (c) different amounts of silt added.

It can be seen from Figure 12a that there is no significant difference in the effect of cement content on the rising section of the stress–strain curve of the desulfurization gypsum–fly ash flow-state lightweight soil. But as the cement content increases, its decreasing section gradually slows down significantly. The opening of the stress–strain curve gradually increases downwards. This indicates that an increase in cement content can significantly improve the ductility of the desulfurization gypsum–fly ash flowable lightweight soil.

It can be seen from Figure 12b that as the mass ratio of fly ash to desulfurization gypsum increases, the slope of the rising section of the stress–strain curve shows a decreasing trend, and its peak strain gradually increases. When the mass ratio of fly ash to desulfurization gypsum is 2:1, the slope of the descending section of the stress–strain curve is the highest. The stress–strain curve has the smallest downward opening. Therefore, its brittleness is maximum when the mass ratio is 2:1.

It can be seen from Figure 12c that as the amount of silt gradually increases, the slope of the rising section of the stress–strain curve significantly decreases. But the impact on the descending segment is not significant. The descending section of the stress–strain curve at each dosage is almost parallel. It is worth mentioning that as the amount of silt increases, the strain corresponding to its peak stress gradually increases. From this, it can be seen that an increase in the amount of silt significantly increases the deformation of the desulfurization gypsum–fly ash flow-state lightweight soil.

3.3.2. Elastic Modulus

The elastic modulus can be calculated from the slope of the stress–strain curve of the material. The stress–strain curves of the desulfurization gypsum–fly ash flowable lightweight soil under different mix ratios in this experiment are shown in Figure 12. The stress–strain relationship of the desulfurization gypsum–fly ash flowable lightweight soil exhibits nonlinear characteristics similar to that of cement soil. And its rising section has a good linear elastic relationship. Therefore, the elastic modulus calculation method for high–plasticity clay proposed by Li et al. [38] can be used. The slope of the line connecting the point corresponding to 0.5 times the peak stress in the rising segment and the origin was taken.

The calculation results of the elastic modulus of the desulfurization gypsum–fly ash flowable lightweight soil under different mix ratios are shown in Table 8. The elastic modulus E_{50} of the desulfurization gypsum–fly ash flowable lightweight soil ranges from approximately 120 MPa to 565 MPa. Its elastic modulus is about 2~18 times that of roadbed soil (the elastic modulus of roadbed soil is about 30~60 MPa). Through research by Feng et al. [39], it was found that roadbed materials with higher elastic moduli can significantly reduce the post-construction settlement and traffic load on the stress of pipelines in trench backfill engineering. Therefore, using a desulfurized gypsum–fly ash flowable lightweight soil for backfilling can reduce post-construction settlement. It can change the stress state of the abutment under traffic loads.

Table 8. Elastic modulus of desulfurization gypsum–fly ash flowable lightweight soil.

Sample	Elastic Modulus $\frac{E_{50}}{MPa}$
Y1	246.2
Y2	321.1
Y3	267.7
Y4	436.7
Y5	150.1
Y6	564.5
Y7	217.3
Y8	121.2

It can be seen from Table 8 that with the increase in cement content and the mass ratio of fly ash to desulfurization gypsum, the elastic modulus E_{50} of the desulfurization gypsum–fly ash flowable lightweight soil first increases and then decreases, reaching its maximum when the cement content is 8% and the mass ratio of fly ash to desulfurization gypsum is 2:1. The elastic modulus E_{50} decreases with the increase in silt content. This further indicates that an increase in the amount of silt can increase the deformation of the desulfurization gypsum–fly ash flowable lightweight soil.

3.3.3. Uniaxial Compressive Constitutive Model

Establishment of Constitutive Model

Based on the analysis above, the stress–strain relationship curve of the desulfurization gypsum–fly ash flowable lightweight soil falls between that of soil and concrete. Unlike soil, this curve exhibits a distinct linear elastic stage. However, it can undergo significant plastic deformation under a load, similar to concrete. Currently, research on the constitutive model of cement–soil primarily focuses on describing the direct or modified constitutive relationship between soil and concrete. In this study, a constitutive model of soil was selected to describe the constitutive relationship of the desulfurization gypsum–fly ash flowable lightweight soil.

Among the existing mathematical models for describing the stress–strain relationship of soil, the most influential one is the Duncan–Zhang model. This model is a nonlinear variable elasticity model. The parameters in the model can be determined and calculated very clearly. The original expression of the Duncan–Zhang model is as follows:

$$(\sigma_1 - \sigma_3) = \frac{\varepsilon}{\frac{1}{E} - \frac{R_f \varepsilon}{(\sigma_1 - \sigma_3)_f}} \tag{2}$$

where $(\sigma_1 - \sigma_3)$ is deviatoric stress, ε is axial strain, E is the initial tangent modulus, R_f is the failure stress ratio, and $(\sigma_1 - \sigma_3)_f$ is the failure deviator stress.

During the uniaxial compression test, due to the fact that $\sigma_3 = 0$, based on the hypothesis of Hooke's Law, Formula (2) can be expressed as follows:

$$\frac{\sigma}{\sigma_p} = \frac{\frac{\varepsilon}{\varepsilon_p}}{a + f\left(\frac{\varepsilon}{\varepsilon_p}\right)^b} \tag{3}$$

where σ_p is peak stress, ε_p is the strain corresponding to peak stress, and f is the stress failure ratio, which is approximately R_f. As long as reasonable values are taken for the parameters a, f, and b in the formula, the stress–strain relationship of the desulfurization gypsum–fly ash flowable lightweight soil can be well described.

Analysis and Verification of Constitutive Models

To verify the accuracy of this constitutive model, the stress–strain curves of the desulfurization gypsum–fly ash flowable lightweight soil under different mix ratios were analyzed. They were calculated according to Equation (3) above. The fitting curve is shown in Figure 13, and the values of the three fitting parameters (a, f, and b) and their correlations are shown in Figure 13.

From Figure 13, it can be seen that the experimental data of the desulfurization gypsum–fly ash flow-state lightweight soil under different mix ratios are in good agreement with the calculated values of this model, and the correlation is greater than 0.96. Therefore, the modified Duncan–Zhang model can well describe the stress–strain relationship of the desulfurization gypsum–fly ash flowable lightweight soil. This has certain theoretical and practical significance.

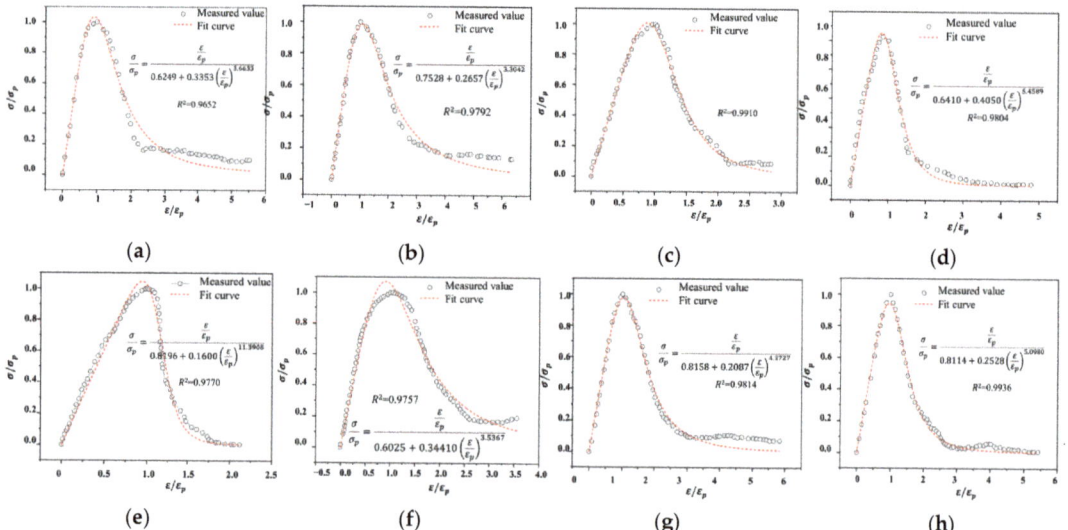

Figure 13. Fitting curve of stress–strain relationship for desulfurization gypsum–fly ash flowable lightweight soil under different mix ratios: (**a**) Sample Y1; (**b**) Sample Y2; (**c**) Sample Y3; (**d**) Sample Y4; (**e**) Sample Y5; (**f**) Sample Y6; (**g**) Sample Y7; (**h**) Sample Y8.

3.4. Mechanism of Strength Formation of Desulfurization Gypsum–Fly Ash Flowable Lightweight Soil

Desulfurized gypsum–fly ash flowable lightweight soil is a high flowable roadbed backfill material made by uniformly mixing Portland cement, fly ash, and desulfurization gypsum as a ternary cementitious system with powder soil. The mechanism of strength formation is actually the complex process of microstructure formation in the system composed of composite cementitious materials and silt. The specific formation process is discussed and analyzed below.

3.4.1. Cement Hydration Reaction

After adding water to mix of the desulfurization gypsum–fly ash fluidized lightweight soil, the Portland cement will immediately start the hydration reaction. The hydration reaction generates gel products such as calcium silicate hydrate and calcium aluminate hydrate, and a large amount of $Ca(OH)_2$ will be generated at the same time.

3.4.2. Activation of Fly Ash Activity

Fly ash has volcanic activity, and its main components are Al_2O_3 and SiO_2. Based on Figure 1, it can be seen that fly ash mainly exists in the form of glass ball particles. In the alkaline environment created by the cement hydration reaction, the chemical bonds of Si-O and Al-O on the surface of the glass particles of fly ash break, leading to the activation of fly ash and enhancing the reaction rate of volcanic ash. The pozzolanic reaction of fly ash mainly involves the reaction of Ca^{2+} in $Ca(OH)_2$ with the active oxides Al_2O_3 and SiO_2 of fly ash, generating cementitious substances such as hydrated calcium aluminate and hydrated calcium silicate. The specific reaction formula is as follows:

$$Al_2O_3 + nCa(OH)_2 + xH_2O \rightarrow nCaO \cdot SiO_2 \cdot xH_2O \qquad (4)$$

$$SiO_2 + mCa(OH)_2 + yH_2O \rightarrow mCaO \cdot Al_2O_3 \cdot yH_2O \qquad (5)$$

$$2SiO_2 + Al_2O_3 + Ca(OH)_2 + 3H_2O \rightarrow CaO \cdot Al_2O_3 \cdot 2SiO_2 \cdot 4H_2O \qquad (6)$$

3.4.3. Promoting Reaction of Desulfurization Gypsum

The primary component of desulfurization gypsum is $CaSO_4 \cdot 2H_2O$. The presence of SO_4^{2-} not only facilitates the breaking of Si-O and Al-O chemical bonds on the surface of fly ash particles but also reacts with AlO^{2-} and Ca^{2+} to form ettringite. This reaction leads to a reduction in the AlO^{2-} content within the structural system, thereby enhancing the disintegration of fly ash particles. In addition, desulfurization gypsum can also provide sufficient Ca^{2+} for the reaction system, promoting the generation of more cementitious substances. The main reactions involved in desulfurization gypsum are as follows:

$$Ca^{2+} + Al_2O_3 + OH^- + SO_4^{2-} \rightarrow 3CaO \cdot Al_2O_3 \cdot 3CaSO_4 \cdot 32H_2O \qquad (7)$$

To gain a more intuitive understanding of the strength formation process of the desulfurization gypsum–fly ash flowable lightweight soil, for the specimens with the first mix ratio (cement content of 8%, fly ash/desulfurization gypsum mass ratio of 2:1, and silt content of 20%), samples were taken for SEM testing at the curing ages of 7 and 28 days, as shown in Figures 14 and 15.

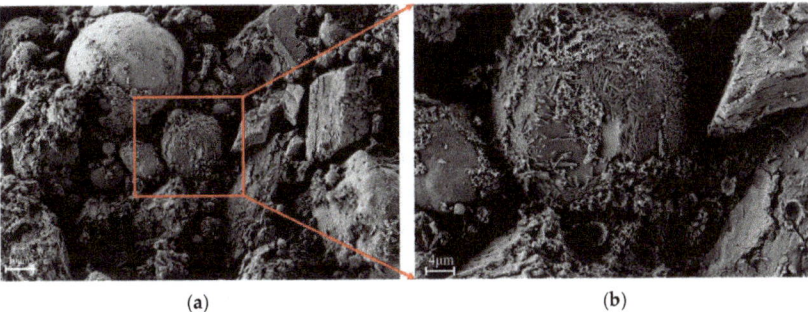

Figure 14. SEM images of specimens cured for 7 d: (**a**) 10 μm; (**b**) 4 μm.

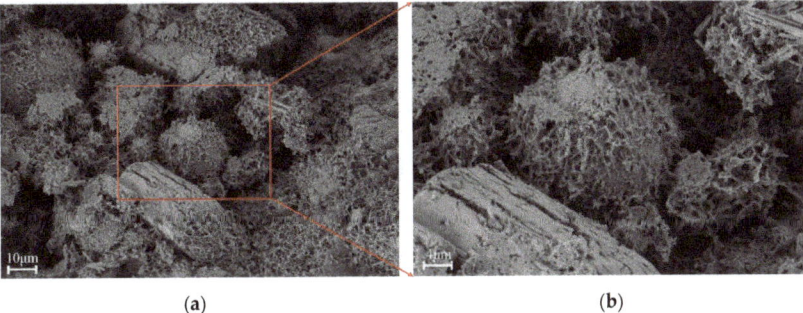

Figure 15. SEM images of specimens cured for 28 d: (**a**) 10 μm; (**b**) 4 μm.

Based on the findings in Figure 14, it is evident that after 7 days of curing, fly ash, desulfurization gypsum, and silt particles have been partially wrapped in flocculent and needle-like substances. However, complete wrapping has not yet occurred. Figure 14b further illustrates that the fly ash pellets are densely packed, with a visible disintegration gap in the glass layer. Notably, a significant amount of flocculent material is present around the gap. These observations indicate that the volcanic ash reaction of fly ash has commenced during the 7-day curing period. The figure primarily showcases the hydration reaction of cement, the interaction between fly ash and $Ca(OH)_2$, and the formation of hydrated calcium silicate and hydrated calcium aluminate.

According to Figure 15, it is evident that at a curing age of 28 days, the fluidized lightweight soil structure of the desulfurization gypsum–fly ash contains more flocculent substances. The fly ash, desulfurization gypsum, and powder soil particles are completely enveloped by a continuous network structure. Additionally, there is a needle-like substance known as ettringite present in the structure, indicating the involvement of desulfurization gypsum in the reaction. The different particles are interconnected and overlapped by flocs, resulting in the formation of a cohesive whole. This enhances the adhesion between particles and improves the structural density.

4. Conclusions

This article first explains the raw materials, testing methods, and mix design scheme. By evaluating the flowability and unconfined compressive strength of the desulfurization gypsum–fly ash flowable lightweight soil, each mix ratio is determined to meet the requirements of roadbed backfilling; in addition, the development process of its stress–strain curve was studied, and a uniaxial compressive constitutive model was proposed. The main conclusions are as follows:

(1) When the cement content ranges from 6% to 10%, the fly ash/desulfurization gypsum mass ratio ranges from 1:1 to 3:1, and the fly ash content ranges from 0% to 60%. The flowability, 7 d compressive strength, and 28 d compressive strength of the desulfurization gypsum–fly ash flowable lightweight soil vary from 150 mm to 240 mm, 0.38 to 2.89 MPa, and 1.2 to 5.13MPa, respectively. It is recommended to select specific mix proportions based on actual engineering requirements.

(2) The increases in cement content and silt content both lead to a decrease in the flowability of desulfurization gypsum–fly ash flowable lightweight soil. However, as the mass ratio of fly ash to desulfurization gypsum increases, the flowability reaches its maximum at a mass ratio of 2:1. The mass ratio of fly ash/desulfurization gypsum has the greatest impact on its fluidity, with a maximum decrease of 27% when it increases from 2:1 to 3:1; The flowability of desulfurized gypsum–fly ash fluidized lightweight soil shows a loss after standing for 30 min, with a decrease range of 1.36% to 5.13%.

(3) The compressive strength of the desulfurization gypsum–fly ash flowable lightweight soil is positively proportional to the cement content and age and inversely proportional to the silt content. When the mass ratio of the fly ash/desulfurization gypsum increases from 1:1 to 2:1, its compressive strength increases significantly. When the mass ratios are 2:1 and 3:1, its compressive strength is very close. Considering production costs, it is recommended to choose a fly ash/desulfurization gypsum mass ratio of 2:1.

(4) The development process of the stress–strain curve of the desulfurized gypsum–fly ash flowable lightweight soil can be divided into four stages: linear elastic stage, plastic yield stage, failure stage, and residual strength maintenance stage. Based on the stress–strain relationship test results, a uniaxial compressive constitutive model of the desulfurization gypsum–fly ash flowable lightweight soil was proposed. The model was fitted and analyzed with the test results, and the correlation was greater than 0.96, indicating a high degree of agreement.

(5) Compared to traditional fluidized fly ash, the desulfurization gypsum in the fluidized lightweight soil of the desulfurization gypsum fly ash can promote the disintegration of fly ash, thereby increasing the specific surface area. This provides more contact points and promotes the hardening process, enhancing the interlocking force between particles and the formation of cementitious substances and further enhancing strength.

Author Contributions: Conceptualization, S.Z. and X.Z.; methodology, N.H. and X.Z.; software, T.Z.; validation, X.Z., J.L. and H.Z.; formal analysis, X.Z.; investigation, S.Z. and X.Z.; resources, S.Z.; data curation, J.L.; writing—original draft preparation, J.L.; writing—review and editing, H.Z.; visualization, S.Z.; supervision, X.Z.; project administration, S.Z. and J.L.; funding acquisition, S.Z. All authors have read and agreed to the published version of the manuscript.

Funding: This research was funded by the Science and Technology Plan of Shandong Provincial Department of Transportation (2021B117).

Institutional Review Board Statement: Not applicable.

Informed Consent Statement: Not applicable.

Data Availability Statement: Not applicable.

Conflicts of Interest: The authors declare no conflict of interest.

References

1. Yuan, Q.; Qian, J.S.; Yang, J.J. Indoor experimental study of flowable fly ash. *Coal Ash.* **2011**, *23*, 1–3.
2. Zhang, S.F.; Wang, J.Y.; Gao, L.Y.; Guo, G.M.; Jiang, W.L. Experimental study on mix design and road performance of fluid fly ash subgrade filler. *J. Highw. Transp. Res. Dev.* **2023**, *40*, 84–92.
3. Yuan, X.Y.; Pu, Y.D.; Gui, Z.Y. Effect of hydroxylated graphene on properties of fly ashcement matrix composites. *Mat. Rep.* **2023**, 1–17.
4. Jia, Y. Research on Application of Liquid Cement Fly Ash Filled in Bridge Abutment. Master's Thesis, Chang'an University, Xi'an, China, 2009.
5. Huang, Z.Q.; Yu, Y.Y. Study on the physical and mechanical properties of fly ash modified red layer filler. *Sci. Technol. Innov.* **2022**, 21–23.
6. An, S.; Wang, B.M.; Chen, W.X. Interaction mechanism of carbide slag activating slag-fly ash composite cementitious materials. *Bull. Chin. Ceram. Soc.* **2023**, *42*, 1333–1343.
7. Liu, Y.; Ceng, X.; Wang, B.W. Preparation and strength mechanism of alkali-activated fly ash-slag-carbide slag based geopolymer. *Bull. Chin. Ceram. Soc.* **2023**, *42*, 1353–1362.
8. Zhang, T.T.; Wang, C.L.; Zhang, Y.X. Effect of fly ash content on performance of high performance concrete with seawater and sea sand. *Bull. Chin. Ceram. Soc.* **2022**, *41*, 1677–1688.
9. Mao, Y.; Zhang, H.; Luo, C.W. Simulation study on compactness of fly ash cement block based on discrete element method. *Ind. Miner. Process.* **2020**, *49*, 36–40.
10. Ma, C.C.; Xu, F.X.; Yang, Y. Study on tensile creep characteristics of high volume fly ash concrete at early ages. *J. Zhejiang Univ. Technol.* **2023**, *51*, 131–138.
11. Liu, J.; Li, Z.L.; Zhang, W.Z. Research advances in cold-bonded artificial lightweight aggregates made from industrial solid waste materials. *Mat. Rep.* **2023**, 1–31.
12. Janardhanatn, R.; Burns, F.; Peindl, R.D. Mix design for flowable fly-ash backfill material. *J. Mater. Civil Eng.* **1992**, *4*, 252–263. [CrossRef]
13. Wang, Y.Z. Application Research of Casting Type of Cement Fly Ash in Backfilling on Back of Bridge Abutments and Culverts. Master's Thesis, Chang'an University, Xi'an, China, 2008.
14. Dapeng, Z.; Dongmin, W.; Hongzhi, C. Hydration characteristics of cement with high volume circulating fluidized bed fly ash. *Constr. Build. Mater.* **2023**, *380*, 131310.
15. He, P.; Zhang, X.; Chen, H. Waste-to-resource strategies for the use of circulating fluidized bed fly ash in construction materials: A mini review. *Powder Technol.* **2021**, *393*, 773–785. [CrossRef]
16. Lin, Y.; Jiang, X.; Zhao, Y. Zeolite greenly synthesized from fly ash and its resource utilization: A review. *Sci. Total Environ.* **2022**, *851*, 158182. [CrossRef] [PubMed]
17. Yang, C.F.; Wang, S.; Sun, J.S. The best water content of liquid fly ash. *J. Hebei Univ. Technol.* **2014**, *43*, 92–95.
18. Qi, Z.L. The Optimization of Liquid Fly Ash Embankment in the Tang-Jin Freeway Rebuid. Master's Thesis, Hebei University of Technology, Tianjin, China, 2015.
19. Wang, S. The Application of Liquid Fly Ash in Tang Jin High-Speed Expansion Project. Master's Thesis, Hebei University of Technology, Tianjin, China, 2015.
20. Wan, C.C.; Jiang, T.H.; Yu, Y. Basic Mechanical Properties Study of Polypropylene Foam Concrete Based on Orthogonal Test. *Bull. Chin. Ceram. Soc.* **2023**, 1–14.
21. Sun, J.S.; Lu, X.; Li, H.L. Study on mix design and properties of foamed fluid fly ash. *J. Guangxi Univ. Nat. Sci. Ed.* **2017**, *42*, 352–358.
22. Zhen, W.K.; Zhao, Y.Y.; Wang, Y.C. The Simulation study on static mixer for foamed concrete preparation. *Mater. Rep.* **2023**, 1–20.
23. Gao, Z.H.; Chen, B.; Chen, J.L. Pore structure and mechanical properties of foam concrete under freeze-thaw environment. *Acta Mater. Compos. Sin.* **2023**, 1–11.

24. Chen, Z.C.; Guo, L.P.; Li, Y.Q. Experimental study on paraffin emulsion phase change foamed concrete. *Bull. Chin. Ceram. Soc.* **2023**, *42*, 1623–1629+1649.
25. Yong, Y.; Bing, C. Potential use of soil in lightweight foamed concrete. *KSCE J. Civ. Eng.* **2016**, *20*, 2420–2427. [CrossRef]
26. Gabr, M.; Bowders, J.J. Controlled low-strength material using fly ash and AMD sludge. *J. Hazard. Mater.* **2000**, *76*, 251–263. [CrossRef]
27. Do, M.T.; Kim, Y. Engineering properties of controlled low strength material (CLSM) incorporating red mud. *Int. J. Geo-Eng.* **2016**, *7*, 7. [CrossRef]
28. Wu, H.; Huang, B.; Shu, X. Utilization of solid wastes/byproducts from paper mills in Controlled Low Strength Material (CLSM). *Constr. Build. Mater.* **2016**, *118*, 155–163. [CrossRef]
29. Do, M.T.; Kim, S.Y.; Kang, O.G. Thermal conductivity of controlled low strength material (CLSM) made entirely from by-products. *Key Eng. Mater.* **2018**, *773*, 244–248. [CrossRef]
30. Türkel, S. Long-term compressive strength and some other properties of controlled low strength materials made with pozzolanic cement and Class C fly ash. *J. Hazard. Mater.* **2006**, *137*, 261–266. [CrossRef]
31. Chittoori, B.; Puppala, A.J.; Raavi, A. Strength and stiffness characterization of controlled low-strength material using native high-plasticity clay. *J. Mater. Civ. Eng.* **2014**, *26*, 04014007. [CrossRef]
32. Bui, A.T.; Pham, H.V.; Nguyen, T.D. Effectiveness of Lubricants and Fly Ash Additive on Surface Damage Resistance under ASTM Standard Operating Conditions. *Coatings* **2023**, *13*, 851. [CrossRef]
33. Ionescu, B.A.; Barbu, A.-M.; Lăzărescu, A.-V.; Rada, S.; Gabor, T.; Florean, C. The Influence of Substitution of Fly Ash with Marble Dust or Blast Furnace Slag on the Properties of the Alkali-Activated Geopolymer Paste. *Coatings* **2023**, *13*, 403. [CrossRef]
34. Liu, C.; Jia, Y. Effect of Redispersible Latex Powder and Fly Ash on Properties of Mortar. *Coatings* **2022**, *12*, 1930. [CrossRef]
35. *ASTM D6103-2004*; Standard Test Method for Flow Consistency of Controlled Low Strength Material (CLSM). ASTM: West Conshohocken, PA, USA, 2004.
36. *ASTM D6103-2017*; Standard Test Method for Flow Consistency of Controlled Low Strength Material (CLSM). ASTM: West Conshohocken, PA, USA, 2017.
37. *JCJ/T70-2009*; Standard for Basic Performance Testing of Building Mortar. Ministry of Housing and Urban-Rural Development of the People's Republic of China: Beijing, China, 2009.
38. Li, Q.; Fan, Y.; Shah, S.P. Effect of nano-metakaolin on establishment of internal structure of fly ash cement paste. *J. Build. Eng.* **2023**, *77*, 107484. [CrossRef]
39. Feng, Z.; Shen, D.; Huang, Q.; Zhang, T. Effect of fly ash on early-age properties and viscoelastic behaviors of supersulfated cement concrete under different degrees of restraint. *Constr. Build. Mater.* **2023**, *401*, 132895. [CrossRef]

Disclaimer/Publisher's Note: The statements, opinions and data contained in all publications are solely those of the individual author(s) and contributor(s) and not of MDPI and/or the editor(s). MDPI and/or the editor(s) disclaim responsibility for any injury to people or property resulting from any ideas, methods, instructions or products referred to in the content.

Article

The Advantages on Using GGBS and ACBFS Aggregate to Obtain an Ecological Road Concrete

Liliana Maria Nicula [1,2,*], Daniela Lucia Manea [1,*], Dorina Simedru [3], Oana Cadar [3], Ioan Ardelean [4] and Mihai Liviu Dragomir [1,*]

1. Faculty of Civil Engineering, Technical University of Cluj-Napoca, 15, C. Daicoviciu Street, 400114 Cluj-Napoca, Romania
2. Faculty of Construction, Cadaster and Architecture, University of Oradea, 4, B.S. Delavrancea Street, 410058 Oradea, Romania
3. INCDO-INOE 2000, Research Institute for Analytical Instrumentation, 67 Donath Street, 400293 Cluj-Napoca, Romania; dorina.simedru@icia.ro (D.S.); oana.cadar@icia.ro (O.C.)
4. Department of Physics and Chemistry, Technical University of Cluj-Napoca, 400114 Cluj-Napoca, Romania; ioan.ardelean@phys.utcluj.ro
* Correspondence: liliana.nicula@infra.utcluj.ro (L.M.N.); daniela.manea@ccm.utcluj.ro (D.L.M.); mihai.dragomir@cfdp.utcluj.ro (M.L.D.)

Citation: Nicula, L.M.; Manea, D.L.; Simedru, D.; Cadar, O.; Ardelean, I.; Dragomir, M.L. The Advantages on Using GGBS and ACBFS Aggregate to Obtain an Ecological Road Concrete. Coatings 2023, 13, 1368. https://doi.org/10.3390/coatings13081368

Academic Editor: Valeria Vignali

Received: 29 June 2023
Revised: 27 July 2023
Accepted: 31 July 2023
Published: 3 August 2023

Copyright: © 2023 by the authors. Licensee MDPI, Basel, Switzerland. This article is an open access article distributed under the terms and conditions of the Creative Commons Attribution (CC BY) license (https://creativecommons.org/licenses/by/4.0/).

Abstract: This work aims to show the advantages of using GGBS (Ground Granulated Blast Furnace Slag) and ACBFS aggregate (Air-Cooled Blast Furnace Slag) on the tensile strength and durability properties of infrastructure concrete at the reference age of 28 days. Three concrete mixes were prepared: the first one was a control sample; the second one had 15% GGBS (instead of Portland cement) and 25% ACBFS (instead of natural sand); and the third had 15% GGBS (instead of Portland cement) and 50% ACBFS (instead of natural sand). The studies on mortars focused on the ratio of compressive strength (CS) in correlation with the specific surface area (obtained by the Blain method). The microstructure of the prepared mortars was examined at the age of 28 days by X-ray diffraction, SEM electron microscopy with an energy-dispersive EDX spectrometer, and NMR nuclear magnetic resonance relaxometry. The results of the tests carried out afterwards on the concretes containing slag (15% GGBS and 25% or 50% ACBFS) showed values that met high-quality criteria for exfoliation ($S_{56} < 0.1$ kg/m^2), carbonation, and gelling G100 (with a loss of resistance to compression $\eta < 25\%$). The slag concretes showed a degree of gelation of G100 (with a loss of compressive strength below 25%), low volume losses below 18,000 mm^3/5000 mm^2 (corresponding to wear class 4, grade I), and moderate penetration of chloride ions (according to the RCPT test). All of these allow the concrete with slag (GGBS/ACBFS) to be recommended as an ecological road concrete. Our study proved that a high-class road concrete of BcR 5.0 can be obtained, with tensile strengths of a minimum 5 MPa at 28 days (the higher road concrete class in Romania, according to national standards).

Keywords: blast furnace slag; hydration activity index; flexural tensile strength; exfoliated mass; penetrability of chloride ions; wear resistance; frost–thaw resistance

1. Introduction

In less than one century, concrete has become the most widely used construction material in the world [1], which is why large amounts of natural resources are consumed in its production. As a measure to ensure the ecological balance, it is necessary to implement the use of various unconventional and renewable resources [2,3]. An important characteristic of cementitious materials is their ability to accommodate large amounts of waste by replacing some or all of the cement and/or natural aggregate. Extensive research work has been carried out to evaluate the suitability of various industrial by-products as supplementary cementitious materials (SCM) and their influence on the rheology and physical and mechanical properties of mortars and concretes [4–10]. In the steel industry, millions of tons of slag are a by-product that has to be neutralized for environmental reasons. The slag obtained is classified according to the origin of the minerals and the solidification process. Slag that cools slowly to atmospheric temperature is called air-cooled blast furnace slag (ACBFS), and slag that is instantaneously cooled under pressurized water jets is called granulated slag; the latter is ground into a powder that turns into ground granulated blast furnace slag. A crystalline character is observed in ACBFS slag by slow cooling in the air [11], while little or no crystallization occurs in slag rapidly cooled in water, where the glass content predominates [12,13].

The durability and mechanical properties of cement composites with GGBS largely depend on the hydraulic activity of the slag. The properties of the slag powder and the hydraulic activity are mainly affected by the glass content, chemical composition, and fineness of the slag powder. The vitreous content and chemical composition are passively determined by the ironmaking process, but the activation performance of GGBS can be changed by the fineness of the slag powder. The fineness of GGBS is usually expressed in terms of specific surface area and particle size distribution. Different levels of fineness can be achieved by grinding cementitious materials using different methods and grinding times [14–16]. By increasing the fineness of GGBS, the cementitious properties of cement containing steel slag are significantly improved [17]. The use of GGBS obtained by wet milling results in improved mechanical strength, reduced porosity, and low $Ca(OH)_2$ content in concrete composites [18]. One of the important uses for GGBS is in concrete structures in marine environments as it has been shown to reduce the penetration of chlorides into concrete. Other positive effects of GGBS in concrete are protection against sulfate attacks and the alkali–silica reaction.

More stringent requirements apply to pavement concretes used on roads, platforms, and/or airfields than those used in hydraulic or civil engineering [19–21]. Atmospheric factors caused by temperature fluctuations and the influence of traffic caused by dynamic loads, as well as the presence of chemical deicing agents, contribute to the demanding operating conditions of pavement concrete [21,22]. The pollutant transport processes through concrete are different and can occur through absorption, when the movement of liquids is due to the capillary forces generated in the capillary pores. Transport occurs by permeability when the fluid movement is due to the action of pressure and by diffusion when the fluid movement is due to a concentration gradient [23]. Finally, the penetration of aggressive substances from the operating environment into the component impairs the durability of road concrete [24]. Due to the limitations related to the percentage of cement substitution, for example, the Swedish standard SS 13 70 03 [25] allows the addition of GGBS up to 25% of the CEM I cement for exposure class XF4, a class which is specific to road concrete. According to the Indian standard IS: 383, the replacement of natural sand with blast furnace slag is allowed up to a limit of 50% for unreinforced concrete [11], and the ACI 213R guidance [26] recommends the range of 20%–60% for the replacement of natural sand with ACBFS. The substitution percentage limitation is based on previous research, which indicated that ACBFS aggregate could be a potential internal source of sulfates released into the pore solution. These sulfates contribute to the filling of air voids in the concrete with ettringite. The filling process reduces the space for water to expand as it freezes, making the concrete microstructure more susceptible to freeze–thaw damage [27]. On the other hand, the study in [28] highlighted that the porous and rough ACBFS aggregate increased the contact area with the concrete matrix; thus, the concave holes and micropores on the surface of the slag aggregate could be filled with mortar and hydrated cement paste. This process can increase the interlocking and mechanical bond between the aggregate and the concrete matrix. ACBFS slag is used as a fine and a coarse aggregate in concrete [11,29,30]. The study in [31] reported that replacing coarse aggregate with up to 40% ACBFS and cement with 10% fly ash increased the concrete's compressive and flexural strength. The study in [32] showed that road concrete with coarse aggregate replacement by ACBFS and fly ash cement developed a 6% higher compressive strength, while the flexural strength decreased by 15% compared to the reference.

Coarse aggregate road concrete ACBFS showed a decrease in wear resistance in a study [33]. Due to the low resistance to the crushing and fragmentation of coarse aggregate in the slag, it is recommended to use them only in the road infrastructure base layer [12]. Regarding the use of ACBFS as a fine aggregate, the study in [34] highlighted better properties in geopolymer mortars than in Portland cement mortar. The influence of the replacement degree of up to 100% natural sand by ACBFS in the concrete composition showed no detrimental effects on the properties studied, such as workability and mechanical strength, or on durability, which recorded a reduction in porosity, water absorption, chloride permeability, etc., at the age of 90 days [11].

Most of the research has examined the individual effects of GGBS replacing cement and ACBFS aggregate replacing sand in concrete. The research results [35–38] show that GGBS does not significantly improve compressive strength up to 28 days but that it does significantly improve it after 56 and 90 days as the pozzolanic reaction of GGBS gradually continues. Moreover, the studies in [38,39] show that by using GGBS with a low reactivity index in mortar mixtures at older ages (90 and 360 days) lower porosity and lower water absorption were achieved. The 0–10 micron particles of the GGBS particle distribution were strongly correlated with the hydration activity index at different curing ages [40].The study in [41] highlights a reduction in the shrinkage of concrete at the age of 150 days when granulated slag was used with a reduced hydration activity index.

This study focuses on the durability of rigid pavements and the environmental benefits of using possible industrial waste as raw material in the mass of road concrete, while ensuring the transition to a circular economy. At least two benefits emerge from this research; the first is the reduction in the energy consumption needed for the additional grinding of slag powder, and the second is the conservation of natural resources. First, the effect of GGBS fineness on compressive strengths was investigated on two mortars with 50% GGBS with specific Blaine surface areas of 360 m^2/kg and 330 m^2/kg. The ratio of compressive strengths between slag mortars and standard mortar was recorded at 7 and 28 days. A detailed study of the microstructure of the mortars was carried out using X-ray diffraction, SEM electron microscopy with an EDX energy dispersive spectrometer, and NMR nuclear magnetic resonance relaxometry.

To further explore the blast furnace slag benefits, two pavement concrete compositions were studied; in these, the cement was replaced with 15% GGBS with a low Blaine specific surface area (330 m^2/kg) equivalent to that of Portland cement, and the natural sand was replaced with 25% and 50% ACBFS. The tensile strength and durability properties were tested, including freeze–thaw resistance with and without deicing agent, abrasion resistance, resistance to carbonation, and chlorine ion permeability by the RCPT rapid test. The analysis of the results sought to establish the optimal percentages of using blast furnace slag in the composition of road concrete with performances equivalent to those of the reference concrete at the age of 28 days.

2. Materials and Methods

2.1. Materials

Portland cement-CEM I 42.5R, supplied by Romcim, CRH Romania Company, Romania, presents technical properties with values falling within the limits recommended by SR EN 197-1 [42]. The granulated blast furnace slag purchased from Steelworks Galati has a glass content of 95% and an oxide composition as given in Table 1, and the physicomechanical properties have been declared according to SR EN 15167-1 [43]. The superunitary ratio (1.07) of calcium oxide to silicon indicates efficient hydraulic activity, and the 70% concentration of oxides of silicon, calcium, aluminum, magnesium, and iron accumulated in GGBS indicates good pozzolanic activity according to standard D 6868 [38,44,45].

Table 1. Oxide analysis of granulated and ground GGBS slag.

(GGBS)	SiO$_2$	Al$_2$O$_3$	MnO	MgO	CaO	Fe$_2$O$_3$	Na$_2$O	K$_2$O	Other
(%)	38.10	9.50	0.23	8.10	40.80	0.56	0.30	0.68	1.73

Air-cooled blast furnace slag aggregate ACBFS, crushed to 0/4 mm size, was purchased from Galati Steelworks. The quality assurance of the slag aggregate (0/4) mm; natural sand NA quality (0/4) mm; and gravel (4/8) mm was controlled according to SR EN 12620 [46]. The geometric and physical–mechanical properties of the crushed quarry chippings of grade (16/25) mm correspond to the requirements of SR 667 [47] and SR EN 13043 [48] for grade (8/16) mm. Figure 1a–e present the granulometric composition of the used aggregate, determined in accordance with SR EN 933-1 [49]. For the ACBFS, the slag aggregates are given in Figure 1f, photographic images before and after crushing.

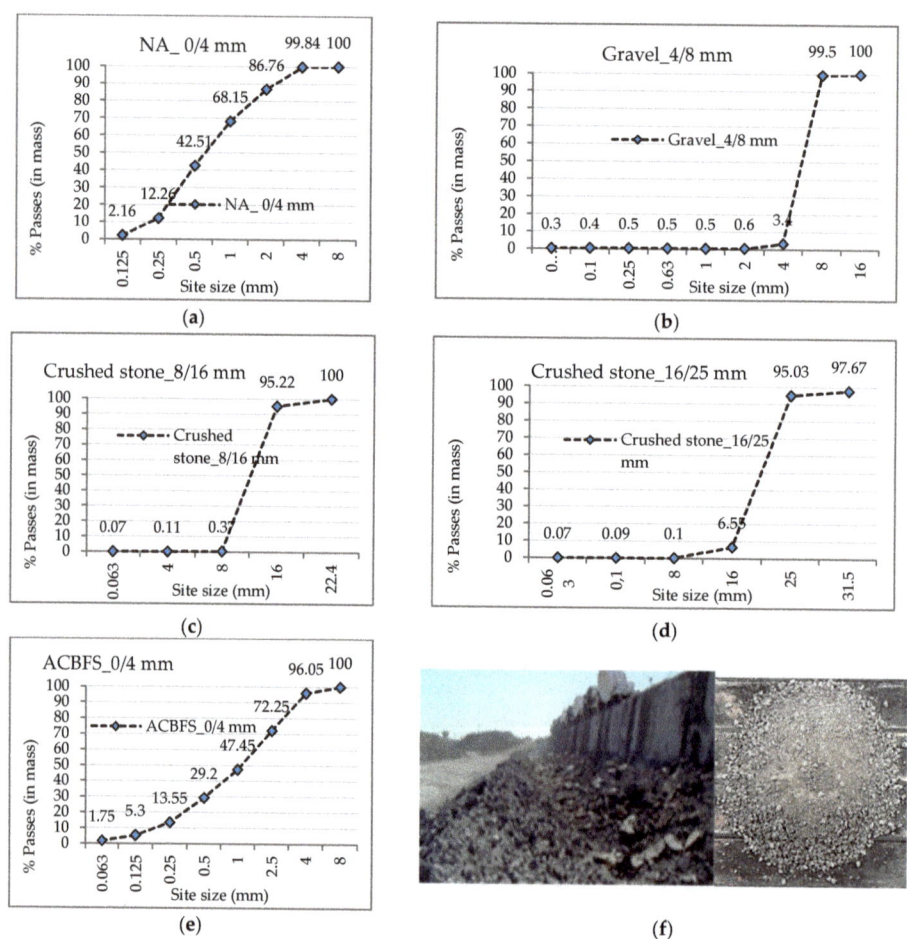

Figure 1. Granulometric composition for aggregate: (**a**) natural sand 0/4 mm; (**b**) gravel 4/8 mm; (**c**) crushed stone 8/16 mm; (**d**) crushed stone 16/25 mm; (**e**) ACBFS 0/4 mm; (**f**) ACBFS slag during the air-cooling process and after crushing (Source: Galati Steel Works and the Technical University of Cluj-Napoca Laboratory).

The additives used were purchased from Master Builders Solutions Romania group, Romania, MasterGlenium SKY 527 superplasticizer additive, and Master Air 9060 air entrainer additive, with characteristics following SR EN 934-2 [50]. The water was taken from the Cluj-Napoca supply system, and the properties corresponded to SR EN 1008 [51].

2.1.1. Preliminary Mortar Mixtures

To assess the compressive strength, preliminary tests were carried out on the mortars with different specific GGBS surfaces. Then, the ratio of the compressive strength (CS) between the mortars with GGBS and the standard mortar was determined to show the influence of the specific surface area on the mortars; these experiments were similar to those of the study in [52]. The recipe for standard mortar M I with 100% Portland cement and two mortar recipes with 50% GGBS M II/360 and M III/330 and different specific surface areas of 360 m^2/kg and 330 m^2/kg were created in the experiment. The prepared mortars were stored in a polyethylene bag at a relative humidity of (90 ± 5)% for 7 days and then in a humidified air box at a temperature of (20 ± 2) °C and a relative humidity of (65 ± 5)%. As curing conditions, water according to SR EN 196-1 [53] or a wet environment according to SR EN 1015-11 [54] can be optionally applied [55]. For this experiment, we chose the method of curing the samples in a wet environment, which is less favorable for compressive strength development and closer to the in situ conditions. The water to binder ratio was 0.5, and the quantities of materials used to make the mortars are listed in Table 2. The prismatic specimens of 40 × 40 × 160 mm were tested according to SR EN 1015-11 [54].

Table 2. The composition of mortars.

Materials	M I	M II/360	M III/330
Specific binder surface (SS)	331 m^2/kg	360 m^2/kg	330 m^2/kg
Portland cement, (g)	450 ± 2	225 ± 1	225 ± 1
(GGBS < 63 µm), (g)	-	225 ± 1	225 ± 1
(NA_0/4 mm), (g)	1350 ± 5	1350 ± 5	1350 ± 5
Water, (ml)	225	225	225

2.1.2. Concrete Mixtures

In the run-up to this work, the optimal proportions of the blast furnace slag material substitution were analyzed, and the material proportions were selected based on the various preliminary recipes in the study in [56]. We then proceeded with three paving mixes, one reference and two that were constantly incorporated with 15% GGBS blast furnace slag and 25% or 50% ACBFS slag aggregate crushed to a (0/4) mm size.

When preparing the concrete, the aggregate was saturated with the SSD dry surface. The initial moisture state of aggregate in concrete containing recycled materials has a major impact on the fresh and hardened state of the concrete. To avoid changes in the effective water–cement ratio and prevent the transfer of water from the mix to the porous aggregate, water balancing is generally suggested [57]. The quantities of material (kg/m^3) for the concrete production are given in Table 3.

The road concrete design parameters were established for resistance class BcR 5.0, which corresponds to very heavy traffic, with casting in the sliding formwork system, according to the requirements of NE 014 [58] and SR EN 206 [59]. The test results for the concrete mix are shown in Table 4. It can be seen that the results were within the limits established by the applicable standards and that the compressive strengths in the preliminary recipes reached the minimum value of 50 MPa from the seventh day which are found in the previous work, [56].

Table 3. Concrete composition kg/m³.

Materials (Kg/m³)	S 0/0 Reference Concrete	S 15/25 (15% GGBS_25% ACBFS)	S 15/50 (15% GGBS_50% ACBFS)
Cement ©	370	314.50	314.50
(GGBS < 63 μm)	-	55.50	55.50
Binder (b)	370	370	370
Water (w)	151.68	154.51	155.92
w/b	0.41	0.418	0.421
(ACBFS_0/4 mm)	-	163.44	326.90
Natural sand (AN_0/4 mm)	654.70	490.33	326.90
Coarse aggregate (AC_4/25 mm)	1215.88	1214.16	1214.18
Total aggregate	1870.58	1867.94	1867.97
Admixtures (SP-SKY 527)	5.55	6.18	6.29
Admixtures (MA 9060)	0.74	0.74	0.81

Table 4. Design parameters of road concrete composites in accordance with NE 014 [58] and SR EN 206 [59].

Parameter Design	Min. Dosage Cement	(w/l)	Compaction (%)	Density (kg/m²)	Occluded Air (%)	fcm 28 days MPa	fcfm 28 days MPa
NE 014 and SR EN 206	360 kg/m³	max. 0.45	1.15 ÷ 1.35	2390 ± 30	5.0 ÷ 6.5	min. 50	min. 5.5
Results	370 kg/m³	0.41 ÷ 0.421	1.26 ÷ 1.33	2392 ÷ 2394	5.1 ÷ 6.4	-	-

2.2. Method

2.2.1. Influence of GGBS on the Ratio of Compressive Strength of Mortars at 7 and 28 Days

The CS ratio of mortars was calculated with Equation (1):

$$\text{CS ratio} = (fc_{GGBS}/fc_C) \times 100, (\%), \quad (1)$$

In the above relationship, (fc_{GGBS}) is the mean compressive strength (MPa) of the mortar made from 50:50 slag and cement, and (fc_C) is the mean compressive strength (MPa) of the mortar made from cement, (Figure 2a,b). Table 5 shows the minimum hydration activity index values determined according to the specifications SR EN 15167:1 [43] and ASTM C989 [60], correlated with the resistance classes for GGBS.

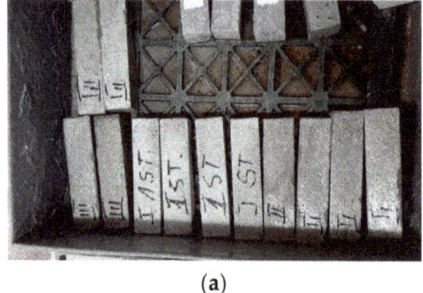

(a)　　　　　　　　　　(b)　　　　　　　　　　(c)

Figure 2. (a) Preservation of mortar specimens; (b) compressive strength test on mortars; (c) mortar samples prepared for the NMR relaxometry measurements.

Table 5. The minimum value of the hydration activity index at 7 and 28 days of slag GGBS.

SR EN 15167:1		ASTM C989 (Class 80)		ASTM C989 (Class 100)		ASTM C989 (Class 120)	
7 days	28 days	7 days	28 days	7 days	28 days	7 days	28 days
45	70	-	75	75	95	95	115

2.2.2. Characterization of the Microstructure of Standard Mortars and Mortars with GGBS

Phase composition analysis by X-ray diffraction (XRD) and microstructure analysis by electron microscopy coupled with energy dispersive spectroscopy (SEM–EDX) were performed on the fragments resulting from the compression test on the standard mortar (M I) and the mortar with 50% GGBS (M II/360, M III/330) at the ages of 7 and 28 days.

X-ray Diffraction on Mortars at 7 and 28 Days

XRD patterns were recorded on the powder samples using a D8 Advance diffractometer (Bruker, Karlsruhe, Germany) with CuK α1 radiation (λ = 1.5418 Å), operating at 40 kV and 35 mA, at room temperature. The semi-quantitative evaluation was performed following the reference intensity ratio (RIR) method [61].

SEM Measurements on Mortars at 7 and 28 Days

SEM–EDX analysis was performed at room temperature using a scanning electron microscope (VEGAS 3 SBU, Tescan, Brno-Kohoutovice, Czech Republic) with a Quantax EDX XFlash (Bruker, Karlsruhe, Germany) detector. The ~4 mm^2 samples were mounted with carbon tape on an SEM stub.

2.2.3. NMR Relaxometry Measurements Made on Mortars, at 28 Days

From the intact prismatic samples prepared with standard mortar and mortar with 50% GGBS, a cylindrical sample with a length of 20 mm and a diameter of 9.5 mm was taken at the age of 28 days. The sample was used to extract the relative distribution of the pore sizes by the NMR relaxometry technique (Figure 2c). After oven drying, the samples were placed and sealed in 10 mm diameter glass tubes, and NMR measurements were performed to reveal the intra-C-S-H pores. To better highlight the inter-C-S-H and capillary pores, a second set of NMR measurements was performed on cyclohexane-saturated samples [62]. The NMR measurements were performed with a MinispecMQ20 low-field device (Bruker, Germany) using the CPMG technique (see Ref. [57] for a description). Before each measurement, the samples were brought to the thermal equilibrium at a temperature of 35 °C. A total of 2000 spin echoes were recorded in each experiment, and the time between two echoes was kept at 0.1 ms to mitigate the effects of internal gradients on the relaxation measurements. The relaxation time distributions were extracted from the CPMG series using a numerical Laplace inversion of the recorded data [63,64].

2.2.4. Road Concrete Mixes with Embedded Slag (GGBS and ACBFS), Tested at 28 Days

In this study, hardened concrete, mechanical strengths (flexural and compression), wear resistance, freeze–thaw resistance, permeable void content, corrosion resistance by chlorine ion penetration, and carbonation were evaluated. For each mixture, three 150 × 150 × 600 mm prismatic samples were cast and kept in air for 24 h and then kept in water at (20 ± 2) °C until the reference age of 28 days. Subsequently, flexural tests were carried out according to SR EN 12390-5 [65], (Figure 3a,b). To determine the corrosion resistance by carbonation according to SR CR 12793, a 1% phenolphthalein solution was sprayed onto the freshly cleaved fragments [66] (Figure 3c). From the remaining prisms, 18 cubes with a side length of 150 mm were cut; with these, the compressive strengths were evaluated according to SR EN 12390-3 [67], and the loss of compressive strength between the water-preserved control samples and the samples tested in 100 freeze–thaw cycles was evaluated according to SR 3518 [68] (Figure 3d). For each mixture, 9 slabs with a cross-section of 150 × 150 mm and a thickness of 50 mm were cut from the remaining fragments to estimate the mass loss by exfoliation in the presence of thawing agents (3% NaCl) after 56 freeze–thaw cycles according to SR assessment CEN/TS 12390-9 [69] (Figure 3e), and 9 cubes with a side length of 71 mm were used to assess the volume loss through wear according to SR EN 1338 [70] (Figure 3f,g). The permeable void content was evaluated in 71 mm cubes according to ASTM C 642 [71] (Figure 3h).

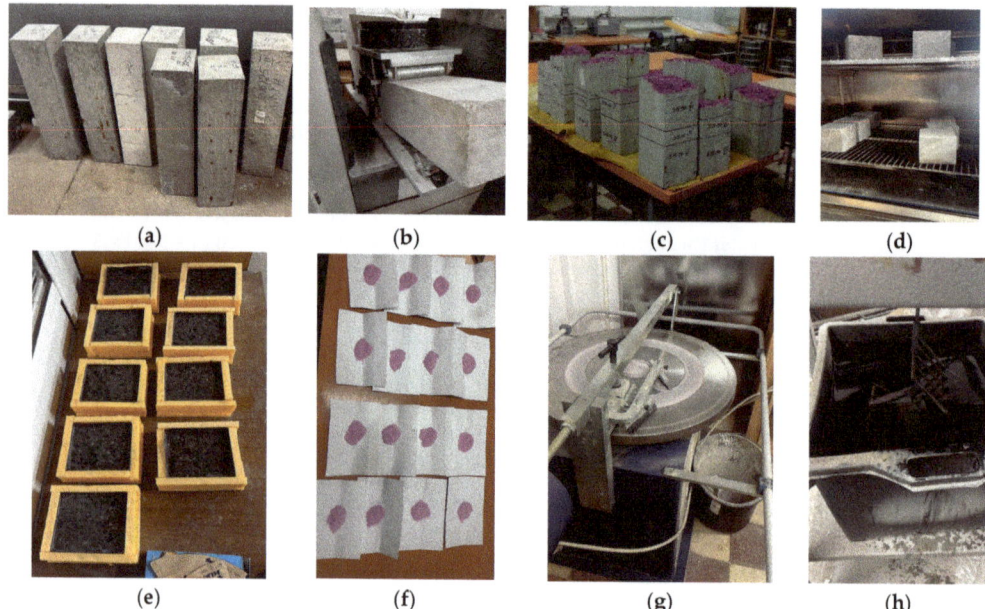

Figure 3. (**a**) Prisms 150 × 150 × 600 mm; (**b**) bending tensile strength test; (**c**) prism fragments sprayed with phenolphthalein solution; (**d**) cubes of 150 mm sides maintained in the thermostatic chamber for 100 freeze–thaw cycles; (**e**) samples prepared to assess loss of mass through exfoliation; (**f**) abrasive material; (**g**) Böhme abrasive disc wear test; (**h**) weighing of samples with hydrostatic balance.

The concrete quality in terms of frost–thaw resistance was assessed against the criteria presented in Table 6, and the conditions for wear resistance are given in Table 7.

Table 6. Acceptance criteria for the frost resistance of concrete SS 13 72 44 [72,73], NE012-1 [74], and SR 3518 [68].

Frost–Thaw Resistance	Criteria SS 13 72 44	Criteria NE 012-1	Criteria SR 3518
Very good	$m_{56} < 0.10$ kg/m^2		
Good	$m_{56} < 0.20$ kg/m^2	$m_{56} < 0.50$ kg/m^2	-
(High)	or $m_{56} < 0.50$ kg/m^2 and $m_{56}/m_{28} < 2$ or $m_{112} < 0.50$ kg/m^2		
Acceptable (Moderate)	$m_{56} < 1.00$ kg/m^2 and $m_{56}/m_{28} < 2$ or $m_{112} < 1.00$ kg/m^2	$m_{56} < 1.0$ kg/m^2	-
Unacceptable (Low)	$m_{56} \geq 1.00$ kg/m^2 and $m_{56}/m_{28} \geq 2$ or $m_{112} \geq 1.00$ kg/m^2	$m_{56} < 2.0$ kg/m^2	-
Degree of gelation G100	-	-	Loss of compressive strength (η) < 25%

Table 7. Wear resistance classes SR EN 1338 [70].

Class	1	3	4
Mark	F	H	I
Criteria	No measured performance	≤20,000 mm³/5000 mm²	≤18,000 mm³/5000 mm²

To assess chlorine ion penetration, cylindrical specimens 100 mm in diameter and approximately 200 mm long were cast, from which 50 mm thick strips were cut for the RPCT test according to ASTM C 1202 [75]. The samples, protected on the side surface with epoxy material and covered with boiled and chilled water, were kept in the vacuum saturation apparatus with a vacuum pump and manometer for 20 h until fully saturated (Figure 4a). Immediately after completion of the RCPT test (Figure 4b), the chlorine ion migration coefficient was then determined on the same samples using the NT BUILD 492 colorimetric method [76]. A silver nitrate solution (AgNO$_3$) was sprayed at 0.1 mol/dm³ (0.1 M) onto axially split samples. After 15 min, a light gray precipitate of silver chloride formed on the surface of the sample on the part of the face where chlorine ions were present. The chlorine penetration depth (X_d) for each sample was obtained from the average of seven measurements taken at different positions across the sample width. Then, the migration coefficients of the chlorine ions (D) were calculated using the Nernst–Planck Equation (2) [76,77]:

$$D = \frac{0.0239(273+T)L}{(V-2)t}\left(x_d - 0.0238\sqrt{\frac{(273+T)Lx_d}{V-2}}\right) \quad (2)$$

under:

D—Migration coefficient in non-equilibrium state ($\times 10^{-12}$ m²/s);
V—Applied voltage (V);
T—Mean value between initial and final temperature in the anolytic solution, NaCl (°C);
L—Sample thickness (mm);
X_d—Average chlorine penetration depth (mm).

(a)   (b)

Figure 4. (a) Vacuum apparatus and pressure gauge; (b) RCPT rapid test.

The correlation between the past electric charge Q and chlorine ion penetrability is shown in Table 8.

Table 8. Interpretation of results according to ASTM C 1202 [75].

Past electrical Charge (Coulomb)	High	Moderate	Low	Very Low
Penetrability of chlorine ions	>4000	2000–4000	1000–2000	100–1000

3. Results and Discussion

3.1. Mortar Mixes with GGBS

3.1.1. Influence of GGBS on the Ratio of Compressive Strength of Mortars at 7 and 28 Days

The mean values of the three specimens of the standard mortar M I and the slag mortars M II/360 and M III/330 that were tested for compressive strength and the CS ratios for the mortars are shown in Table 9.

Table 9. Compressive strength fc determined for mortars (MPa) and CS ratio (%).

Age	fc_MI	fc_M II/360	fc_M III/330	CS Ratio-MII/360	CS Ratio-MIII/330
7 days	34.40	22.16	20.34	64	59
28 days	43.62	40.26	35.14	92	81

The compressive strengths of the standard mortar and the test mortars are shown in Figure 5a. It can be seen that the compressive strength of the mortar samples increased with age; this was due to the continued duration of the hydration process and the reactions of the cement with the water in the pores and capillaries of the system [78]. The hydration reactions led to the formation of new hydration products, of which C-S-H occupied 50% ÷ 60% of the volume of cement paste after 28 days and was the main product contributing to the development of mechanical strength [79]. The CS ratio of the standard and GGBS mortars with the two specific surface areas at 7 and 28 days are shown in Figure 5b. The results recorded for the CS ratio in the two test mortars with GGB, as associated with the requirements specified in SR EN 15167:1 [43], classified the GGBS slag powder as being in resistance class 80 according to ASTM C989 [60]. With the specific GGBS area increase from 330 to 360 m^2/kg, the CS ratio increased by 8.47% after 7 days and by 13.58% after 28 days. The larger the specific surface area, the stronger the reaction capacity and the faster the increase in the hydration rate [14,40,80], which increased the strength of M II/360 slag mortar more compared to M III/330. The linear equations in Figure 4b show that the CS ratio could reach 100 for the specific area of 382 m^2/kg at the age of 28 days and the specific area of 576 m^2/kg at the age of 7 days under the conditions of the linear increase in the specific surface area, according to the results of the study in [40].

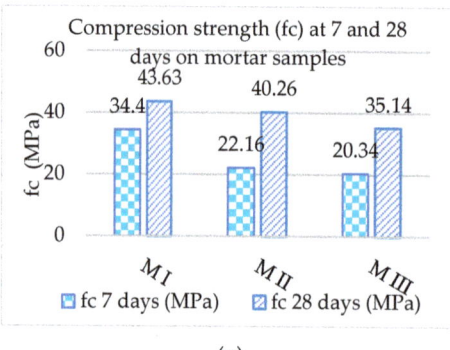
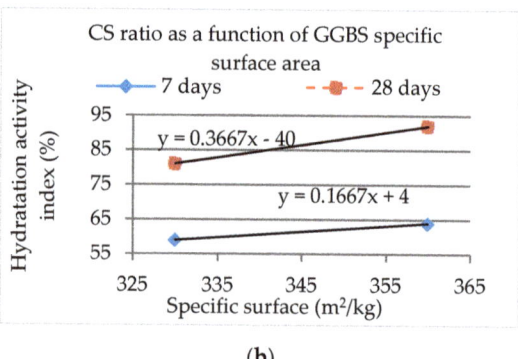

(a)   (b)

Figure 5. (a) Compressive strengths of mortars; (b) evolution of CS ratio by specific surface area of (GGBS) at the ages of 7 and 28 days.

3.1.2. XRD Measurements with X-ray Diffractometer Made on Mortars at 7 and 28 Days

The XRD patterns of the mortar samples at the ages of 7 and 28 days are presented in Figure 6.

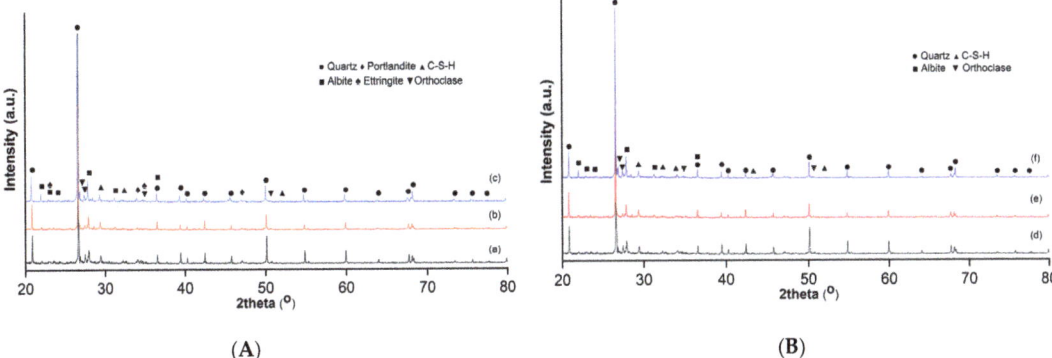

Figure 6. (**A**) Diffraction patterns of M I; (**B**) Diffraction patterns of M II; Diffraction patterns of (a) M I, (b) M II/360, (c) M III/330 mortar samples at 7 days and (d) M I, (e) M II/360, (f) M III/330 mortar samples at 28 days.

The semi-quantitative analysis of the crystalline phases identified in the investigated samples is presented in Table 10. At the age of 7 days, quartz SiO_2 was the main phase observed; this was followed by albite $NaAlSi_3O_8$, portlandite $Ca(OH)_2$, ettringite $Ca_6Al_2(SO_4)_3(OH)_{12} \cdot 26H_2O$, calcium silicate hydrate CaH_2O_4Si, and orthoclase $K(AlSi_3O_3)$ (Figure 5). At the age of 28 days, the diffraction peaks of portlandite and ettringite, which are characteristic of early cement hydration [81], were no longer observed.

Table 10. Phase fraction (%) identified by RIR method of mortar samples investigated at the ages of 7 and 28 days.

Mortar Mixture	M I		M II/360		M III/360	
	7 Days	28 Days	7 Days	28 Days	7 Days	28 Days
Quartz	+++	+++	+++	+++	+++	+++
Albite (Ab)	++	++	++	++	++	++
Orthoclase	+	+	+	+	+	+
Calcium-silicate (C-S-H)	+	+	+	+	+	+
Portlandite (CH)	+		+		+	
Ettringite (C-A-S-H)	+		+		+	

+++ major phases (>20%), ++ minor phases (5%–10%), + phases in traces (<5%).

The crystallinity degree of the investigated samples at the ages of 7 and 28 days is presented in Figure 7. At the age of 7 days, the highest value was obtained for MI. Replacing the cement with 50% GGBS led to more disorder in the early-stage structure of the mortars. At the age of 28 days, the degree of crystallinity of MI decreased, probably due to the change in structure leading to the disappearance of the portlandite and ettringite crystals. For the mortars with the addition of GGBS, the degree of crystallinity increased. A possible explanation for this behavior could be the higher proportion of calcium silicate hydrates found in the concrete with GGBS [82] than in the concrete made with Portland cement only. Moreover, the higher degree of crystallinity in slag mortar M II/360 could be attributed to the increased pozzolanic activity of the GGBS particles with higher SS [83–85].

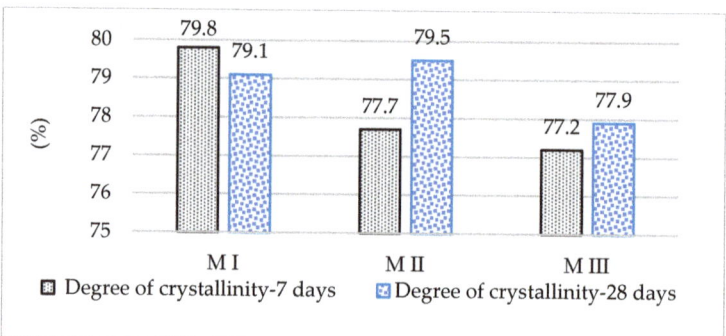

Figure 7. Degree of crystallinity of mortar samples at the ages of 7 days and 28 days.

3.1.3. SEM Measurements with Scanning Electron Microscopy and EDX with Energy Dispersive X-ray Spectroscopy Performed on Mortars Aged 7 and 28 Days

The pore size and morphology of the mortar samples M I, M II/360, and M III/330 at 7 and 28 days were studied by SEM and are shown in Figure 8. As can be seen, their surfaces were irregular and inhomogeneous, with pore sizes larger than capillary pores, which was characteristics of air holes (>a few μm) [86,87].

Table 11 shows the sizes of the pores measured as well as the pore spacing on the samples studied at 7 and 28 days.

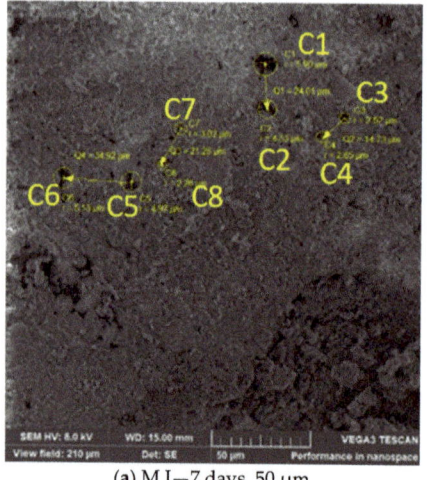
(a) M I—7 days, 50 μm

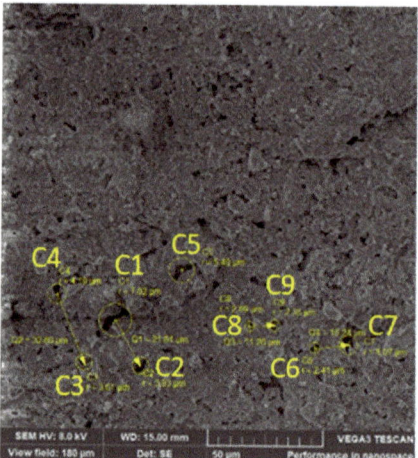
(b) M II/360—7 days, 50 μm

Figure 8. *Cont.*

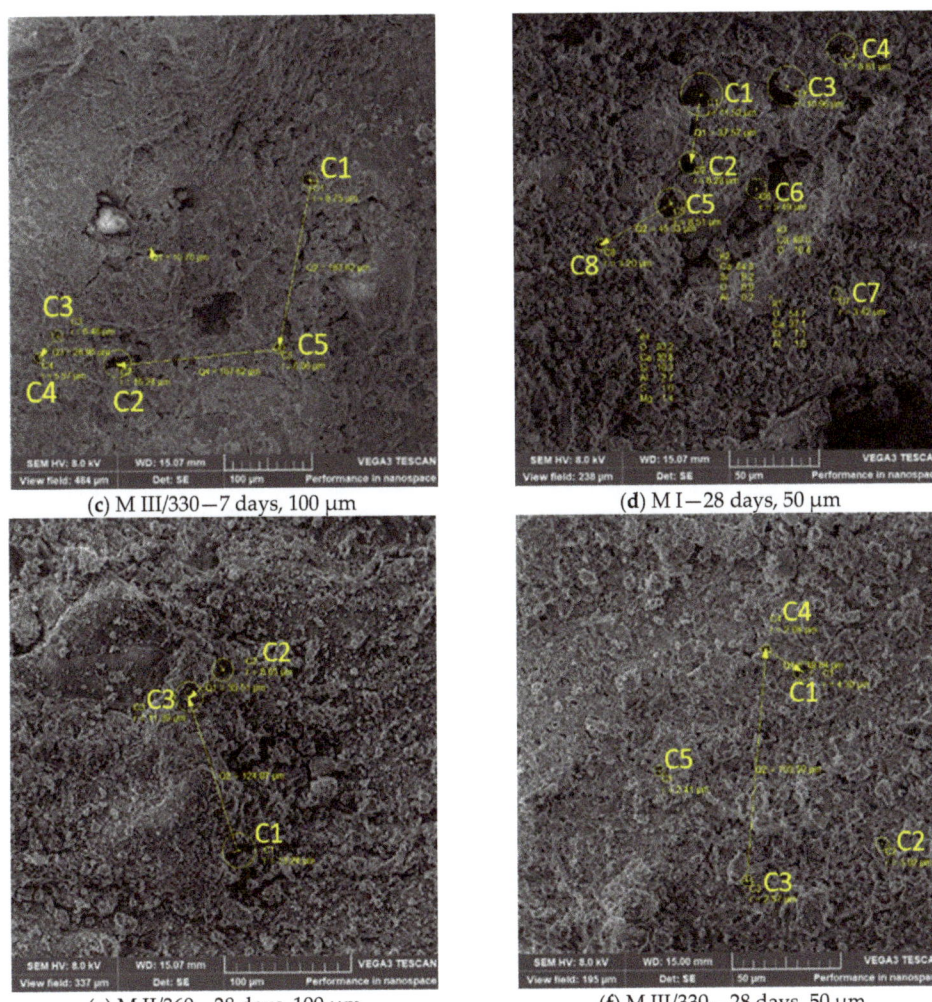

Figure 8. Pore structure of (**a**) M I, (**b**) M II/360, (**c**) M III/330 mortar samples at 7 days, and (**d**) M I, (**e**) M II/360, (**f**) M III/330 mortar samples at 28 days.

Table 11. Pore measurements of mortar samples M I, M II/360, and M III/330.

Sample	Pore Identification Code	Pore Radius (μm)	Distance Code $Q_i(C_i - C_{i+n})$	Distance (μm)	Pore Identification Code	Pore Radius (μm)	Distance Code $Q_i(C_i - C_{i+n})$	Distance (μm)
			7 Days				28 Days	
M I	C1	5.90	Q1(C1 − C2)	24.01	C1	11.52	Q1(C1 − C2)	37.57
	C2	4.53			C2	6.28		
	C3	2.52			C3	10.96		
	C4	2.65	Q2(C3 − C4)	14.73	C4	8.61		
	C5	4.97			C5	8.51		
	C6	5.13	Q4(C5 − C6)	34.92	C6	5.49		
	C7	3.03			C7	3.42		
	C8	2.26	Q3(C7 − C8)	21.28	C8	3.20	Q2(C5 − C8)	45.33

Table 11. Cont.

Sample	Pore Identification Code	Pore Radius (μm)	Distance Code Qi(Ci − Ci + n)	Distance (μm)	Pore Identification Code	Pore Radius (μm)	Distance Code Qi(Ci − Ci + n)	Distance (μm)
			7 Days				28 Days	
M II/360	C1	7.02			C1	13.24		
	C2	3.83	Q1(C1 − C2)	21.81	C2	8.65	Q1(C2 − C3)	35.51
	C3	3.61	Q2(C3 − C4)	32.60	C3	11.39	Q2(C1 − C3)	124.04
	C4	4.19						
	C5	6.57						
	C6	2.41	Q4(C6 − C7)	15.24				
	C7	4.07						
	C8	2.09	Q3(C8 − C9)	11.26				
	C9	2.35						
M III/330	C1	8.75	Q1	10.70	C1	4.30	Q1(C1 − C4)	19.84
	C2	15.24	Q4(C2 − C5)	187.62	C2	3.02		
	C3	6.48	Q3(C3 − C4)	28.96	C3	2.57	Q2(C3 − C4)	103.50
	C4	5.57			C4	2.05		
	C5	6.06	Q2(C1 − C5)	187.62	C5	2.41		

Larger air holes at smaller intervals indicate a more porous surface, and conversely, small pore sizes spaced at larger intervals indicate a more compact surface [69,70].

At 7 days, the standard mortar sample M I (Figure 8a) had the most compact surface, with air holes (from 2.26 to 5.90 μm) at intervals (from 14.73 to 34.92 μm). In comparison, the slag sample M II/360 (Figure 8b) had larger air holes (from 2.09 to 7.02 μm) which were identified at smaller intervals (from 11.26 to 32.60 μm). Sample M III/330 (Figure 8c) was characterized by the largest pore ranges (from 10.70 to 187.62 μm) and sizes (r ranges from 5.57 to 15.24 μm). The different morphology of the slag mortar surfaces at the early age of 7 days is explained by the slower hydration reaction and lower heat release rate in GGBS than in Portland cement [13]. The SEM results were in correlation with the compressive strengths; an increase in porosity in the M II/360 and M III/330 samples caused a decrease in the compressive strengths [88].

At 28 days, the M III/330 slag mortar sample (Figure 8f) showed the most compact surface with smaller air holes (from 2.05 to 4.30 μm) at larger spacings (from 19.84 to 103.50 μm). In comparison, for the sample standard mortar sample M I (Figure 8d), the air holes were larger (r ranges from 3.20 to 11.52 μm), and the distances at the intervals were smaller (from 37.57 to 45.33 μm). For the slag sample M II/360, (Figure 8e) we observed air holes at larger spacings (from 35.51 to 124.04 μm) and pore sizes (from 8.65 to 13.24 μm) close to the standard sample M I. The surface morphology of the M II/360 sample was consistent with the slightly lower compressive strength compared to the M I, which was justified by the fineness of the GGBS. But the denser surface in sample M III/330 contrasted with the compressive strength, which decreased the most compared to the standard mortar. This behavior may be due to the filling of the free pores and the ability of GGBS to reduce the pore volume of the matrix [89] or to the inhomogeneity of the mortar.

The EDX images are presented in Figures 9 and 10, and the results for the concentration of the identified elements are given in Table 12.

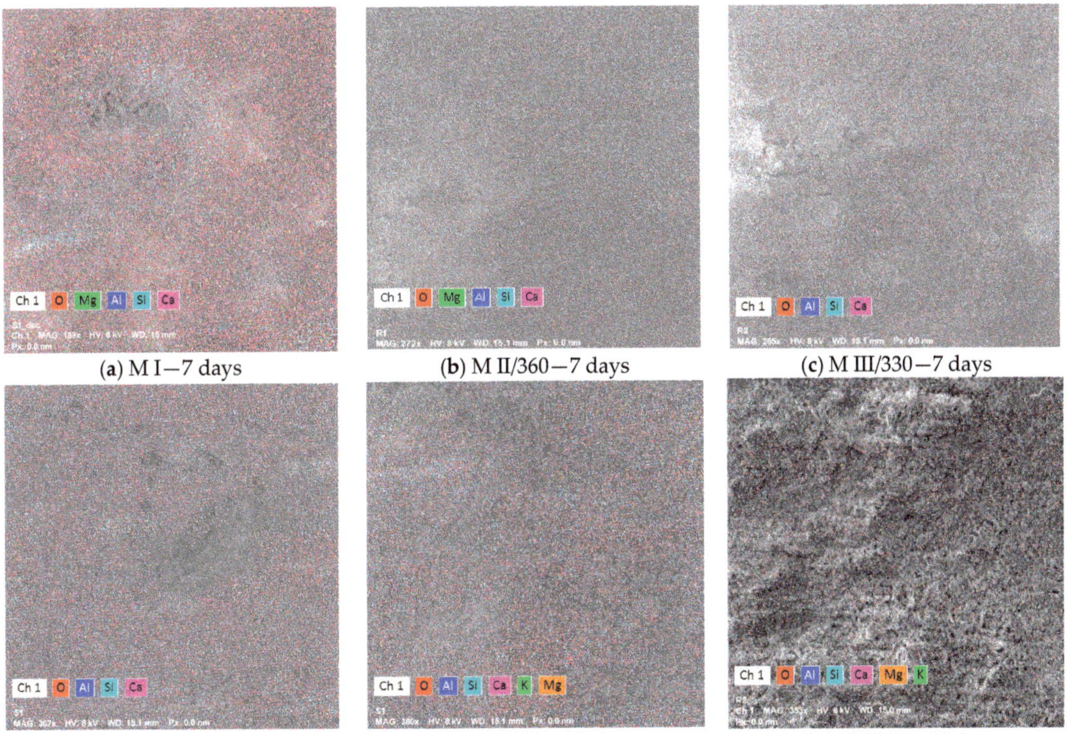

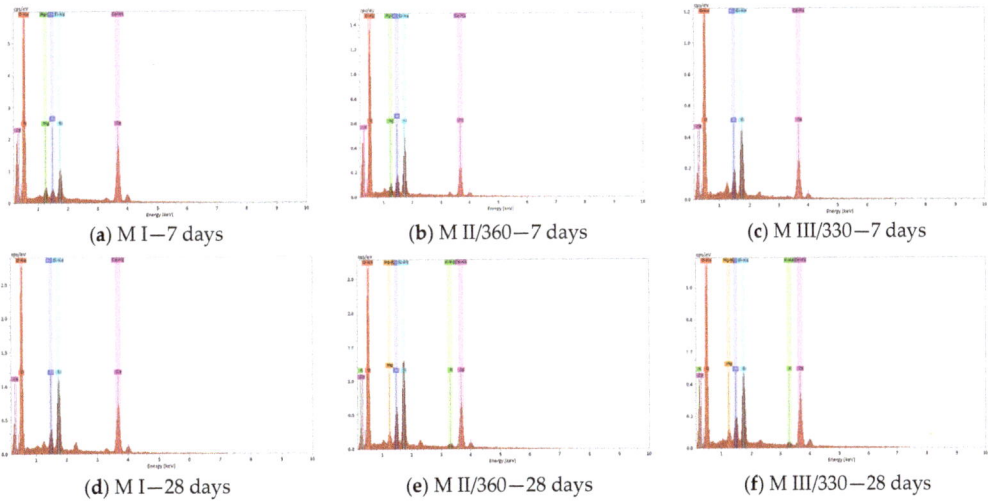

(a) M I—7 days (b) M II/360—7 days (c) M III/330—7 days

(d) M I—28 days (e) M II/360—28 days (f) M III/330—28 days

Figure 9. Surface mapping of (**a**) M I, (**b**) M II/360, (**c**) M III/330 mortar samples at 7 days, and (**d**) M I, (**e**) M II/360, (**f**) M III/330 mortar samples at 28 days.

(a) M I—7 days (b) M II/360—7 days (c) M III/330—7 days

(d) M I—28 days (e) M II/360—28 days (f) M III/330—28 days

Figure 10. *Cont.*

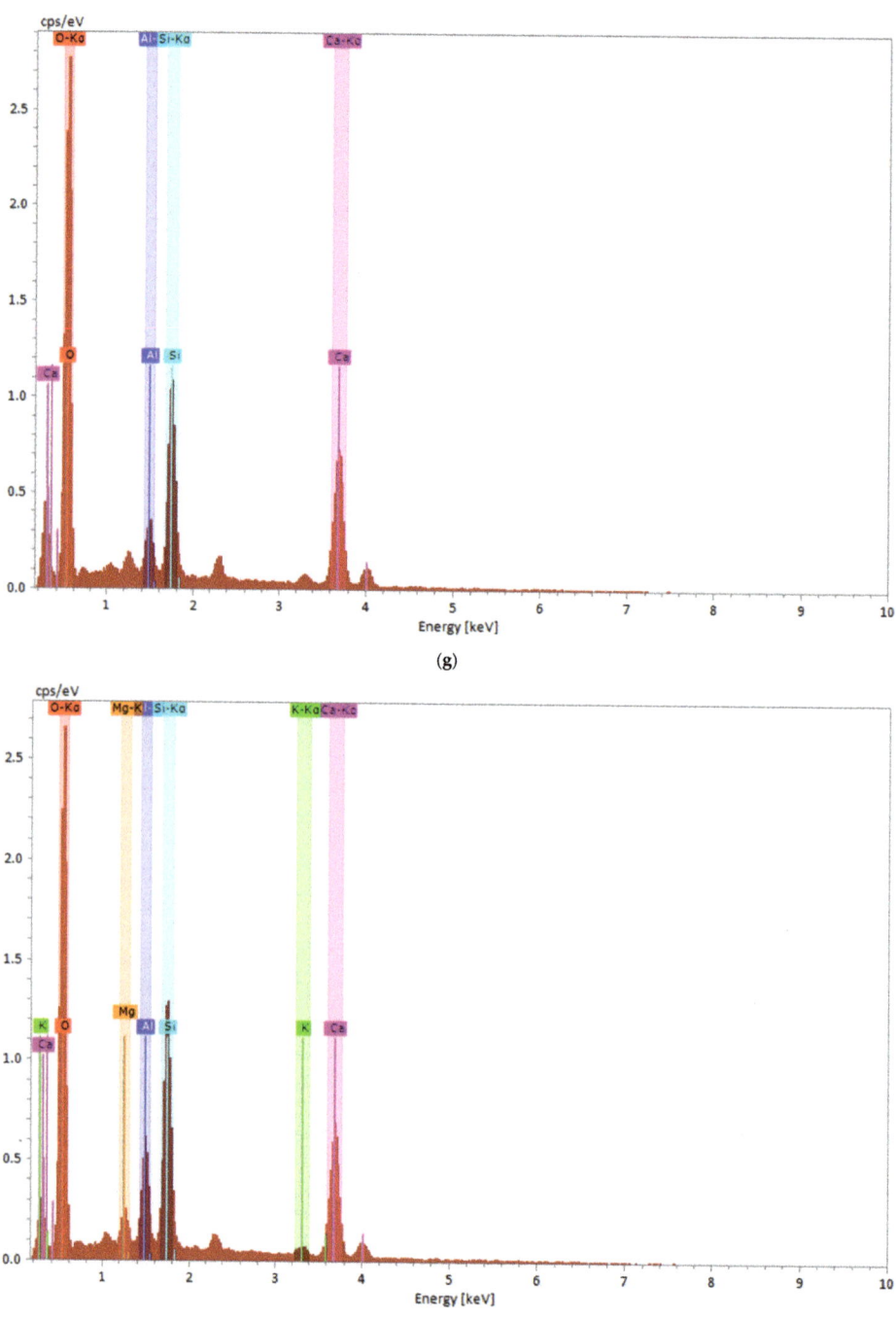

Figure 10. *Cont.*

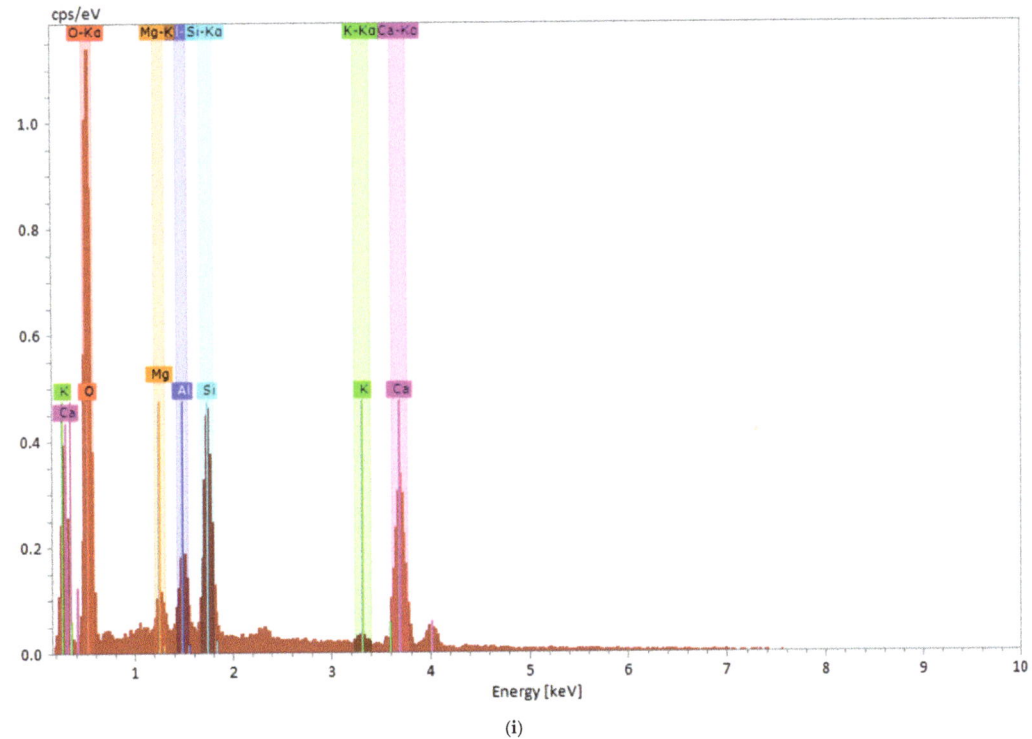

(i)

Figure 10. Surface mapping of (**a**) M I, (**b**) M II/360, (**c**) M III/330 mortar samples at 7 days, and (**d**) M I, (**e**) M II/360, (**f**) M III/330 mortar samples at 28 days; (**g**) M I, (**h**) M II/360, (**i**) M III/330 mortar samples at 28 days.

Table 12. Element concentrations (%) in M I, M II/360, M III/330 obtained by mapping the mortar sample surface.

Mix	O		Ca		Si		Al		Mg		Ca/Si	
	7 Days	28 Days	7 Days	28 Days	7 Days	28 Days	7 Days	28 Days	7 Days	28 Days	7 Days	28 Days
M I	51.69	49.51	38.87	33.47	6.90	14.60	1.15	2.42	1.39	-	5.63	2.29
M II/360	53.44	41.49	28.00	27.04	14.63	15.39	2.7	3.87	1.23	1.15	1.92	1.85
M III/330	53.63	46.96	28.44	32.63	14.69	14.66	3.24	3.25	-	1.47	1.94	2.23

In the three samples analyzed, the predominant element, along with oxygen, was calcium, followed by silicon. The calcium-rich gel formation mechanisms imparted strength to the cementitious materials [90]. Hydrated calcium silicate C-S-H contributed significantly to the macro-properties of the concrete, such as strength and durability [91].

In the slag mortars, these high calcium concentrations can be attributed to the formation of C-S-H calcium silicate hydrate gel in addition to the hydraulic and pozzolanic reactivity of GGBS [89]. In addition, Al and Mg compounds were observed in samples M II/360 and M III/330 since the GGBS contained higher amounts of aluminum and magnesium than the Portland cement [92,93].

Different variations in the Ca/Si ratio characterized the composition of the gels C-S-H [94]:

- For standard mortar M I with 100% cement, the value of 5.63 at 7 days old indicated the presence of most of the calcium-based structure, and the value of 2.29 at 28 days old indicated the presence of different forms of C-S-H there. As C-S-H gels are

amorphous [94], the EDX result was in accordance with the XRD data obtained for the crystallinity degree.
- For M II/360 and M III/330 with 50% GGBS, the values obtained in the range (1.85–2.23) at 7 and 28 days indicated the presence of different forms of C-S-H in the slag mortar samples [91,92].

The EDX measurements correlated with the 28-day SEM measurements, showing a more compact surface in the MIII/330 sample with 50% GGBS with smaller air holes and larger spacings compared to the standard mortar.

3.1.4. NMR Relaxometry Data at 28 Days

The evolution of the CPMG echo series in the empty mortar samples and cyclohexane-saturated samples at 28 days is shown in Figures 10 and 11.

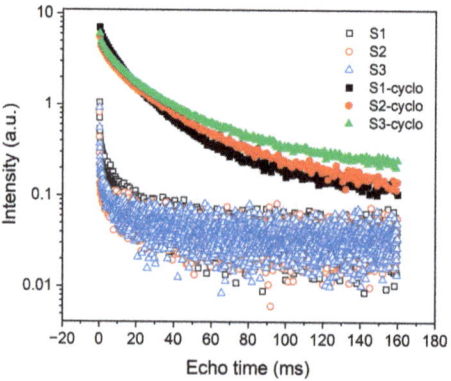

Figure 11. CPMG echo series for cyclohexane-saturated bare mortar samples M I, M II/360, M III/330 at 28 days (note: S1 = M I; S2 = M II/360; S3 = M II/330).

A clearer interpretation of these results comes from using the inverse Laplace transform to reveal the distribution of the relaxation times extracted from the CPMG series [63,64], as shown in Figure 11. The distribution of the relaxation times in the bare mortar samples after oven drying and in the cyclohexane-saturated samples obtained by the inverse Laplace numerical analysis is shown in Figure 12.

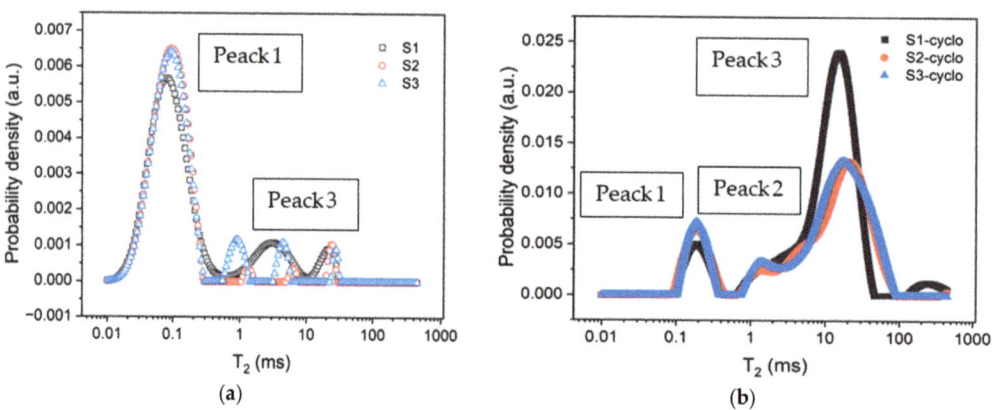

Figure 12. (a) Distribution of relaxation times in empty samples; (b) distribution of relaxation times in cyclohexane-saturated samples at 28 days (note: S1 = M I; S2 = M II/360; S3 = M II/330).

The values for the transverse relaxation time T_2 plotted on the horizontal axis are proportional to the relative pore size, and the values plotted on the vertical axis represent the probability density that such relaxation times exist in the sample (here, a.u. stands for arbitrary units). The probable density indicates the probability of a certain relaxation time in the measured sample. In Figure 12, three peaks can be distinguished, the position of which can be assigned to the three types of pores: intra-gel pores C-S-H (Peack 1) with dimensions up to 2 nm; inter-gel pores C-S-H (Peack 2) with dimensions up to 10 nm; and capillary pores (Peack 3) with dimensions between 50 nm and 1 µm [62].

The peak area is proportional to the amount of water or cyclohexane that is absorbed into the pores. The highest amount of water in the gel pores penetrating C-S-H (Peack 1, Figure 12a) was recorded in the slag mortar samples M II/360 and M III/330. But the gel pores in the cementitious materials had a protective effect due to their small size; the water in the pores of the gel did not freeze during the operation and was characterized by minimal permeability to liquids and gases. The C-S-H inter-gel pores, (Peack 2, Figure 12b) showed a similar relaxation time distribution in the three measured samples. The macro-capillaries with a radius in the range of 0.1 ÷ 10 µm were filled with water only in direct contact with water but had the ability of capillary condensation of moisture, making the cementitious materials hygroscopic. The main pores that caused damage to the structure of the cement stone were capillary pores [21,90].

The surface area of the Peack 3 in Figure 12b, corresponding to the capillary pores saturated with cyclohexane, shows that the test samples M II/360 and M III/330 with 50% GGBS absorbed the least amount of cyclohexane, and the sample of standard MI absorbed the greatest amount. These results indicated a lower capillary porosity compared to the standard mortar MI. The region of the transverse relaxation time distribution T_2 (on the Ox axis) moved slightly to the right at higher values (100 ms) for samples M II/360 and M III/330 compared to sample M I, which had a larger radius pore but a greater decreasing probability density below 0.01. In addition, it was observed that the maximum intensity (Peack 3) in sample M III/330 was slightly shifted to the left at the lower values compared to sample M II/360 which had smaller radius pores; the NMR results were consistent with those obtained by SEM–EDX.

3.2. Road Concrete Mixes with Embedded Slag (GGBS and ACBFS), Tested at 28 Days

3.2.1. Mechanical, Mechanical Wear, Freeze–Thaw Resistance, and Permeable Pore Content

The flexural tensile strength evolution at 28 days, the compressive strengths of the blanks and samples maintained in the thermostatic chamber, and the loss of compressive strength after 100 freeze–thaw cycles are shown in Figure 13. It was observed in all mixtures that the value of the tensile strength by bending in (Figure 13a) was above the limit of 5.5 MPa, which allowed the classification of the road concrete as class BcR 5.0, corresponding to very heavy traffic according to NE 014 [58]. However, the tensile strengths decreased slightly with the increasing dosage of ACBFS aggregate, by up to 4.1% for composite S15/25 and 10% for S15/50 compared to the reference mixture S 0/0. Regarding compressive strength, the S15/50 composite performed better than the S15/25 composite although it was up to 7.58% lower than the S0/0 reference mixture, (Figure 13b). It was also observed that the two slag mixtures had lower compressive strength losses at 100 freeze–thaw cycles than the reference composite S0/0 (Figure 13c), with values below the limit of 25% set in SR 3518 [68].

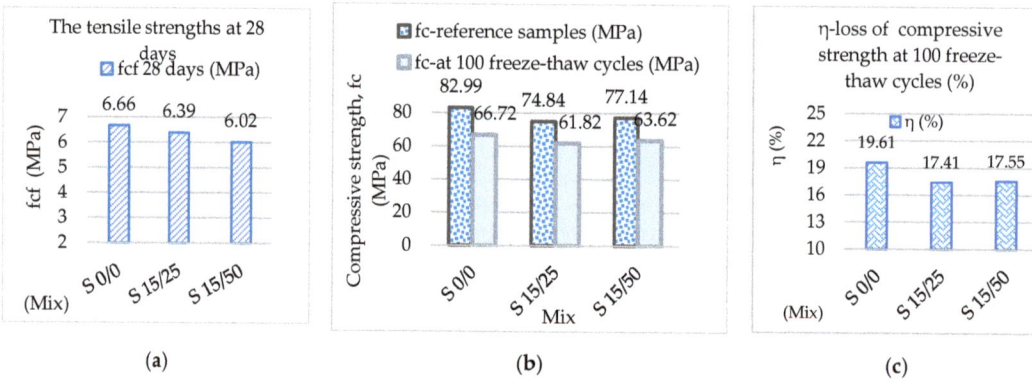

Figure 13. (a) The tensile strengths at 28 days; (b) the compressive strength of the reference samples and the compressive strength at 100 freeze–thaw cycles, (c) loss of compressive strength.

The total exfoliated mass S at 56 freeze–thaw cycles, the volume loss from attrition, and the permeable void content gave the results shown in Figure 14a. It shows an 8.9% increase in exfoliated cumulative mass in the S15/50; Figure 14b shows a wear volume loss of 5.5% in the S15/50 composite compared to the reference S0/0 composite. For the slag mixtures, however, the total value of the exfoliated mass S56 was <0.1 kg/m^2, which corresponds to a very good exfoliation resistance according to SS 13 72 44 [72,73] and falls into the same wear resistance class as the reference concrete (class 4, marking I, with values ≤18,000 mm^3/5000 mm^2) [70]. In Figure 14c, the permeable pore content of slag compositions shows values above the reference composite S0/0.

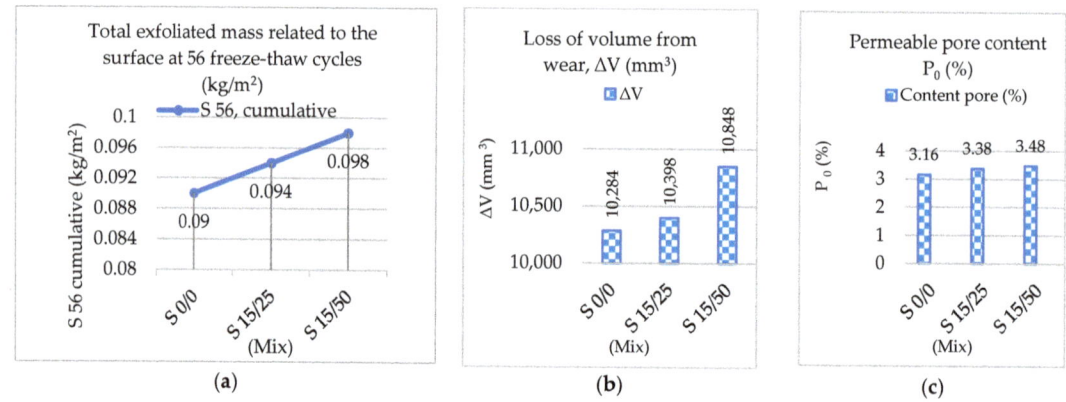

Figure 14. (a) Total mass exfoliated after 56 freeze–thaw cycles; (b) volume loss after mechanical wear test; (c) permeable pore content.

Figure 14 shows the relationship between the permeable void content and compressive strength, the total mass exfoliated after 56 freeze–thaw cycles, and the volume loss through abrasion and tensile strength. It was found that a second-order polynomial relationship developed between these features with a very good correlation coefficient (R-value), with decreasing permeable void content, flexural strengths, and freeze–thaw resistance and a wear resistance increase (Figure 15a–c). The porosity increase justified the reduction in performance parameters since more porous concrete usually results in a lower degree of hydration and lower compressive and tensile strengths [24].

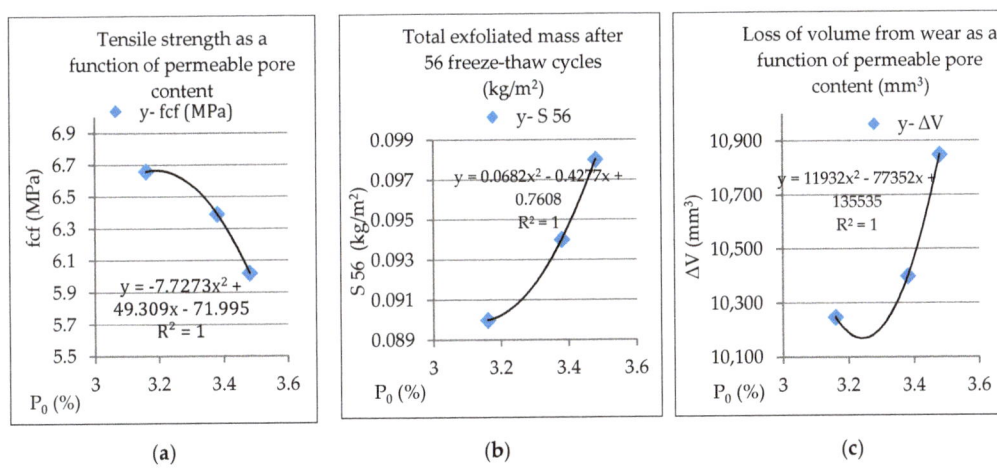

Figure 15. Relationship between pore permeable content and (**a**) flexural tensile strengths; (**b**) the total mass exfoliated after 56 freeze–thaw cycles; (**c**) loss of volume from wear.

3.2.2. Corrosion Resistance from Chlorides and Carbonation

The passing electric charges recorded values between 2273 and 3650 Coulombs for all the mixtures made (Figure 16a), which placed them in the moderate class of chlorine ion permeation, in accordance with ASTM C 1202 [75]. A discrepancy in RCPT values was observed with the S15/50 blend, which increased by 61%, while the S15/25 blend increased by only 16%. This observation needs to be investigated in a future study.

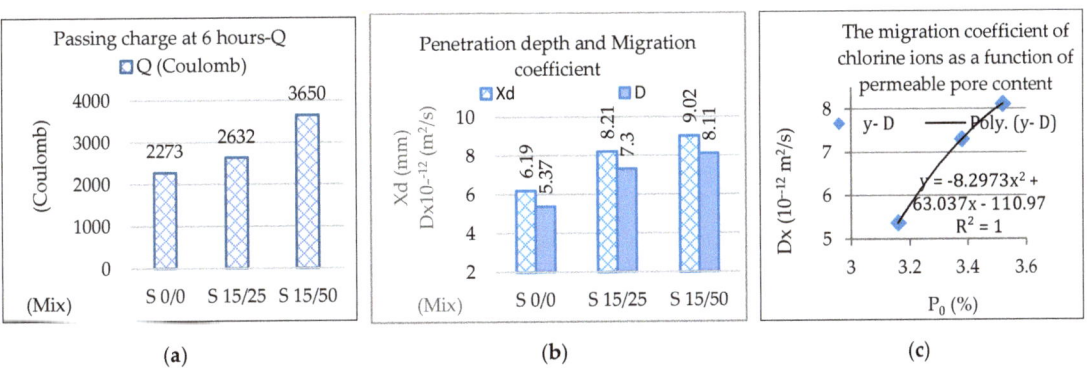

Figure 16. (**a**) Electric charge Q recorded at 6 h; (**b**) penetration depth Xd and migration coefficient Dx (10^{-12} m^2/s) of chlorine ions; (**c**) the bonding relationship between the migration coefficient of chlorine ions and the content of permeable pores.

It can be seen that as the degree of substitution by ACBFS aggregate increases, the penetration depth of the chlorine ions and the migration coefficient calculated with the Nernst–Planck equation increased (Figure 16b). Due to the higher water absorption of the ACBFS aggregate, the water requirement to maintain specific workability increases at higher levels of sand substitution [11]. A higher amount of water led to an increase in the porosity of the S15/50 composite and the highest migration coefficient of chlorine ions compared to the reference mixture. The two features developed a directly proportional relationship, as shown in Figures 16c and 17, with images of the chlorine ion migration front.

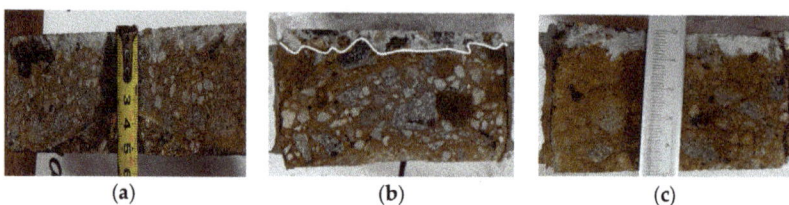

Figure 17. (a) Chlorine ion migration front for sample S 0/0; (b) for sample S 15/25; (c) for sample S 15/50.

3.2.3. Corrosion Resistance from Carbonation

Both the reference samples (Figure 18a) and the slag samples (Figure 18b,c) were unaffected by corrosion by carbonation at the age of 28 days. The color of the indicator solution (red-purple) on the contour areas of the surfaces was kept identical to the interior of the test specimens. The diffusion of carbon dioxide from the cement matrix was impeded by the increased compactness of the concrete cement stone [95,96].

Figure 18. Carbonation progress one hour after spraying phenolphthalein solution: (a) sample S0/0; (b) sample S15/25; (c) per sample S15/50.

The reported results represented the average of the three samples tested for each mixture with the CoV coefficient of variation (%), as shown in Table 13.

Table 13. Coefficient of CoV variation (%) of physic mechanical characteristics.

Mix Coefficient of Variation	S0/0 CoV (%)	S15/25 CoV (%)	S15/50 CoV (%)
The tensile strength—fcf (MPa)	12.4	7.3	1.7
Reference compressive strength—fc (MPa)	5.0	9.0	1.0
Compressive strength at 100 freeze–thaw—fc (MPa)	11	8.0	5.0
The exfoliated mass at 56 freeze–thaw cycles (kg/m^2)	8	7.8	12.2
Loss of volume from wear (mm^3)	5.5	3.6	7.7
Permeable pore content (%)	11.3	3.3	11.2
Passing charge at 6 h (Q)	1.1	1.5	0.8
Penetration depth Xd (mm)	1.9	8.6	0.9
Migration coefficient of chlorine ions (D× (10^{-12} m^2/s))	2.1	9.3	1.0

Note that the ratio of the standard deviation to the mean value obtained was below the accepted limit of 15% [97], with a reasonable quality in the range of 0.8% to 12%.

4. Conclusions

In this work, the influence of blast furnace slag (GGBS and ACBFS) on the physico-mechanical properties at 28 days was studied in three road concrete mixes, as follows: S0/0, a blank sample (exclusively Portland cement), and two other mixes S15/25 and S15/50, in which the Portland cement was replaced by 15% GGBS and the natural sand by 25% ACBFS and 50% ACBFS.

First of all, before making concrete, studies were made on mortars. This method was very useful in finding the best direction for the study of concrete with different percentages of slag (GGBS/ACBFS).

The main results of the physico-mechanical and durability properties of the road concretes and mortars showed the following:

The report on the compressive strengths classified the GGBS powder as resistance class 80, relative to the criteria established by ASTM C989.

At 7 days, the XRD measurements showed the highest level of crystallinity in the standard mortar M I.

After 28 days, mortar M II/360 showed a level of crystallinity above the level of the standard mortar.

At 7 days, the SEM measurements indicated the most compact surface for the standard MI mortar.

At 28 days, the slag mortar M III/330 showed a compact surface. The surface morphology of the M II/360 was similar to that of the standard sample; results were also confirmed by the compressive strength and degree of crystallinity determined by XRD.

The EDX results complemented the SEM assessment and showed the presence of C-S-H in all the samples analyzed.

At 28 days, in M III/330, the Ca/Si ratio value was the closest to the standard M I.

The NMR results showed that the slag mortars M II/360 and M III/330 had the largest proportion of C-S-H intra-pores, which had a protective effect against the freeze–thaw strength due to their small size.

Due to the evolution of the CS ratio and a similar microstructure in both the M II/360 and M III/330 mortars, it was decided to use GGBS slag with a low specific surface, to create an ecological pavement concrete.

The physical and mechanical properties and the durability values of slag concrete S15/25 and S15/50 were:

Above the limit of 5.5 MPa, which allowed the classification of the road concrete as class BcR 5.0, corresponding to very heavy traffic according to NE 014 [58].

The tensile strengths decreased slightly with the increasing dosage of ACBFS aggregate, by up to 4.1% for composite S15/25 and 10% for S 15/50 compared to the reference mixture S0/0.

In terms of compressive strength, the S15/50 composite performed better than the S15/25 composite, although it was up to 7.58% lower than the S0/0 reference mixture.

In general, the results were 10% lower, and they were up to 16% in the RCPT test. However, they were at the same level as the quality criteria of the reference concrete. The low reactivity of GGBS and the increased porosity of ACBFS explain the reduction in resistances: mechanical, freeze–thaw resistance, exfoliation, wear, and chlorine ion penetration.

The pavement mixes with 15% GGBS and 25% or 50% ACBFS met the criteria for the assessment of the concrete quality as being very good in terms of exfoliation (S56 is <0.1 kg/m^2) and carbonation, which corresponds to a G100 degree of gelation (with losses η < 25%), full placement into wear resistance class 4, brand I (volume loss values ≤ 18,000 mm^3/5000 mm^2), and a moderate level of chlorine ion penetration.

This research shows that blast furnace slag containing 15% GGBS and 25% or 50% ACBFS is an effective substitute for cement and natural sand in road concrete for all traffic classes.

Furthermore, from an environmental and carbon footprint perspective, the sustainability of GGBS with a low specific surface area and an ACBFS fine aggregate is evident.

Although the current focus is on mechanical strengths at 28 days, as a reference for determining the strength class of concrete further investigations aim to investigate the influence of blast furnace slag on road concrete composition at the age of 90 days since the pozzolanic activity of GGBS slag lasts longer than that of Portland cement.

The authors are currently working on publishing results on the quality and sustainability of ecological BcR in terms of strength and exposure class.

Author Contributions: Conceptualization, L.M.N., D.L.M., D.S., O.C., I.A. and M.L.D.; methodologies, L.M.N., D.S., O.C. and I.A.; investigations, L.M.N., D.S., O.C. and M.L.D.; writing—preparation of the original draft, L.M.N.; writing—review and editing, L.M.N., D.L.M., D.S., O.C., I.A. and M.L.D.; supervision, D.L.M., D.S., I.A. and M.L.D. All authors have read and agreed to the published version of the manuscript.

Funding: This research was funded by Project "Network of excellence in applied research and innovation for doctoral and postdoctoral programs/InoHubDoc", a project co-funded by the European Social Fund financing agreement No. POCU/993/6/13/153437. This research was also funded by the Ministry of Research, Innovation, and Digitization through Program 1—Development of the national research & development system, Subprogram 1.2—Institutional performance—Projects that finance the RDI excellence, Contract no. 18PFE/30.12.2021.

Institutional Review Board Statement: Not applicable.

Informed Consent Statement: Not applicable.

Data Availability Statement: All the required data that support the finding are presented in the manuscript.

Acknowledgments: The authors acknowledge the financial support by the Project "Network of excellence in applied research and innovation for doctoral and postdoctoral programs/InoHubDoc", a project co-funded by the European Social Fund financing agreement No. POCU/993/6/13/153437. D.S. and O.C. acknowledge the financial support by the Ministry of Research, Innovation, and Digitization through Program 1—Development of the national research & development system, Subprogram 1.2—Institutional performance—Projects that finance the RDI excellence, Contract no. 18PFE/30.12.2021.

Conflicts of Interest: The authors declare no conflict of interest.

References

1. Aitcin, P.-C. Cements of yesterday and today: Concrete of tomorrow. *Cem. Concr. Res.* **2000**, *30*, 1349–1359. [CrossRef]
2. Prakash, R.; Thenmozhi, R.; Raman, S.N. Mechanical characterisation and flexural performance of eco-friendly concrete produced with fly ash as cement replacement and coconut shell coarse aggregate. *Int. J. Environ. Sustain. Dev.* **2019**, *18*, 131–148. [CrossRef]
3. Burlacu, A.; Racanel, C. Reducing Cost of Infrastructure Works Using New Technologies. Available online: https://www.oecd-ilibrary.org/development/road-and-rail-infrastructure-in-asia_9789264302563-en (accessed on 27 February 2023).
4. Corbu, O.; Toma, I.O. Progress in Sustainability and Durability of Concrete and Mortar Composites. *Coatings* **2022**, *12*, 1024. [CrossRef]
5. Ahmad, J.; Martínez-García, R.; Szelag, M.; De-Prado-gil, J.; Marzouki, R.; Alqurashi, M.; Hussein, E.E. Effects of steel fibers (Sf) and ground granulated blast furnace slag (ggbs) on recycled aggregate concrete. *Materials* **2021**, *14*, 7497. [CrossRef] [PubMed]
6. Dimulescu, C.; Burlacu, A. Industrial Waste Materials as Alternative Fillers in Asphalt Mixtures. *Sustainability* **2021**, *13*, 8068. [CrossRef]
7. Popescu, D.; Burlacu, A. Considerations on the Benefits of Using Recyclable Materials for Road Construction. *Rom. J. Transp. Infrastruct.* **2017**, *6*, 43–53. [CrossRef]
8. Forton, A.; Mangiafico, S.; Sauzéat, C.; Di Benedetto, H.; Marc, P. Behaviour of binder blends: Experimental results and modelling from LVE properties of pure binder, RAP binder and rejuvenator. *Road Mater. Pavement Des.* **2021**, *22*, S197–S213. [CrossRef]
9. Cadar, R.D.; Boitor, R.M.; Dragomir, M.L. An Analysis of Reclaimed Asphalt Pavement from a Single Source—Case Study: A Secondary Road in Romania. *Sustainability* **2022**, *14*, 7075. [CrossRef]
10. Forton, A.; Mangiafico, S.; Sauzéat, C.; Di Benedetto, H.; Marc, P. Properties of blends of fresh and RAP binders with rejuvenator: Experimental and estimated results. *Constr. Build. Mater.* **2020**, *236*, 117555. [CrossRef]
11. Jain, S.; Santhanam, M.; Rakesh, S.; Kumar, A.; Gupta, A.K.; Kumar, R.; Sen, S.; Ramna, R.V. Utilization of Air-Cooled Blast Furnace Slag As a 100% Replacement of River Sand in Mortar and Concrete. *Indian Concr. J.* **2022**, *96*, 6–21.
12. Smith, K.D.; Morian, D.A.; Van Dam, T.J. *Use of Air-Cooled Blast Furnace Slag as Coarse Aggregate in Concrete Pavements—A Guide to Best Practice*; Report No. FHWA-HIF-12-009; Federal Highway Administration: Philadelphia, PA, USA, 2012.
13. ACI-233R-03; Slag Cement in Concrete and Mortar. American Concrete Institute: Farmington Hills, MI, USA, 2003.
14. Khan, K.; Amin, M.N. Influence of fineness of volcanic ash and its blends with quarry dust and slag on compressive strength of mortar under different curing temperatures. *Constr. Build. Mater.* **2017**, *154*, 514–528. [CrossRef]
15. Liu, S.; Li, Q.; Xie, G.; Li, L.; Xiao, H. Effect of grinding time on the particle characteristics of glass powder. *Powder Technol.* **2016**, *295*, 133–141. [CrossRef]
16. Liu, S.; Li, Q.; Song, J. Study on the grinding kinetics of copper tailing powder. *Powder Technol.* **2018**, *330*, 105–113. [CrossRef]

17. Yen, T.; Hsu, T.H.; Liu, W.Y.; Chen, S. Influence of class F fly ash on the abrasion–erosion resistance of high-strength concrete. *Constr. Build. Mater.* **2007**, *21*, 458–463. [CrossRef]
18. Wang, Y.; He, X.; Su, Y.; Yang, J.; Strnadel, B.; Wang, X. Efficiency of wet-grinding on the mechano-chemical activation of granulated blast furnace slag (GBFS). *Constr. Build. Mater.* **2019**, *199*, 185–193. [CrossRef]
19. Petrova, T.; Chistyakov, E.; Makarov, Y. Methods of road surface durability improvement. *Transp. Res. Procedia* **2018**, *36*, 586–590. [CrossRef]
20. Garipov, A.; Makarov, D.; Khozin, V.; Stepanov, S.; Ayupov, D. Cement concrete modified by fine-dispersed anionactive bitumen emulsion for road construction. *IOP Conf. Ser. Mater. Sci. Eng.* **2020**, *890*, 12107. [CrossRef]
21. Smirnov, D.; Stepanov, S.Y.; Garipov, R.; Garayev, T.; Sungatullin, T. Influence of the porosity structure of road concrete on its durability. In Proceedings of the E3S Web of Conferences, Virtual Event, 6–7 September 2021.
22. Tolmachov, S. Research of the reasons of frost destruction of road concrete. *Key Eng. Mater.* **2020**, *864*, 175–179. [CrossRef]
23. Bjegović, D.; Serdar, M.; Oslaković, I.S.; Jacobs, F.; Beushausen, H.; Andrade, C.; Monteiro, A.V.; Paulini, P.; Nanukuttan, S. Test methods for concrete durability indicators. *RILEM State-Art Rep.* **2016**, *18*, 51–105. [CrossRef]
24. Correia, V.; Ferreira, J.G.; Tang, L.; Lindvall, A. Effect of the addition of GGBS on the frost scaling and chloride migration resistance of concrete. *Appl. Sci.* **2020**, *10*, 3940. [CrossRef]
25. SS 13 70 03; Concrete-Usage of EN 206-1 in Sweden. Swedish Standard Institute: Stockholm, Sweden, 2008.
26. ACI 213R-03; Guide for Structural Lightweight Aggregate Concrete. American Concrete Institute: Farmington Hills, MI, USA, 2003.
27. Panchmatia, P.; Olek, J.; Kim, T. The influence of air cooled blast furnace slag (ACBFS) aggregate on the concentration of sulfates in concrete's pore solution. *Constr. Build. Mater.* **2018**, *168*, 394–403. [CrossRef]
28. Wang, A.; Deng, M.; Sun, D.; Li, B.; Tang, M. Effect of crushed air-cooled blast furnace slag on mechanical properties of concrete. *J. Wuhan Univ. Technol. Sci. Ed.* **2012**, *27*, 758–762. [CrossRef]
29. Patra, R.K.; Mukharjee, B.B. Influence of incorporation of granulated blast furnace slag as replacement of fine aggregate on properties of concrete. *J. Clean. Prod.* **2017**, *165*, 468–476. [CrossRef]
30. Singh, G.; Das, S.; Ahmed, A.A.; Saha, S.; Karmakar, S. Study of Granulated Blast Furnace Slag as Fine Aggregates in Concrete for Sustainable Infrastructure. *Procedia-Soc. Behav. Sci.* **2015**, *195*, 2272–2279. [CrossRef]
31. Sandhu, R.S.; Singh, J.; Dhanoa, G.S. Use of Air Cooled Blast Furnace Slag (ACBFS) as Coarse Aggregates—A Case Study. *Int. J. Innov. Eng. Res. Technol.* **2015**, *2*, 1–10.
32. Verian, K.P.; Panchmatia, P.; Olek, J.; Nantung, T. Pavement concrete with air-cooled blast furnace slag and dolomite as coarse aggregates: Effects of deicers and freeze-thaw cycles. *Transp. Res. Rec.* **2015**, *2508*, 55–64. [CrossRef]
33. Kılıç, A.; Atiş, C.D.; Teymen, A.; Karahan, O.; Özcan, F.; Bilim, C.; Özdemir, M. The influence of aggregate type on the strength and abrasion resistance of high strength concrete. *Cem. Concr. Compos.* **2008**, *30*, 290–296. [CrossRef]
34. Luna-Galiano, Y.; Leiva Fernández, C.; Villegas Sánchez, R.; Fernández-Pereira, C. Development of Geopolymer Mortars Using Air-Cooled Blast Furnace Slag and Biomass Bottom Ashes as Fine Aggregates. *Processes* **2023**, *11*, 1597. [CrossRef]
35. Ahmad, J.; Martínez-García, R.; De-Prado-gil, J.; Irshad, K.; El-Shorbagy, M.A.; Fediuk, R.; Vatin, N.I. Concrete with Partial Substitution of Waste Glass and Recycled Concrete Aggregate. *Materials* **2022**, *15*, 430. [CrossRef] [PubMed]
36. Ahmad, J.; Aslam, F.; Martinez-Garcia, R.; de-Prado-Gil, J.; Qaidi, S.M.A.; Brahmia, A. Effects of waste glass and waste marble on mechanical and durability performance of concrete. *Sci. Rep.* **2021**, *11*, 21525. [CrossRef] [PubMed]
37. Ahmad, J.; Tufail, R.F.; Aslam, F.; Mosavi, A.; Alyousef, R.; Javed, M.F.; Zaid, O.; Khan Niazi, M.S. A step towards sustainable self-compacting concrete by using partial substitution of wheat straw ash and bentonite clay instead of cement. *Sustainability* **2021**, *13*, 824. [CrossRef]
38. Ahmad, J.; Kontoleon, K.J.; Majdi, A.; Naqash, M.T.; Deifalla, A.F.; Ben Kahla, N.; Isleem, H.F.; Qaidi, S.M.A. A Comprehensive Review on the Ground Granulated Blast Furnace Slag (GGBS) in Concrete Production. *Sustainability* **2022**, *14*, 8783. [CrossRef]
39. Hadj-sadok, A.; Kenai, S.; Courard, L.; Darimont, A. Microstructure and durability of mortars modified with medium active blast furnace slag. *Constr. Build. Mater.* **2011**, *25*, 1018–1025. [CrossRef]
40. Dai, J.; Wang, Q.; Xie, C.; Xue, Y.; Duan, Y.; Cui, X. The Effect of Fineness on the Hydration Activity Index of Ground Granulated Blast Furnace Slag. *Materials* **2019**, *12*, 2984. [CrossRef] [PubMed]
41. Nicula, L.M.; Manea, D.L.; Simedru, D.; Cadar, O.; Becze, A.; Dragomir, M.L. The Influence of Blast Furnace Slag on Cement Concrete Road by Microstructure Characterization and Assessment of Physical-Mechanical Resistances at 150/480 Days. *Materials* **2023**, *16*, 3332. [CrossRef] [PubMed]
42. SR EN 197; Standard Cement—Part 1: Composition, Specification, and Conformity Criteria Common Cements. ASRO: Bucharest, Romania, 2011.
43. SR EN 15167; Ground Granulated Blast Furnace Slag for Use in Concrete, Mortar and Grout Part 1: Definitions, Specifications and Conformity Criteria. ASRO: Bucharest, Romania, 2007.
44. D6868 Standard; Specification for Biodegradable Plastics Used as Coatings on Paper and Other Compostable Substrates. ASTM International: West Conshohocken, PA, USA, 2017.
45. Lothenbach, B.; Scrivener, K.; Hooton, R.D. Supplementary cementitious materials. *Cem. Concr. Res.* **2011**, *41*, 1244–1256. [CrossRef]
46. SR EN 12620+A1; Aggregates for Concrete. ASRO: Bucharest, Romania, 2008.
47. SR 667; Natural Aggregates and Processed Stone for Roads. ASRO: Bucharest, Romania, 2001.
48. SR EN 13043; Aggregates for Bituminous Mixtures. ASRO: Bucharest, Romania, 2004.

49. SR EN 933-1; Tests to Determine the Geometric Characteristics of the Aggregates. Part 1: Determination of Granularity. Granulometric Analysis by Sieving. ASRO: Bucharest, Romania, 2012.
50. SR EN 934-2+A1; Concrete Additives. ASRO: Bucharest, Romania, 2002.
51. SR EN 1008; Mixing Water for Concrete. ASRO: Bucharest, Romania, 2003.
52. Corbu, O.; Ioani, A.M.; Bakri, A.; Meiță, V.; Szilaghyi, H.; Sandu, A.V. The Pozzoolanic Activity Level of Powder Waste Glass in Comparisons with other Powders. *Key Eng. Mater.* **2015**, *660*, 237–243. [CrossRef]
53. SR EN 196-1; Methods of Testing Cement Part 1: Determination of Strength. ASRO: Bucharest, Romania, 2016.
54. SR EN 1015-11; Methods of Testing Masonry Mortars. Part 11: Determination of Flexural and Compressive Strength of Hardened Mortar. ASRO: Bucharest, Romania, 2020.
55. Sajedi, F.; Razak, H.A. Comparison of different methods for activation of ordinary Portland cement-slag mortars. *Constr. Build. Mater.* **2011**, *25*, 30–38. [CrossRef]
56. Nicula, L.M.; Manea, D.L.; Simedru, D.; Dragomir, M.L. Investigations Related to the Opportunity of Using Furnace Slag in the Composition of Road Cement Concrete. In Proceedings of the International Conference on Innovative Research, Iasi, Romania, 11–12 May 2023; p. 85.
57. Sosa, M.E.; Villagrán Zaccardi, Y.A.; Zega, C.J. A critical review of the resulting effective water-to-cement ratio of fine recycled aggregate concrete. *Constr. Build. Mater.* **2021**, *313*, 125536. [CrossRef]
58. NE 014; The Norm for the Execution of Cement Concrete Road Pavements in a Fixed and Sliding Formwork System. Matrix ROM: Bucharest, Romania, 2007; ISBN 978-973-755-185-6.
59. SR EN 206; Standard for Concrete-Part 1: Specification, Performance, Production and Conformity. ASRO: Bucharest, Romania, 2021.
60. ASTM C989:C989/C989M Standard Specification for Slag Cement for Use in Concrete and Mortars. Available online: https://img.antpedia.com/standard/files/pdfs_ora/20210202/ASTM%20C989-18a.pdf (accessed on 28 June 2023).
61. Jenkins, R.; Snyder, R.L. *Introduction to X-ray Powder Diffractometry*; Wiley Online Library: Hoboken, NJ, USA, 1996; Volume 138.
62. Bede, A.; Scurtu, A.; Ardelean, I. NMR relaxation of molecules confined inside the cement paste pores under partially saturated conditions. *Cem. Concr. Res.* **2016**, *89*, 56–62. [CrossRef]
63. Venkataramanan, L.; Song, Y.-Q.; Hurlimann, M.D. Solving Fredholm integrals of the first kind with tensor product structure in 2 and 2.5 dimensions. *IEEE Trans. Signal Process.* **2002**, *50*, 1017–1026. [CrossRef]
64. Provencher, S.W. CONTIN: A general purpose constrained regularization program for inverting noisy linear algebraic and integral equations. *Comput. Phys. Commun.* **1982**, *27*, 229–242. [CrossRef]
65. SR EN 12390; Test on Hardened Concrete. Part 5: Bending Tensile Strength of Specimens. ASRO: Bucharest, Romania, 2019.
66. SR CR 12793; Determination of the Depth of the Carbonation Layer of Hardened Concrete. ASRO: Bucharest, Romania, 2002.
67. SR EN 12390; Standard for Test-Hardened Concrete—Part 3: Compressive Strength of Test Specimens. ASRO: Bucharest, Romania, 2019.
68. SR 3518; Tests on Concrete: Determination of the Freeze-Tawing Resistance by Measuring the Variations of the Resistance Strength and/or of the Dynamic Relative Elastics Modulus. ASRO: Bucharest, Romania, 2009.
69. SR CEN/TS 12390; Test on Hardened Concrete. Part 9: Resistance to Freeze-Thaw Using De-icing Salts Exfoliating. ASRO: Bucharest, Romania, 2017.
70. SR EN 1338:2006; Concrete Pavers. Test Conditions and Methods. ASRO: Bucharest, Romania, 2006.
71. ASTM C 642; Standard Test Method for Density, Absorption and Voids in Hardened Concrete. ASTM International: West Conshohocken, PA, USA, 2006.
72. SS 13 72 44; Concrete Testing—Hardened Concrete—Scaling at Freezing. Swedish National Testing and Reasearch Institute: Borras, Sweden, 2005.
73. Nicula, L.M.; Corbu, O.; Iliescu, M. Methods for assessing the frost-thaw resistance of road concrete used in our country and at European level. *IOP Conf. Ser. Mater. Sci. Eng.* **2020**, *877*, 012025. [CrossRef]
74. Ne 012 *Normative for the Production of Concrete and the Execution of Works in Concrete, Reinforced Concrete and Prestressed Concrete Part 1: Production of Concrete*; Ministry of Development, Public Works and Administration, Technical University of Construction: Bucharest, Romania, 2022.
75. ASTM C 1202; Electrical Indication of Concrete's Ability to Resist Chloride Ion Penetration. American Society for Testing and Materials, Annual Book of ASTM Standards: Philadelphia, PA, USA, 2000.
76. NT BUILD 492: 1999 Chloride Migration Coefficient from Nonsteady-State Migration Experiments. Nord. Finl. Available online: https//www.betonconsultingeng.com/services/concrete-testing/nt-build-492/ (accessed on 28 June 2023).
77. Bassuoni, M.T.; Nehdi, M.L.; Greenough, T.R. Enhancing the Reliability of Evaluating Chloride Ingress in Concrete Using the ASTM C 1202 Rapid Chloride Penetrability Test. *J. ASTM Int.* **2006**, *3*, 3.
78. Jercan, S. *Concrete Roads*; Corvin Publishing House: Deva, Romania, 2002.
79. Raki, L.; Beaudoin, J.; Alizadeh, R.; Makar, T.S. Cement and concrete nanoscience and nanotechnology. *Materials* **2010**, *3*, 918–942. [CrossRef]
80. Liu, S.; Li, L. Influence of fineness on the cementitious properties of steel slag. *J. Therm. Anal. Calorim.* **2014**, *117*, 629–634. [CrossRef]
81. Jakob, C.; Jansen, D.; Ukrainczyk, N.; Koenders, E.; Pott, U.; Stephan, D.; Neubauer, J. Relating Ettringite Formation and Rheological Changes during the Initial Cement Hydration. *Materials* **2019**, *12*, 2957. [CrossRef]
82. What Are the Advantages and Applications of GGBS? Available online: https://constrofacilitator.com/what-are-the-advantages-and-applications-of-ggbs/ (accessed on 26 June 2023).

83. Hosseini, P.; Abolhasani, M.; Mirzaei, F.; Kouhi Anbaran, M.R.; Khaksari, Y.; Famili, H. Influence of two types of nanosilica hydrosols on short-term properties of sustainable white portland cement mortar. *J. Mater. Civ. Eng.* **2018**, *30*, 4017289. [CrossRef]
84. Khaloo, A.; Mobini, M.H.; Hosseini, P. Influence of different types of nano-SiO_2 particles on properties of high-performance concrete. *Constr. Build. Mater.* **2016**, *113*, 188–201. [CrossRef]
85. Madani, H.; Bagheri, A.; Parhizkar, T. The pozzolanic reactivity of monodispersed nanosilica hydrosols and their influence on the hydration characteristics of Portland cement. *Cem. Concr. Res.* **2012**, *42*, 1563–1570. [CrossRef]
86. Song, Y.; Zhou, J.; Bian, Z.; Dai, G. Pore Structure Characterization of Hardened Cement Paste by Multiple Methods. *Adv. Mater. Sci. Eng.* **2019**, *2019*, 3726953. [CrossRef]
87. Scurtu, D.A.; Kovacs, E.; Senila, L.; Levei, E.A.; Simedru, D.; Filip, X.; Dan, M.; Roman, C.; Cadar, O.; David, L. Use of Vine Shoot Waste for Manufacturing Innovative Reinforced Cement Composites. *Appl. Sci.* **2023**, *13*, 134. [CrossRef]
88. Nicula, L.M.; Corbu, O.; Ardelean, I.; Sandu, A.V.; Iliescu, M.; Simedru, D. Freeze–Thaw Effect on Road Concrete Containing Blast Furnace Slag: NMR Relaxometry Investigations. *Materials* **2021**, *14*, 3288. [CrossRef] [PubMed]
89. Ganesh, P.; Murthy, A.R. Tensile behaviour and durability aspects of sustainable ultra-high performance concrete incorporated with GGBS as cementitious material. *Constr. Build. Mater.* **2019**, *197*, 667–680. [CrossRef]
90. Yang, K.; White, C.E. Modeling of aqueous species interaction energies prior to nucleation in cement-based gel systems. *Cem. Concr. Res.* **2021**, *139*, 106266. [CrossRef]
91. Chu, D.C.; Kleib, J.; Amar, M.; Benzerzour, M.; Abriak, N.-E. Determination of the degree of hydration of Portland cement using three different approaches: Scanning electron microscopy (SEM-BSE) and Thermogravimetric analysis (TGA). *Case Stud. Constr. Mater.* **2021**, *15*, e00754. [CrossRef]
92. Richardson, I.G. The nature of C-S-H in hardened cements. *Cem. Concr. Res.* **1999**, *29*, 1131–1147. [CrossRef]
93. Lee, J.; Choi, S. Case Studies in Construction Materials Effect of replacement ratio of ferronickel slag aggregate on characteristics of cementitious mortars at different curing temperatures. *Case Stud. Constr. Mater.* **2023**, *18*, e01882. [CrossRef]
94. Thermodynamic Description of the Solubility of C-S-H Gels in Hydrated Portland Cement. Available online: https://inis.iaea.org/collection/NCLCollectionStore/_Public/43/063/43063335.pdf (accessed on 26 June 2023).
95. Pimienta, P.; Albert, B.; Huetb, B.; Dierkens, M.; Francisco, P.; Rougeaud, P. Durability performance assessment of non-standard cementitious materials for buildings: A general method applied to the French context, Fact sheet 1—Risk of steel corrosion induced by carbonation. *RILEM Tech. Lett.* **2016**, *1*, 102–108. [CrossRef]
96. Medeiros, R.A.; Lima, M.G.; Yazigi, R.; Medeiros, M.H.F. Carbonation depth in 57 years old concrete structure. *Steel Compos. Struct.* **2015**, *19*, 953–966. [CrossRef]
97. Badr, A.; Ashour, A.F.; Platten, A.K. Statistical variations in impact resistance of polypropylene fibre-reinforced concrete. *Int. J. Impact Eng.* **2006**, *32*, 1907–1920. [CrossRef]

Disclaimer/Publisher's Note: The statements, opinions and data contained in all publications are solely those of the individual author(s) and contributor(s) and not of MDPI and/or the editor(s). MDPI and/or the editor(s) disclaim responsibility for any injury to people or property resulting from any ideas, methods, instructions or products referred to in the content.

Article

Study on the Oil Well Cement-Based Composites to Prevent Corrosion by Carbon Dioxide and Hydrogen Sulfide at High 2Temperature

Chunqin Tan [1,2], Mu Wang [1,2], Rongyao Chen [3,4,*] and Fuchang You [3,4,*]

1. Sinopec Key Laboratory of Cementing and Completion, Beijing 102206, China
2. Sinopec Research Institute of Petroleum Engineering Co., Ltd., Beijing 102206, China
3. School of Petroleum Engineering, Yangtze University, Wuhan 430100, China
4. National Engineering Research Center for Oil and Gas Drilling and Completion Technology, Yangtze University, Wuhan 430100, China
* Correspondence: 2021720483@yangtzeu.edu.cn (R.C.); yfc81@yangtzeu.edu.cn (F.Y.)

Abstract: Complex wells with high temperature and the presence of carbon dioxide and hydrogen sulfide acid gas require the use of high-temperature and high-density anti-corrosion cement slurry for cementing operations, and conventional cement slurry does not have the advantages of high density, high-temperature resistance, or corrosion resistance. In order to avoid the severe corrosion of cement slurry by carbon dioxide and hydrogen sulfide at high temperatures, solid phase particles with different particle sizes are combined with polymer materials to form a dense, high-density, high-temperature- and corrosion-resistant cement slurry. In this paper, we consider the use of manganese ore powder weighting agent, composite high-temperature stabilizer, inorganic preservative slag and organic preservative resin to improve the corrosion resistance of cement slurry, design a high-density cement slurry that is resistant to high temperature and carbon dioxide and hydrogen sulfide corrosion, and evaluate the performances of the cement slurry at 180 °C. The results show that the manganese ore powder weighting agent effectively improves the density of the cement slurry. Using composite silica fume with different particle sizes as a high-temperature stabilizer can ensure the rheology of the cement slurry and improve the ability of the cement sample to resist high-temperature damage. The use of slag and resin as preservatives can effectively reduce the corrosion degree in cement slurry. The high-temperature corrosion-resistant cement slurry systems with different densities designed using these materials exhibit good rheological properties, with water loss of less than 50 mL and a thickening time of more than four hours. The compressive strength decreased by less than 5.8% after 28 days at high temperatures. After being corroded by hydrogen sulfide and carbon dioxide (total pressure 30 MPa, 16.7% hydrogen sulfide and 6.7% carbon dioxide) under high temperature (180 °C) for 30 days, the corrosion depth of the cement sample was less than 2 mm, the reduction of compressive strength was low, and the corrosion resistance was strong. These research results can be used for cementing operations of high-temperature oil and gas wells containing hydrogen sulfide and dioxide.

Keywords: high temperature; corrosion; hydrogen sulfide; carbon dioxide; oil well cement

Citation: Tan, C.; Wang, M.; Chen, R.; You, F. Study on the Oil Well Cement-Based Composites to Prevent Corrosion by Carbon Dioxide and Hydrogen Sulfide at High Temperature. *Coatings* **2023**, *13*, 729. https://doi.org/10.3390/coatings13040729

Academic Editors: Ionut Ovidiu Toma and Ofelia-Cornelia Corbu

Received: 19 February 2023
Revised: 22 March 2023
Accepted: 28 March 2023
Published: 3 April 2023

Copyright: © 2023 by the authors. Licensee MDPI, Basel, Switzerland. This article is an open access article distributed under the terms and conditions of the Creative Commons Attribution (CC BY) license (https://creativecommons.org/licenses/by/4.0/).

1. Introduction

With the continuous development of oil and gas, there are more and more high-temperature and high-pressure oil and gas wells. The high temperature at the bottom of oil and gas wells makes cementing operations difficult, and the sealing quality of the cement sheath cannot be guaranteed. In particular, there are a lot of acid gases in the humid environment, which react easily with the cement slurry. If the cement sample is corroded, the structure of the cement slurry will be damaged, causing fluid channeling, and reducing the production and service life of oil and gas wells [1–3]. On the one hand, the cement slurry needs to be resistant to high temperature. On the other hand, for some oil and gas wells with high temperature, high pressure and acid gas, the design of cement slurry is very difficult [4,5]. This is because conventional oil well cement-based composites are prone to a decline in strength at high temperatures and accelerated corrosion reaction. At present, the common corrosive gases in oil and gas wells are hydrogen sulfide and carbon dioxide, which form sulfuric acid and carbonic acid when combined with formation water [3,6]. As the conventional oil well cement is Portland cement, its hydrated products mainly include calcium hydroxide, hydrated calcium silicate, etc. The cement sample is alkaline, and the acid gas dissolves in water and has strong corrosivity to it, so it has high requirements in terms of the corrosion resistance of the oil well cement-based composites. There have been some studies on the corrosion of cement-based composites by a single acid gas [7–10]. However, due to the synergistic corrosion of carbon dioxide and hydrogen sulfide dissolved in water, the synergistic corrosion of carbon dioxide and hydrogen sulfide will damage the hydration products (mainly calcium hydroxide and calcium silicate) of the cement sample, destroy the structure, increase the pore volume, cause the strength to decrease, and make the cement slurry unable to effectively support the formation and casing pipe [11–13]. The research of O.A. Omosebi [14] showed that under high temperature, the acid gas exerts a high degree of corrosion on the cement sheath and causes serious corrosion damage. Therefore, it is necessary to design a high-density-resistant high-temperature anti-corrosion cement slurry for cementing operations in high-temperature formations containing carbon dioxide and hydrogen sulfide.

In order to ensure the sealing integrity of the cement sheath, the type of additives in cement slurry is relatively fixed. For the design of high-temperature and high-density anti-corrosion cement slurry, dispersant, retarder, fluid loss reducer, and other conventional additives have been studied in high-temperature cement slurry and can be used directly [15–18]. However, the weighting agents, high-temperature stabilizers and anti-corrosion materials need to be specially designed, mainly because the content of these materials is very large, which has a great impact on the performances of cement samples; in particular there has been little research performed to date on materials that can improve corrosion resistance to carbon dioxide and hydrogen sulfide at high temperature. A. Abdulmalek et al. [19] and A. Ahmed et al. [20] studied the use of barite and iron ore powder in the design of high-density cement slurry. The density of the cement slurry designed with these weighting materials reached 2.4 kg/m^3, making them suitable for cementing operations in high-pressure formations. The research of J.K. Qin et al. [21], B.L.D. Costa et al. [22], and H.J. Liu et al. [23] showed that silica fume can be used as a high-temperature stabilizer of oil well cement to improve the performance of cement samples under high temperature. In order to ensure the quality of cementing cement sheath in acidic gas environments, B. Yuan et al. [24], Z.G. Peng et al. [25], and M.X. Bai et al. [26] studied the corrosion resistance of modified cement with the addition of the corrosion-resistant additive CRA in an H_2S-CO_2 environment. The research results showed that cement with polymer preservatives has excellent acid corrosion resistance due to its film-forming and filling effects, reducing permeability and alkalinity. The results of these studies can serve as a reference for the design of high-temperature and high-density carbon-dioxide-emitting and hydrogen-sulfide-corrosive cement slurry. However, these materials cannot be used simultaneously in high-temperature, high-density, hydrogen sulfide and carbon dioxide environments, so these key materials need to be further developed.

To meet the particular requirements of high-temperature and high-density anti-corrosion cement slurry, in this paper, high-temperature-resistant solid materials are designed with various particle sizes to improve the high-temperature stability of cement slurry, adjust the density of cement slurry, and improve the anti-corrosion performance of cement slurry. At the same time, polymer liquid materials and solid materials are combined to synergize and improve the performance of cement slurry through different mechanisms of action. In this paper, the influence of manganese ore powder weighting agent in the performance of cement slurry was studied under high temperature and high pressure to construct high-density cement slurry. In view of the fact that the high-temperature environment makes the cement prone to failure and instability, a high-temperature stabilizer is studied to improve the mechanical property stability of the cement sample under high temperatures. Inorganic and organic anti-corrosion materials are studied and used together to improve the corrosion resistance of cement slurry. Based on the admixture materials studied, a high-temperature, high-density and anti-corrosion cement slurry system was designed, and its performance was evaluated. The research results provide technical support for the cementing of high-temperature and high-pressure oil and gas wells containing carbon dioxide and hydrogen sulfide.

2. Experimental Section

2.1. Experimental Materials

Class G oil well cement is used as the primary material of the cement slurry, and the chemical composition of the oil well cement is shown in Table 1. In this study, fluid loss reducer, dispersant and retarder are used to adjust the performance of cement slurry. The fluid loss reducer is a 2-Acrylamido-2-methylpropane sulfonic acid (AMPS)-type water-soluble polymer that can reduce the permeability of filter cakes of oil well cement slurry. In this study, it is mainly used to reduce the water loss of the cement slurry. The dispersant is an aldehyde ketone condensation polymer that is mainly used to reduce the water loss of cement slurry by adjusting the surface charge of the cement particles to obtain cement slurry with appropriate rheological properties. The retarder consists of polymers with carboxylic and sulfonic acid groups that regulate the setting time of cement slurry through adsorption, chelation, dispersion, and wetting. It is mainly used to regulate the thickening time of cement slurry. The manganese ore powder, silicon powder and slag were purchased from the market. The main component of manganese ore powder is manganese tetroxide, with a particle size of less than 6 microns. It is mainly used to adjust the density of cement slurry. The main component of silicon powder is silicon dioxide, and the particle sizes used in this study are mainly 100 and 300 mesh. Silicon powder is mainly used to improve the high-temperature resistance of oil well cement slurry. The main components of the silicon powder are calcium oxide, silicon dioxide, and aluminum trioxide, with a particle size of less than 30 μm. It is mainly used to improve the corrosion resistance of oil well cement slurry. The polymer resin was obtained from the laboratory, and is a bisphenol A type epoxy resin, which is mainly used as an organic preservative to improve the corrosion resistance of cement paste.

Table 1. Chemical composition of oil well cement.

Component	CaO	SiO_2	Fe_2O_3	Al_2O_3	MgO	$Na_2O + K_2$	Others
Content (%)	64.2	22.5	4.4	4.1	1.6	0.38	2.82

2.2. Experimental Methods

2.2.1. Preparation of Cement Slurry

Depending on the proportions in the cement slurry composition, cement and other solid materials (silica fume, weighting agent, slag) were mixed to form dry mixed ash, water and liquid admixture materials (fluid loss reducer, retarder, dispersant, resin), via wet mixing to form mixed solution; dry mixed ash and the mixed solution were placed in a constant-speed mixer (TG-3060A, Shenyang Taige Petroleum Instrument Equipment Co., Ltd., Shenyang, China), and stirred evenly at 4000 r/min. When the cement slurry was mixed, its evenness was observed, bubbles were eliminated, and the uniformly mixed cement slurry was taken as the sample. The composition of cement slurries with high density is shown in Table 2.

Table 2. Composition of high-density cement slurries. Unit: wt.%.

Density (kg/m³)	Cement	Water	Fluid Loss Reducer	Retarder	Dispersant	Silica Fume	Weighting Agent	Slag	Resin
2.0	100	41	7	2.5	2	35	16	20	8
2.1	100	42	7	2.5	3	35	32	18	8
2.2	100	44	6	2.2	4	35	52	18	8
2.3	100	45	6	2	4.5	35	79	16	10

2.2.2. Construction Performance Tests

The performance testing of the cement slurry was carried out in accordance with the provisions of the Chinese standard GB/T 19139-2012 "Test method for oil well cement". The specific test steps were as follows:

(1) Density test

After the preparation of the cement slurry is completed, the cement slurry is poured into the sample cup of the densimeter, and pushing the cup cover downward into the cup mouth. After the excess cement slurry in the system has been cleaned, it is placed on a balanced rack to test the density of the cement slurry. The density of the cement slurry is tested with a densimeter (YM-3, Qingdao Haitongda Special Instrument Co., Ltd., Shandong, China) under normal temperature and pressure.

(2) Thickening time test

Before the experiment, the sample cup was installed and prepared. After completion of the preparation of the cement slurry sample, the cement slurry is poured into the sample cup, which is sealed. The test cup containing the sample is placed into the thickener. After the kettle is filled with hydrocarbon oil, a temperature sensor is inserted, and the kettle is sealed. The specified temperature and pressure are set on the instrument and the test is started. The thickening time is evaluated with a high-temperature and high-pressure thickener (TG-8040DA, Shenyang Taige Petroleum Instrument Equipment Co., Ltd., Shenyang, China). The temperature and pressure are set to 180 °C and 60 MPa, respectively. When the consistency of the cement slurry reaches 100 Bc, the time is recorded as the thickening time.

(3) Water loss test

The temperature of the high-temperature and high-pressure water loss meter is set to the specified temperature, the tested cement slurry is poured into a water loss test cup, the water loss of the cement slurry is tested under a pressure of 6.9 MPa, the filtration loss is recorded for 30 min, and the water loss is calculated. The water loss of the cement slurry is evaluated using a high-temperature and high-pressure water loss instrument (TG-71, Shenyang Taige Petroleum Instrument Equipment Co., Ltd., Shenyang, China).

(4) Rheological test

During the rheological properties test, the prepared cement slurry is placed in a constant-temperature mixer and cured for 20 min to simulate the flow of cement slurry in the well. Subsequently, the sample is poured into a test cup, and the rheological properties of the cement slurry are tested using a six-speed rheometer (ZNN-D6, Qingdao Chuangmeng Instrument Co., Ltd., Shandong, China). The rheology of the cement slurry is analyzed by evaluating the readings of a six-speed rotational viscometer at different rotational speeds (600 r/min, 300 r/min, 200 r/min, 100 r/min, 6 r/min, 3 r/min).

2.2.3. Compressive Strength Test

After the cement slurry has been prepared, the cement slurry is poured into a sample mold with a compressive strength standard, and the mold containing the cement slurry is cured in a high-temperature and high-pressure curing kettle (TG-7370D, Shenyang Taige Petroleum Instrument Equipment Co., Ltd., Shenyang, China) at 180 °C for a specified time. Subsequently, the cement stone mold is taken out and the mold is removed to obtain a 50.8 mm cube sample. When it was necessary to evaluate the mechanical properties of corroded cement stones, the compressive strength test was performed after the cement samples are corroded according to the corrosion testing process. The universal mechanical testing machine (HY-20080, Shanghai Hengyi Precision Instrument Co., Ltd., Shanghai, China) is used to evaluate the compressive strength of the cement sample. Compressive strength is the maximum stress of the cement slurry in the process of compression failure. The loading rate during compression is 1.2 kN/s. The sample used is a cube with a side length of 50.8 mm. Three cement stones are tested in each group, and the average value is calculated as the experimental result.

2.2.4. Corrosion Depth Test

The prepared cement slurries are poured into a cylindrical mold with a diameter of 25 mm and a height of 25 mm, and they are placed in a pressurized curing kettle at a temperature of 180 °C and a pressure of 21 MPa for 72 h and then demolded to form an uncorroded cement sample. The cement sample is placed in a high-temperature and high-pressure corrosion tester (TL-3, Jingzhou Tallin Machinery Co., Ltd., Jingzhou, China) made of Hastelloy alloy that is resistant to acid gas corrosion, and its appearance is shown in Figure 1. The total pressure of the corrosion test is 30 MPa, in which the partial pressures of H_2S, CO_2 and N_2 are 4 MPa, 2 MPa and 24 MPa, respectively. After reaching the specified corrosion time, the cement samples are taken out to obtain corroded cement samples.

Figure 1. Photo of high-temperature and high-pressure corrosion tester.

In order to test the corrosion degree of the cement sample, the corrosion depth is taken as the evaluation standard. Due to the alkalinity of the cement sample formed during the hydration of oil well cement, it will turn red or purplish red when encountering phenolphthalein. After the cement stone has been corroded, this alkalinity disappears, causing the corroded area to encounter phenolphthalein without discoloration. Therefore, based on this principle, it is possible to test the corrosion degree of cement stone. After the cement stone has been corroded, the cement sample is cut and the corrosion area is demarcated on the basis of the turning red characteristic of phenolphthalein meeting alkali. The four boundary thickness values of the sample that do not turn red are measured with a vernier caliper, and the average value is taken as the corrosion depth of the cement sample. A schematic diagram of the test is shown in Figure 2.

Figure 2. Process of corrosion depth test.

2.2.5. Micromorphology Test

The prepared cement sample was crushed, and the smooth part in the middle was taken as the sample. The hydration of the sample is terminated by soaking in alcohol for 24 h. Then, the samples are dried in an oven at 60 °C for 24 h, and the microscopic morphologies of the cement samples are observed using a scanning electron microscope (SU8010, Hitachi, Tokyo, Japan).

3. Design of High-Temperature-Resistant High-Density Anti-Corrosion Cement Slurry

3.1. Some Considerations for the Design of Cement Slurry

Since many studies have been performed on high-temperature retarders, fluid loss reducers and dispersants, this paper mainly focuses on the weighting agents, high-temperature stabilizers and preservatives with high content in cement slurry. The high-temperature retarder, high-temperature fluid loss reducer and high-temperature dispersant used were purchased from the market. The high-temperature retarder mainly consists of a multi-polymer, which can form a chelate with calcium ions on the surface of cement particles, preferentially adsorb onto the surface of hydration products, slow down their hydration speed, and show a high-temperature retarding effect. The high-temperature fluid loss reducer mainly consists of an AMPS terpolymer. It increases the viscosity of cement slurry and enhances its filtration resistance when dissolved in water. At the same time, it can be adsorbed on cement particles to form a cohesive state with viscoelasticity, block the pores of the filter cake, improve the compactness of the filter cake, and reduce the water loss of the cement slurry. The dispersant can improve the dispersion of cement particles in cement slurry and improve the rheology of the cement slurry. These additives are added according to the requirements of the experiment to ensure that the cement slurry has good performance.

In the CO_2/H_2S mixed gas environment, CO_2 dissolves in water to form an acid-soluble solution of H_2CO_3, and the cement slurry is corrodes under the action of carbonic acid to form $CaCO_3$ crystals [27–30]. At the same time, calcium-silicate-hydrate(CSH) also reacts with carbonic acid to form calcium carbonate. With the continuous action of CO_2 in the aqueous solution, calcium carbonate is transformed into water-soluble calcium bicarbonate, which continues to consume calcium hydroxide inside the cement sample, leading to the degradation of the compressive strength, the increase in the permeability, and an impact on the performance of the cement sample. The reaction process between carbon dioxide and hydration products in an aqueous solution is shown in Formulas (1)–(5) [31,32]. When H_2S is dissolved in water, it forms H_2SO_4, which reacts with calcium hydroxide and CSH in the hydration products to generate expansive $CaSO_4 \cdot 2H_2O$. The internal stress generated by the expansion causes the internal generation of the cement sample to be listed, further aggravating the corrosion. At the same time, tricalcium aluminate and tetra calcium ferroaluminate in the cement slurry also react with sulfate in the aqueous solution, producing expansive ettringite, destroying the internal structure of the cement sample. The reaction process between carbon dioxide and hydration products in an aqueous solution is shown in Formulas (6)–(12) [33,34].

$$CO_2 + H_2O \rightarrow H_2CO_3 \rightarrow H^+ + HCO_3^- \tag{1}$$

$$Ca(OH)_2 + H^+ + HCO_3^- \rightarrow CaCO_3 + 2H_2O \tag{2}$$

$$CSH + H^+ + HCO_3^- \rightarrow CaCO_3 + SiO_2 + H_2O \tag{3}$$

$$CO_2 + H_2O + CaCO_3 \rightarrow Ca(HCO_3)_2 \tag{4}$$

$$Ca(HCO_3)_2 + Ca(OH)_2 \rightarrow 2CaCO_3 + 2H_2O \tag{5}$$

$$H_2S \rightarrow 2H^+ + S^{2-} \tag{6}$$

$$H^+ + OH^- \rightarrow H_2O \tag{7}$$

$$Ca(OH)_2 + H_2S \rightarrow CaS + 2H_2O \tag{8}$$

$$CSH + H_2S + H_2O \rightarrow CaSO_4 \cdot 2H_2O + C_{(m)}S_{(n)}H_{(x)} \tag{9}$$

$$CSH + H_2S+ \rightarrow SiO_2 + CaS + H_2O \tag{10}$$

$$3CaO \cdot Al_2O_3 \cdot 6H_2O + 3Ca^{2+} + 3SO_4^{2-} \rightarrow 3CaO \cdot Al_2O_3 \cdot 3CaSO_4 + 6H_2O \tag{11}$$

$$4CaOAl_2O_3 \cdot Fe_2O_3 \cdot 3H_2O + 3Ca^{2+} + 3SO_4^{2-} + 29H_2O \rightarrow \\ 3CaO \cdot Al_2O_3 \cdot 3CaSO_4 \cdot 32H_2O + 2Fe^{3+} + 6OH^- \tag{12}$$

In an environment of high-temperature CO_2/H_2S mixed gas, the corrosion process of the cement sample is more complex, seriously affecting the compressive strength, permeability and other properties of the cement sample. The main reason for the corrosion of cement is the reaction of hydration products and acid fluids. In view of the corrosion damage, the corrosion resistance of cement can be improved by reducing the alkali substances and improving the compactness of cement slurry to prevent acidic liquid from flowing into the cement sample and forming an acid-resistant cement structure. We consider combining a variety of mineral admixtures with organic substances to add cement, giving full play to their respective advantages and the synergistic effect of their combination, so as to maximize the corrosion resistance of the cement slurry. At the same time, the preservative used needs to allows for high-temperature resistance.

3.2. High-Temperature-Resistant Manganese Ore Powder Weighting Agent

High-temperature oil and gas wells are usually characterized by high pressure, which requires high-density cement slurry in order to perform cementing operations. When the density of the cement slurry is low, oil, gas and water channeling, and even blowout accidents, can easily be caused. At the same time, higher-density cement slurry can also improve displacement efficiency. The commonly used method for increasing the density of cement slurry is to add weighting agent materials [35,36]. A manganese ore powder with high-temperature resistance and a regular shape is selected as the weighting material for the high-temperature anti-corrosion cement slurry. Its main component is Mn_3O_4, which is considered to be one of the weighting materials for oil and gas well-working fluid that offers the best performance at present. The microscopic morphology of manganese ore powder is shown in Figure 3. In order to evaluate the performance of manganese ore powder, different amounts of manganese ore powder were added to the cement slurry to study the impact of manganese ore powder addition in the performance of the cement slurry at 180 °C. The results are shown in Figure 4.

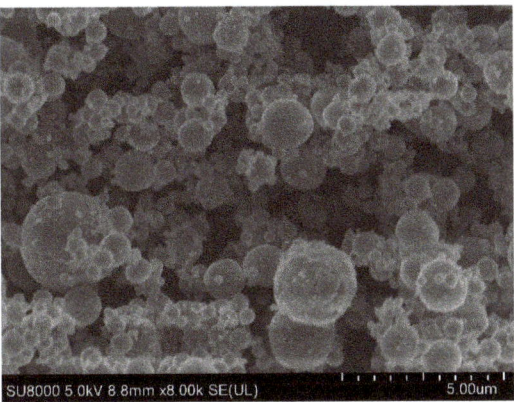

Figure 3. Particle size distribution of manganese ore powder.

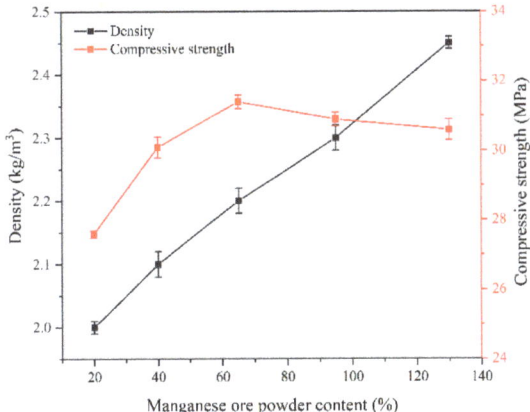

Figure 4. Effect of manganese ore powder in the performance of cement slurry.

Figure 3 shows the micromorphology of the manganese ore powder. The regular particle shape of manganese ore powder is spherical, which helps manganese ore powder to be embedded into the pores of the cement slurry. At the same time, the particle size of manganese ore powder is very small. Mineral materials with small particle size usually have a particular activity, and can improve the compactness and microstructure of cement sample, which is beneficial in the design of high-density anti-corrosion cement slurry. The performances of cement slurries formed using different contents of manganese ore powder are shown in Figure 4. Considering that the cement slurry requires better anti-corrosion performance under high-temperature and acid gas environments, manganese ore powder can effectively improve the density of cement slurry, and the prepared high-density cement slurry possesses high compressive strength. The main reason for this is that the manganese ore powder with small particle size forms a closely packed structure inside the cement sample, which improves the density of the cement slurry and prevents the acidic fluid from entering the cement sample to cause corrosion. This is the advantage of using manganese ore powder over other weighting agents.

3.3. Multi-Particle High-Temperature Stabilizer

The strength of oil well cement decreases after curing for a long time at temperatures higher than 110 °C. One study found that adding 35%~40% silica fume could effectively ameliorate the reduction in the high-temperature strength of the cement sample [37]. Currently, the high-temperature stabilizer used to resist the strength deterioration of cement sample is usually 100 mesh (150 μm) silica fume. Due to its large particle size, the stability of high-temperature cement slurry may worsen when designed. Smaller silica fume is more easily suspended in the cement slurry and filled into the cement sample, improving the compactness of the cement sample. However, smaller silica fume has a more significant impact on the rheology of the cement slurry. In order to study high-temperature stabilizers with good performance, 100 mesh and 300 mesh (48 μm) of composite silicon powder are considered. On the one hand, the pore structure of the cement sample is improved through the combination of different particle sizes. On the other hand, the influence of materials with small particle sizes on the rheology of cement slurry is decreased. By analyzing the influence of different silica fume additions on the performance of the cement slurry, the performance and effect of silica fume can be evaluated.

Table 3 shows the influence of the proportion of silica fume with different meshes on the rheology and stability of the cement slurry. Meanwhile, the influence of silica fume particle size on the permeability and the high-temperature stability of the composite silica fume are evaluated. With increasing content of 300 mesh silica fume, the rheological test reading of the cement slurry increases. When the proportion of 100 mesh silica fume is less than 30%, the rheology cannot be tested at 300 r/min, which indicates that the rheology is very poor. Moreover, when the content of 300 silica fume is greater than 50%, there is no free liquid in the cement slurry, which indicates an improvement in the stability of the cement slurry. Figure 5a shows the influence of the proportion of silica fume on permeability. The smaller the particle size of the silica fume, the lower the permeability of the cement slurry, improving the corrosion resistance of cement sample. The difference between the permeability of 75% 300 mesh silica fume and that of 100% 300 mesh silica fume is slight. Comprehensively considering the influence of composite silica fume on the performance of the cement slurry, a ratio of 100 mesh: 300 mesh silica fume equal to 25:75 is determined to be the best proportion. The effect of composite silicon powder on high-temperature strength stability is shown in Figure 5b. The compressive strength of the cement sample increases throughout curing for three days. When the curing time is greater than three days under high temperature, the strength of the cement slurry fluctuates, but the range of change is small. This shows that the composite silica fume is helpful for improving the high-temperature stability of cement sample.

3.4. High-Temperature-Resistant Preservative Mixed with Organic and Inorganic Materials

The corrosion degree of the cement samples is easily aggravated under high temperature, which makes designing anti-corrosion materials more difficult. In order to design high-temperature-resistant anti-corrosion cement slurry, we consider the addition of organic preservatives and inorganic preservatives to the cement slurry to improve the corrosion resistance in different ways. Inorganic preservatives can be used to reduce the corrosion degree of the cement samples by acid gas, mainly by reducing the content of Portland cement in unit volume and improving the compactness of the cement sample. Organic preservatives mainly form a polymer structure in the cement slurry, organizing acid gas to come into contact with the hydration products, thus reducing the permeability and corrosion degree of the cement slurry [38]. In this study, slag was selected as an inorganic preservative and resin as an organic preservative. Preservatives of different quality were added to the cement slurry to evaluate the degree of corrosion of the cement slurry containing the preservatives at 180 °C. The evaluation results of slag and resin are shown in Figures 6 and 7, respectively.

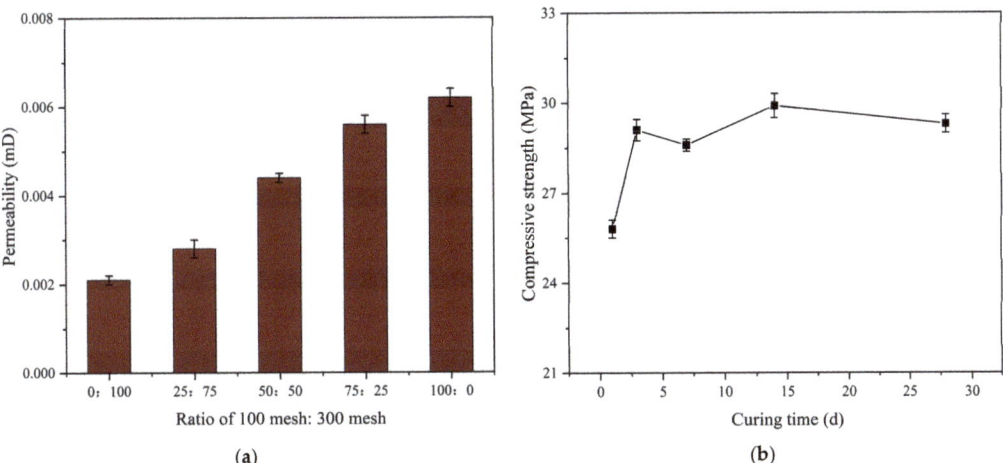

Figure 5. Performances of mixed silica fume cement slurry. (**a**) Permeability; (**b**) Compressive strength.

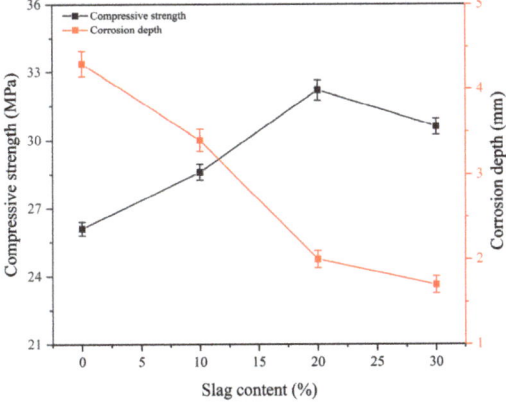

Figure 6. Effect of slag on cement slurry performance.

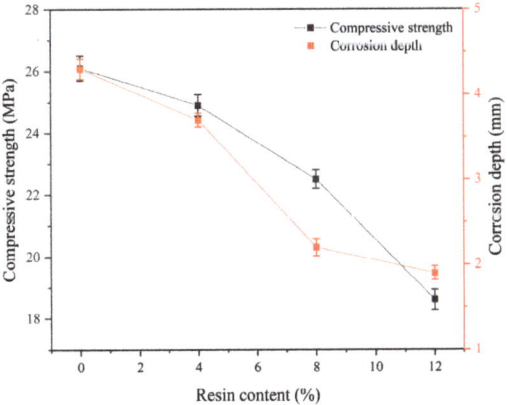

Figure 7. Effect of polymer resin on cement slurry performance.

Table 3. Effect of composite silica fume on the rheology and free liquid of cement slurry.

100 Mesh:300 Mesh	300 r/min	200 r/min	100 r/min	6 r/min	3 r/min	Free Liquid (%)
0:100	—	205	114	12	9	0
25:75	237	165	89	9	5	0
50:50	221	158	86	8	5	0
75:25	205	142	80	7	4	0.1
100:0	186	123	66	4	2	0.2

As a cementitious material to improve the performance of cement slurry, slag can be added in large amounts to the cement slurry, and is helpful in improving the performance of cement slurry at normal temperatures [39–41]. An inorganic preservative is considered in order to improve the high-temperature corrosion resistance of the cement slurry. The influence of the slag on the performance of the cement sample is shown in Figure 6. When the content is lower than 20%, the strength of the slag cement sample increases to a certain extent, thus improving the performance of the cement slurry. At the same time, the addition of slag effectively improves the corrosion resistance, and the corrosion depth of cement sample with 20% slag addition is 60.4% lower than that of pure cement slurry. The hydration products of slag contain almost no Ca (OH)$_2$. The main impact on the corrosion resistance of the cement sample is that slag can reduce the alkali content and improve the structure of the cement sample. The cement with the addition of slag has a strong anti-corrosion effect.

In addition to inorganic anti-corrosion materials, organic polymer anti-corrosion materials are also employed to form a synergistic effect, improving the corrosion resistance. The resin is demonstrated to improve the elasticity and toughness of cement slurry. It has the same effect as latex, while having less foaming than latex. Figure 7 shows the influence of the addition of resin on the mechanical properties and corrosion resistance of the cement sample. With the addition of resin, the compressive strength continues to degrade. The corrosion depth of the cement sample with 8% resin dosage is only 1.9 mm, which is obviously lower than that of the pure cement sample. The resin polymer particles form a film with high adhesion between the cement slurry and the aggregate. Some polymers with reactive groups can form an interpenetrating network structure and improve the adhesion between particles. A covering film is formed inside the cement sample, which prevents corrosive gases from invading the cement sample, protecting the structure from corrosion damage, and enhancing the corrosion resistance of the cement sample.

4. Evaluation of High-Temperature-Resistant High-Density Anti-Corrosion Cement Slurry

4.1. Construction Performances

Through the study of high-temperature resistant additives, the types and dosage of additives were determined, and a high-temperature-resistant high-density anti-corrosion cement slurry system was constructed. The composition of the cement slurry is shown in Table 2. The performances of the cement slurry are shown in Table 4 and Figure 8.

It can be seen from Table 4 and Figure 8 that the designed high-density anti-corrosion cement slurries have good rheology, thickening time is more than four hours, and water loss is less than 50 mL. All properties meet the requirements of cementing operations, ensuring the construction quality of the cement slurry.

Table 4. Rheology of cement slurries with different densities.

Density (kg/m^3)	300 r/min	200 r/min	100 r/min	6 r/min	3 r/min
2.0	223	160	87	9	5
2.1	243	178	93	6	4
2.2	277	197	105	7	5
2.3	253	189	110	9	6

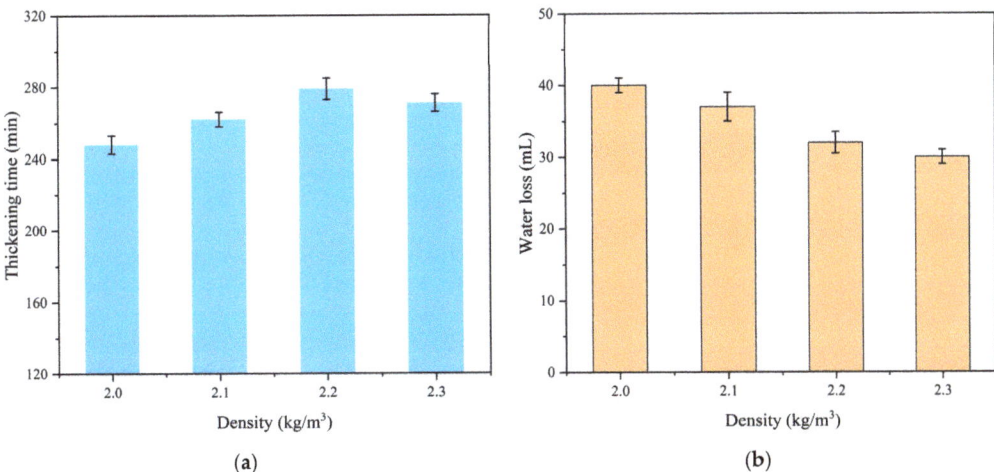

Figure 8. Thickening time and water loss of cement slurries with different densities.(**a**) Thickening time; (**b**) Water loss.

4.2. Compressive Strength at High Temperature

For cement slurry used in high-temperature environments, construction performance is closely related to construction quality, and the stability of the mechanical properties of the cement sample under high temperature is closely related to the long-term cementing quality. If the high-temperature mechanical properties are poor, the cement slurry will not be able to effectively seal the oil, gas and water, which may result in the abandonment of oil and gas wells. For oil and gas wells containing carbon dioxide and hydrogen sulfide, in particular, sealing failure can further lead to acid gas channeling, threatening the safety of oil and gas wells. By evaluating the compressive strength of cement paste cured at high temperatures for different times and analyzing the changes in compressive strength, it is possible to study the stability of the mechanical properties of cement paste at high temperatures. The stability of cement slurry at high temperature is shown in Figure 9.

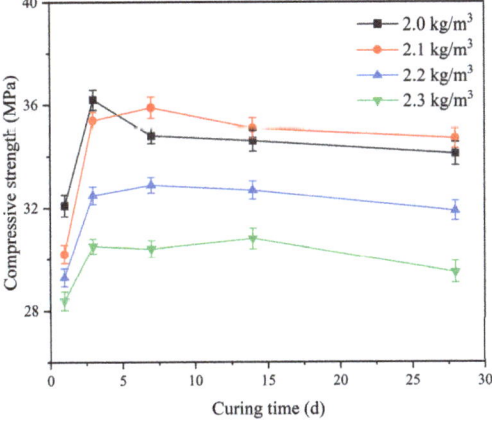

Figure 9. Compressive strength of high-temperature-resistant and high-density anti-corrosion cement slurry.

It can be seen from Figure 9 that the compressive strength of the cement sample increases for curing times of up to three days and declines after three days of curing, but this decline is slight. This is mainly due to the influence of high temperature on the hydration products of oil well cement, and the high-temperature stabilizer is able to ameliorate the decrease in the strength of the cement sample. When the density of the cement slurry is higher, the compressive strength of the cement sample decreases for curing times of less than three days, but the compressive strength is less affected by temperature during long-term curing. After 28 days of curing, the compressive strength of cement sample decreases by less than 5.8% compared with the maximum strength of the cement sample. It can be considered that, through the design of the cement slurry, the high-density cement slurry has stable mechanical properties at high temperature. When the cement paste does not contain a high-temperature stabilizer of silica fume, and when it is subjected to temperatures higher than 110 °C, the painting product $C_3S_2H_3$ of the cement paste no longer remains stable, and will be transformed into a hydration product of lower-strength hydrated dicalcium silicate. At higher temperature, the crystallization degree of the hydration products of the cement slurry are limited, and the dehydration of the crystals further weakens the compressive strength of the cement paste. When adding a silicon powder stabilizer to the cement slurry, the silicon powder can react with the hydration products formed by the cement to obtain tobermorite and xonotlite with a C/S ratio close to 1, which changes the chemical composition of the cement paste and improves its strength, thereby preventing the decline in the strength of the cement paste.

4.3. Corrosion Performance

To evaluate the resistance of different high-density cement slurries to carbon dioxide and hydrogen sulfide, the corrosion performance of cement samples subjected to different lengths of time at high temperature was evaluated. Figure 10 shows the appearance of the corrosion depth of the cement slurry after being corroded for 30 days. As more manganese ore powder is added to the cement slurry, the cement slurry becomes reddish brown. When the cement slurry encounters phenolphthalein after being corroded, the color of the cement slurry becomes deeper. The corrosion depth of the cement sample can be measured from the corrosion morphology. Figure 11 shows the corrosion degree of the cement slurries under high temperature. With the increase in density, due to the increase in spherical manganese ore powder weighting agent and the decrease in alkaline Portland cement slurry, the corrosion depth of the cement slurry decreases, which is related to the density of the cement slurry. With increasing corrosion time, the corrosion depth increases. However, the growth range of the corrosion depth after 7 days is far greater than that after 30 days. This is mainly because the early stage of the corrosion mainly occurs on the surface of the cement slurry, and the acid gas can penetrate easily. With the extension of the corrosion time, the penetration depth of the acid gas increases, reducing the corrosion rate. The corrosion depth of cement slurries with different densities after being corroded for 30 days is less than 2 mm. Figure 12 shows the change in compressive strength of high-density cement slurry following different corrosion times. As the cement slurry is corroded under high temperature, the compressive strength of the cement slurry degrades. During the early stage of corrosion, the compressive strength of cement slurry decreases rapidly, which is related to the degree of corrosion. The compressive strength of the cement slurry with a density of 2.0 kg/m^3 decreases the most after being corroded. When the cement slurry is corroded for 30 days, the compressive strength of 2.0 kg/m^3 cement slurry decreases by 16.6% compared with that of non-corroded cement sample. This is the result of combining the decrease in strength of the cement sample under high temperature with the influence of corrosion on compressive strength. The designed high-density cement slurry has excellent corrosion resistance. The cement stone was corroded to a low degree, helping it to maintain good sealing quality in the well.

Figure 10. Corroded appearance of cement sample.

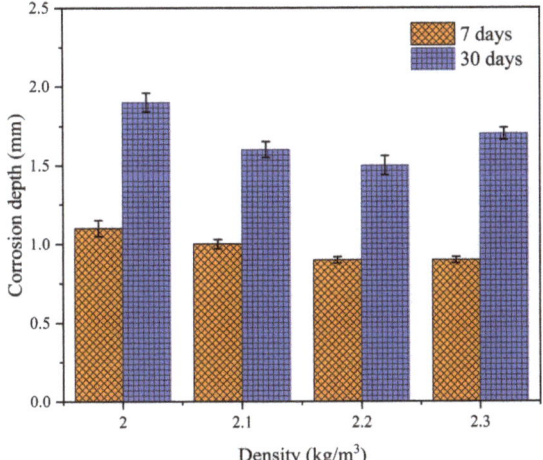

Figure 11. Corrosion depth of high-density anti-corrosion cement slurry.

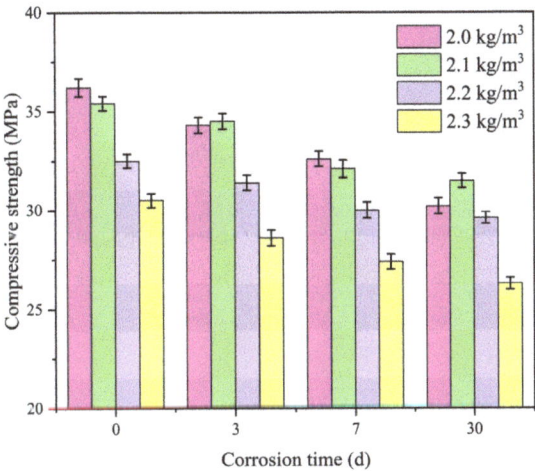

Figure 12. Compressive strength of cement samples with different corrosion times.

4.4. Micromorphology

In order to better analyze the anti-corrosion performance of the high-temperature and high-density cement slurry system, SEM was used to observe the microscopic morphology of the highest- and lowest-density cement slurry with and without a corrosion layer. As can be observed in Figures 13 and 14, there is no obvious crack defect in the cement slurry before corrosion, and the filling of ultra-fine weighted granular materials can be observed in the high-density cement slurry. The manganese ore powder weighting agent inside the cement slurry can still be observed after corrosion, but the structure of the cement slurry becomes obviously less dense than before corrosion. The internal pores of the cement sample increase, and the complete bonding structure of the cement stone is damaged. This is mainly due to the influence of corrosion on the hydration products of cement slurry, resulting in changes in the structure of the cement slurry and defects in the microstructure of cement slurry, which is not conducive to maintaining its integrity. The internal structure of the cement slurry with less corrosion is dense, and the performance of the cement sample is better.

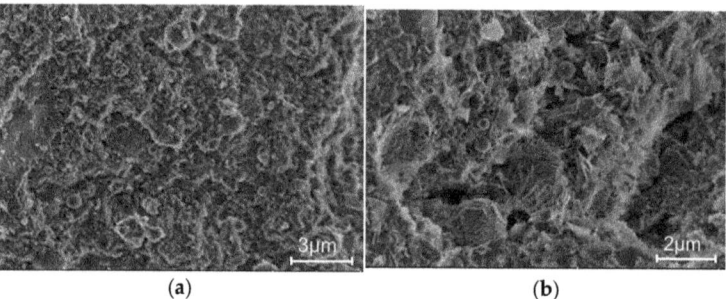

Figure 13. Morphology of 2.0 kg/m^3 cement sample: (**a**) non-corroded; (**b**) corroded.

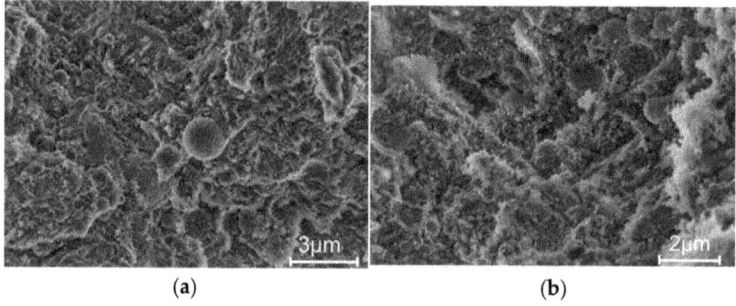

Figure 14. Morphology of 2.3 kg/m^3 cement sample: (**a**) non-corroded; (**b**) corroded.

5. Conclusions

(1) The manganese ore powder weighting agent is spherical in shape, which can not only effectively improve the density of cement slurry, but also has little effect on the compressive strength. The high-temperature stabilizer is determined to be a mixture of silica fume with different particle sizes, and the performance of the high-temperature stabilizer is better when the proportion of 100 mesh: 300 mesh is equal to 25:75.

(2) The designed slag and resin can effectively reduce the corrosion depth of cement slurry. Mixing inorganic material slag and organic polymer resin as the preservative of cement slurry can reduce the corrosion degree of cement slurry.

(3) The rheology, water loss and thickening time of the designed high-temperature-resistant high-density anti-corrosion cement slurry are excellent, the water loss is

(4) After being corroded by hydrogen sulfide and carbon dioxide at high temperature, the designed oil well cement-based composite has low corrosion depth and strong corrosion resistance. The corrosion resistance of the cement slurry is effectively guaranteed by using manganese ore powder, multi-particle silicon powder, slag, and resin preservatives.

less than 50mL, and the thickening time is greater than four hours. The compressive strength of different high-density cement slurries is stable under high temperatures.

Author Contributions: Conceptualization, R.C. and F.Y.; Data curation, C.T., M.W. and F.Y.; Formal analysis, M.W. and R.C.; Investigation, C.T., M.W., R.C. and F.Y.; Methodology, C.T. and M.W.; Project administration, F.Y.; Supervision, R.C.; Validation, C.T.; Writing—original draft, C.T.; Writing—review and editing, M.W., R.C. and F.Y. All authors have read and agreed to the published version of the manuscript.

Funding: This research was funded by Sinopec Key Laboratory of Cementing and Completion (grant number 35800000-22-ZC0607-0004).

Institutional Review Board Statement: Not applicable.

Informed Consent Statement: Not applicable.

Data Availability Statement: The data is contained within the article.

Conflicts of Interest: The authors declare no conflict of interest.

References

1. Zhang, X.G.; Zheng, Y.Z.; Guo, Z.M.; Ma, Y.; Wang, Y.; Gu, T.; Jiao, L.B.; Liu, K.Q.; Hu, Z.Z. Effect of CO_2 solution on Portland cement paste under flowing, migration, and static conditions. *J. Nat. Gas Sci. Eng.* **2021**, *95*, 104179. [CrossRef]
2. Hoa, L.; Bassler, R.; Bettge, D.; Buggisch, E.; Schiller, B.N.; Beck, M. Corrosion Study on Wellbore Materials for the CO_2 Injection Process. *Process.* **2021**, *9*, 115. [CrossRef]
3. Zhou, C.Y.; Zeng, L.H.; Sun, Y.; Zhou, M.; Lei, M.Y.; Wan, W.; Feng, Y. Corrosion behavior and mechanism analysis of oil well cement under CO_2 and H_2S conditions. *Petrophysics* **2022**, *63*, 642–651. [CrossRef]
4. Omosebi, O.; Maheshwari, H.; Ahmed, R.; Shah, S.; Osisanya, S.; Santra, A.; Saasen, A. Investigating temperature effect on degradation of well cement in HPHT carbonic acid environment. *J. Nat. Gas Sci. Eng.* **2016**, *26*, 1344–1362. [CrossRef]
5. Liu, H.T.; Qin, J.N.; Zhou, B.; Liu, Z.F.; Yuan, Z.T.; Zhang, Z.; Ai, Z.Q.; Pang, X.Y.; Liu, X.L. Effects of Curing Pressure on the Long-Term Strength Retrogression of Oil Well Cement Cured under 200 degrees C. *Energies* **2022**, *15*, 6071. [CrossRef]
6. Lian, J.; Yue, J.; Xing, X.S.; Wu, Z.Q. Design and Evaluation of the Elastic and Anti-Corrosion Cement Slurry for Carbon Dioxide Storage. *Energies* **2023**, *16*, 435. [CrossRef]
7. Yin, S.H.; Yang, Y.F.; Zhang, T.S.; Guo, G.F.; Yu, F. Effect of carbonic acid water on the degradation of Portland cement paste: Corrosion process and kinetics. *Constr. Build. Mater.* **2015**, *91*, 39–46. [CrossRef]
8. Lin, Y.H.; Zhu, D.J.; Zeng, D.Z.; Yang, Y.G.; Shi, T.H.; Deng, K.H.; Ren, C.Q.; Zhang, D.P.; Wang, F. Experimental studies on corrosion of cement in CO_2 injection wells under supercritical conditions. *Corros. Sci.* **2013**, *74*, 13–21. [CrossRef]
9. Silva, H.G.C.; Terradillos, P.G.; Zornoza, E.; Mendoza-Rangel, J.M.; Castro-Borges, P.; Alvarado, C.A.J. Improving Sustainability through Corrosion Resistance of Reinforced Concrete by Using a Manufactured Blended Cement and Fly Ash. *Sustainability* **2018**, *10*, 2004. [CrossRef]
10. Tittarelli, F.; Carsana, M.; Bellezze, T. Corrosion behavior of reinforced no-fines concrete. *Corros. Sci.* **2013**, *70*, 119–126. [CrossRef]
11. Wang, D.C.; Noguchi, T.; Nozaki, T. Increasing efficiency of carbon dioxide sequestration through high temperature carbonation of cement-based materials. *J. Clean. Prod.* **2019**, *238*, 117980. [CrossRef]
12. Mei, K.Y.; Cheng, X.W.; Pu, Y.; Ma, Y.; Gao, X.S.; Yu, Y.J.; Zhang, C.M.; Zhuang, J.; Guo, X.Y. Evolution of silicate structure during corrosion of tricalcium silicate (C3S) and dicalcium silicate (C2S) with hydrogen sulphide (H_2S). *Corros. Sci.* **2020**, *163*, 108301. [CrossRef]
13. Mei, K.Y.; Cheng, X.W.; Gu, T.; Zheng, Y.Z.; Gong, P.; Li, B.; Zhang, C.M.; Zhang, L.W.; Dai, B.B. Effects of Fe and Al ions during hydrogen sulphide (H_2S)-induced corrosion of tetracalcium aluminoferrite (C(4)AF) and tricalcium aluminate (C(3)A). *J. Hazard. Mater.* **2021**, *403*, 123928. [CrossRef] [PubMed]
14. Omosebi, O.; Maheshwari, H.; Ahmed, R.; Shah, S.; Osisanya, S. Experimental study study of the effects of CO_2 concentration and pressure at elevated temperature on the mechanical integrity of oil and gas well cement. *J. Nat. Gas Sci. Eng.* **2017**, *44*, 299–313. [CrossRef]
15. Xu, Y.; Hu, M.M.; Chen, D.; Liu, Z.X.; Yu, Y.J.; Zhang, H.; Guo, J.T. Performance and working mechanism of amphoteric polycarboxylate-based dispersant and sulfonated acetone formaldehyde polycondensate-based dispersant in oil well cement. *Constr. Build. Mater.* **2020**, *233*, 117147. [CrossRef]

16. Yan, S.M.; Wang, Y.J.; Wang, F.H.; Yang, S.; Wu, Y.N.; Yan, S.D. Synthesis and mechanism study of temperature-resistant fluid loss reducer for oil well cement. *Adv. Cem. Res.* **2017**, *29*, 183–193. [CrossRef]
17. Tiemeyer, C.; Plank, J. Working mechanism of a high temperature (200 degrees C) synthetic cement retarder and its interaction with an AMPS (R)-based fluid loss polymer in oil well cement. *J. Appl. Polym. Sci.* **2012**, *124*, 4772–4781. [CrossRef]
18. Tian, Y.; Yu, R.G.; Zhang, N.; Yang, Y.T. Preparation of high-temperature-resistant and high-strength low-density oil well cement. *Emerg. Mater. Res.* **2020**, *9*, 163–167. [CrossRef]
19. Abdulmalek, A.; Salaheldin, E.; Adjei, A.S. Influence of weighting materials on the properties of oil-well cement. *ACS Omega* **2020**, *5*, 27618–27625. [CrossRef]
20. Ahmed, A.; Mahmoud, A.A.; Elkatatny, S.; Chen, W.Q. The effect of weighting materials on oil-well cement properties while drilling deep wells. *Sustainability* **2019**, *11*, 6776. [CrossRef]
21. Qin, J.K.; Pang, X.Y.; Cheng, G.D.; Bu, Y.H.; Liu, H.J. Influences of different admixtures on the properties of oil well cement systems at HPHT conditions. *Cem. Concr. Compos.* **2021**, *123*, 104202. [CrossRef]
22. Costa, B.L.D.; de Souza, G.G.; Freitas, J.C.D.; Araujo, R.G.D.; Santos, P.H.S. Silica content influence on cement compressive strength in wells subjected to steam injection. *J. Pet. Sci. Eng.* **2017**, *158*, 626–633. [CrossRef]
23. Liu, H.J.; Bu, Y.H.; Zhou, A.N.; Du, J.P.; Zhou, L.W.; Pang, X.Y. Silica sand enhanced cement mortar for cementing steam injection well up to 380 degrees C. *Constr. Build. Mater.* **2021**, *308*, 125142. [CrossRef]
24. Yuan, B.; Wang, Y.Q.; Yang, Y.G.; Xie, Y.Q.; Li, Y. Wellbore sealing integrity of nanosilica-latex modified cement in natural gas reservoirs with high H_2S contents. *Constr. Build. Mater.* **2019**, *192*, 621–632. [CrossRef]
25. Peng, Z.G.; Lv, F.L.; Feng, Q.; Zheng, Y. Enhancing the CO2-H2S corrosion resistance of oil well cement with a modified epoxy resin. *Constr. Build. Mater.* **2022**, *326*, 126854. [CrossRef]
26. Bai, M.X.; Zhang, Z.C.; Fu, X.F. A review on well integrity issues for CO_2 geological storage and enhanced gas recovery. *Renew. Sustain. Energy Rev.* **2016**, *59*, 920–926. [CrossRef]
27. Grandclerc, A.; Gueguen-Minerbe, M.; Nour, I.; Dangla, P.; Chaussadent, T. Impact of cement composition on the adsorption of hydrogen sulphide and its subsequent oxidation onto cementitious material surfaces. *Constr. Build. Mater.* **2017**, *152*, 576–586. [CrossRef]
28. Liu, J.L.; Jia, Y.M. Experimental investigation on durability of cement-steel pipe for wellbores under CO_2 geological storage environment. *Constr. Build. Mater.* **2020**, *236*, 117589. [CrossRef]
29. Carey, J.W. Geochemistry of Wellbore Integrity in CO_2 Sequestration: Portland Cement-Steel-Brine-CO_2 Interactions. *Rev. Mineral. Geochem.* **2013**, *77*, 505–539. [CrossRef]
30. Li, X.; O'Moore, L.; Song, Y.R.; Bond, P.L.; Yuan, Z.G.; Wilkie, S.; Hanzic, L.; Jiang, G. The rapid chemically induced corrosion of concrete sewers at high H_2S concentration. *Water Res.* **2019**, *162*, 95–104. [CrossRef]
31. Bao, H.; Xu, G.; Wang, Q.; Peng, Y.Z.; Liu, J.Y. Study on the deterioration mechanism of cement-based materials in acid water containing aggressive carbon dioxide. *Constr. Build. Mater.* **2020**, *243*, 118233. [CrossRef]
32. Savija, B.; Lukovic, M. Carbonation of slurrys: Understanding, challenges, and opportunities. *Constr. Build. Mater.* **2016**, *117*, 285–301. [CrossRef]
33. Xu, B.H.; Yuan, B.; Wang, Y.Q. Anti-corrosion cement for sour gas (H_2S-CO_2) storage and production of HTHP deep wells. *Appl. Geochem.* **2018**, *96*, 155–163. [CrossRef]
34. Zhang, Y.; Xu, M.; Song, J.; Wang, C.; Wang, X.; Hamad, B.A. Study on the corrosion change law and prediction model of cement stone in oil wells with CO_2 corrosion in ultra-high-temperature acid gas wells. *Constr. Build. Mater.* **2022**, *323*, 125879. [CrossRef]
35. Wang, J.P.; Xiong, Y.M. Research and Application of High-Density Cementing Slurry Technology under the Condition of Oil-based Drilling Fluid in Salt Formation. *Arab. J. Sci. Eng.* **2021**, *47*, 7069–7079. [CrossRef]
36. Zhou, S.M.; Li, G.S.; Wang, Q.C. Research and preparation of ultra-heavy slurry. *Pet. Explor. Dev.* **2013**, *40*, 115–118. [CrossRef]
37. Abdelmelek, N.; Lubloy, E. Effects of Elevated Temperatures on the Properties of High Strength Cement Paste Containing Silica Fume. *Period. Polytech.-Civ. Eng.* **2022**, *66*, 127–137. [CrossRef]
38. Xu, B.H.; Yuan, B.; Wang, Y.Q.; Zeng, S.P.; Yang, Y.H. Nanosilica-latex reduction carbonation-induced degradation in cement of CO_2 geological storage wells. *J. Nat. Gas Sci. Eng.* **2019**, *65*, 237–247. [CrossRef]
39. Wu, Z.Q.; Liu, H.J.; Qu, X.; Wu, G.A.; Xing, X.S.; Cheng, X.W.; Ni, X.C. Improvement of Calcium Aluminate Cement Containing Blast Furnace Slag at 50 degrees C and 315 degrees C. *Front. Mater.* **2022**, *8*, 807596. [CrossRef]
40. Gao, X.; Yao, X.; Wang, C.Y.; Geng, C.Z.; Yang, T. Properties and microstructure of eco-friendly alkali-activated slag cements under hydrothermal conditions relevant to well cementing applications. *Constr. Build. Mater.* **2022**, *318*, 125973. [CrossRef]
41. Martin-Del-Rio, J.J.; Marquez, G.; Flores-Ales, V.; Romero, E.; Rey, O. Comparison of polymer-based slag-mud slurries used for drilling jobs of steam stimulated wells in the Lagunillas oilfield. *Int. J. Oil Gas Coal Technol.* **2021**, *26*, 1–11. [CrossRef]

Disclaimer/Publisher's Note: The statements, opinions and data contained in all publications are solely those of the individual author(s) and contributor(s) and not of MDPI and/or the editor(s). MDPI and/or the editor(s) disclaim responsibility for any injury to people or property resulting from any ideas, methods, instructions or products referred to in the content.

Article

Effect of a New Multi-Walled CNT (MWCNT) Type on the Strength and Elastic Properties of Cement-Based Mortar

Sergiu-Mihai Alexa-Stratulat [1], George Stoian [2], Iulian-Adrian Ghemeș [2], Ana-Maria Toma [1], Daniel Covatariu [1] and Ionut-Ovidiu Toma [1,*]

[1] Faculty of Civil Engineering and Building Services, The "Gheorghe Asachi" Technical University of Iasi, No. 1, Prof. dr.doc. D. Mangeron Street, 700050 Iasi, Romania

[2] Magnetic Materials and Devices Department, National Institute of Research and Development for Technical Physics, No. 47, Prof. dr.doc. D. Mangeron Street, 700050 Iasi, Romania

* Correspondence: ionut.ovidiu.toma@tuiasi.ro; Tel.: +40-232-701455

Abstract: Creating new construction materials with improved strength, elasticity, and durability properties represent the focus of many research works. Significant research effort has been invested in investigating the use of carbon nanotubes (CNTs) in cementitious materials, especially multi-walled carbon nanotubes (MWCNTs) which consist of a series of concentric graphite tubes. The use of MWCNTs is closely related to the use of surfactants and ultra-sonication procedures which may alter their properties and the properties of cement-based materials. The paper presents the preliminary results of an experimental investigation on the suitability of using a new, modified, MWCNT type aimed at eliminating the need of using surfactants and ultrasonication. The modified MWCNTs have a much lower surface energy compared to "classical" ones which would result in a decreased tendency of self-aggregation. A comparison was carried out from the point of view of density, flexural and compressive strength as well as dynamic modulus of elasticity of the obtained mortars. The mortar mix incorporating the modified MWCNTs showed improved mechanical properties even for a low percentage of CNT addition (0.025% by mass of cement). The results are discussed based on the material structure determined from a series of scanning electron microscopy (SEM) and X-ray diffraction (XRD) analyses.

Keywords: modified multi-walled carbon nanotubes; mechanical properties; uniform dispersion; dynamic modulus of elasticity

Citation: Alexa-Stratulat, S.-M.; Stoian, G.; Ghemeș, I.-A.; Toma, A.-M.; Covatariu, D.; Toma, I.-O. Effect of a New Multi-Walled CNT (MWCNT) Type on the Strength and Elastic Properties of Cement-Based Mortar. *Coatings* **2023**, *13*, 492. https://doi.org/10.3390/coatings13030492

Academic Editor: Michał Kulka

Received: 28 January 2023
Revised: 19 February 2023
Accepted: 21 February 2023
Published: 23 February 2023

Copyright: © 2023 by the authors. Licensee MDPI, Basel, Switzerland. This article is an open access article distributed under the terms and conditions of the Creative Commons Attribution (CC BY) license (https://creativecommons.org/licenses/by/4.0/).

1. Introduction

The need to build taller and more durable structures resulted in a growing demand for construction materials with improved elastic, mechanical, and durability properties. A lot of effort has been invested into creating new materials with improved elastic and mechanical characteristics. At the same time, the developed materials would also have to be competitive from the point of view of manufacturing cost vs. improved performance. Another goal would be the applicability of the already existing design guidelines, without the need for major revisions.

The use of nanomaterials in the construction industry was quickly embraced by researchers and practitioners alike due to the added value in terms of improved material properties and behavior [1–5]. Significant research effort has been invested in investigating the use of carbon nanotubes (CNTs) in cementitious materials [3,6–8]. It is generally agreed upon that their use leads to improvements in the physical and mechanical properties. Based on conducted research works, 0.075% was proposed as being optimal for obtaining the lowest sorptivity, the lowest porosity, and a significant improvement in the values of the mechanical strengths [9] of cementitious matrices. Percentages between 0.01%–0.07% of CNTs used together with nano-silica were reported to improve the microstructure of cement mortar in a synergistic effect which resulted in better corrosion resistance and improved durability [10]. Nano-silica contributed towards enhancing the pozzolanic activity whereas the CNT acted as fiber reinforcement between the aggregates and the paste, at the level of the interfacial transition zone (ITZ), bridging the micro-cracks and arresting their development.

There are two distinct categories of CNTs that are currently investigated in terms of their beneficial effect on cementitious materials. Single-walled carbon nanotubes (SWCNTs), are tubes of graphite, sometimes capped at their ends, having the wall made of a single layer of molecules. On the other hand, there are multi-walled carbon nanotubes (MWCNTs), which consist of two or more open concentric graphite tubes [11]. The main advantage of SWCNTs resides in their greater flexibility in terms of being twisted, flattened, or bent around sharp edges (e.g., of aggregates) compared to MWCNTs. On the other hand, taking into account the large amounts of CNTs that would be necessary for large-scale concrete elements, the unit price of SWCNTs vs. MWCNTs and the corresponding production capacity, the use of MWCNTs seems the better choice, even though their structure is not as well understood and are prone to defects compared to their SWCNT counterparts [11,12].

Although MWCNTs show very high values for moduli of elasticity and high aspect ratios coupled with excellent thermal and electrical conductivity, they tend to bundle and coalesce [13]. There are currently two widely used approaches to overcome this issue: the mechanical and the chemical approach. The former involves sonication and high shear mixing which may break the nanotubes, thus reducing their aspect ratio. It is also a time-consuming procedure and not always very efficient. The latter approach implies the use of a dispersive agent, in the form of a surfactant or plasticizer, in order to overcome the strong bonding forces on the surface of CNTs [13]. There have been many attempts to obtain the ideal combination of surfactant type and sonication duration and input energy, most often ending with reporting conflicting results [14]. While some researchers noticed a beneficial contribution of using surfactants together with sonication [15], other studies reported a lack of improvement [16]. Moreover, there is a high probability that, due to their high surface energy, MWCNTs will still bundle together after being mixed with cement and aggregates.

The downside of using surfactants resides in the fact that high concentrations of surfactants are needed to obtain stable and uniform CNT dispersions. Moreover, the use of dispersants/surfactants was previously reported to negatively impact the properties of cementitious matrices due to inhibition of cement hydration and air entrapment [17,18] resulting in lower values for the mechanical properties of cement-based materials [19]. It has been suggested that the use of antifoaming agents could be applied as a countermeasure to the undesired effect of surfactants [20,21].

The paper presents the preliminary results of an experimental investigation on the suitability of using a new, modified, MWCNT type aimed at solving the above-mentioned shortcomings of "traditional" MWCNTs when used in cement-based materials, namely the use of surfactants together with sonication procedures to ensure their uniform dispersion. The modified MWCNTs were obtained using the same procedure as "traditional" MWCNTs with the key difference consisting in the fact that they are capped at their ends. In this aspect, the new MWCNTs are very similar to SWCNTs. A comparison was carried out from the point of view of density, flexural and compressive strength as well as dynamic modulus of elasticity of the obtained mortars. The results are discussed based on the material structure determined from a series of SEM and XRD analyses. The mortar mix incorporating the new MWCNTs exhibited improved elastic and mechanical properties compared to the reference mix, even though no surfactant was used, and no sonication procedure was applied. The statistical analysis of the results leads to small values for the coefficient of variation for each investigated material property. This supports the conclusion that the new, modified, MWCNTs were uniformly distributed within the mortar mix.

2. Materials and Methods

2.1. Materials

The chosen mortar mix, by volume, was 1:3:0.6 for cement, sand, and water, respectively. A rapid hardening CEM II B-M (S-LL) 42.5R cement complying with standard specifications [22] was used. It is classified as a composite cement consisting of 65% ÷ 79% cement clinker and 21% ÷ 35% a mixture between ground-granulated blast furnace slag (GGBS) and limestone.

The natural sand, readily available on the market, had a particle diameter range from 0 to 4 mm. Tap water was used in the mortar mix.

The new, modified, MWCNTs were produced in the laboratory of the National Institute of Research and Development for Technical Physics based on the chloride-mediated chemical vapor deposition (CVD) method [23]. The structure of the obtained MWCNTs, after 20 min of growth, is shown in Figure 1. The length of the MWCNTs was about 300 μm, as seen in Figure 1a, whereas the outer diameter was around 30 nm, according to the HR-TEM image presented in Figure 1b.

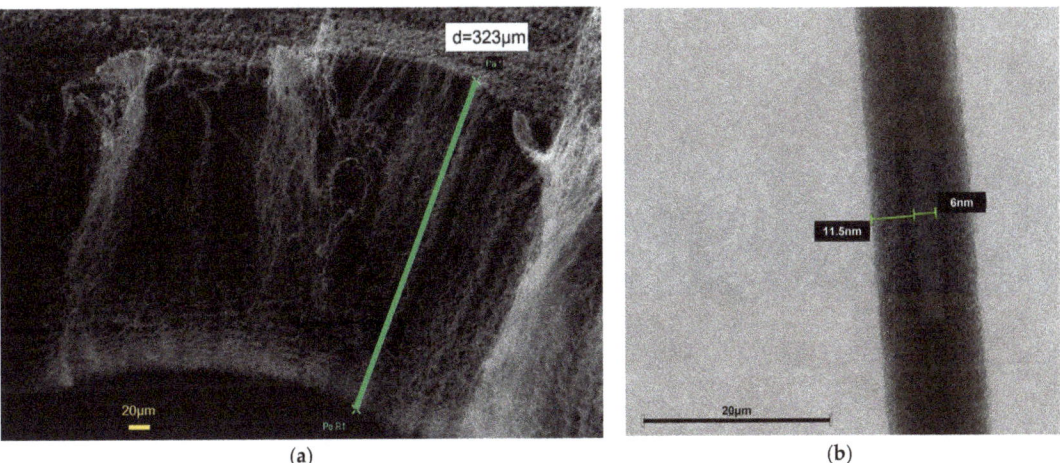

Figure 1. Structure of MWCNTs: (**a**) SEM image of vertically aligned MWCNT arrays; (**b**) HR-TEM image of a single nanotube.

The new MWCNT used in this research as a possible solution to mitigate the issues related to using surfactants and sonication procedures was based on the same CVD method. These new carbon-based nanotubes (CBN) had a diameter of 30–40 nm and a length in the range of 200–400 nm, as measured from the SEM image shown in Figure 2. Anhydrous iron chloride ($FeCl_2$) was introduced in the center of a quartz tube electric furnace which was then evacuated to a pressure of about 1 mTorr using a rotary pump. The furnace was then heated up to a temperature of 900 °C with ramping rate of 40 °C/min. When the synthesis temperature was reached, acetylene (C_2H_2) was introduced into the reaction chamber at a flow rate of 0.5 L/min. The growth pressure was kept constant around the value of 10 Torr for a growth time of 60 min. Finally, the chamber was naturally cooled down to room temperature under vacuum and particles were collected using a permanent magnet.

These new, modified, MWCNTs (CBN) can be considered as precursors of "traditional" MWCNTs. They were obtained using the same procedure with the main difference consisting in the fact that they are capped at both of their ends. A Fe particle (white end in Figure 2) closes one end of the new MWCNT whereas the other end is capped by a C particle [24]. This results in a lower surface energy compared to regular MWCNTs and therefore they would be much easier to disperse and avoid self-aggregation, even in the absence of surfactants and/or sonication procedure.

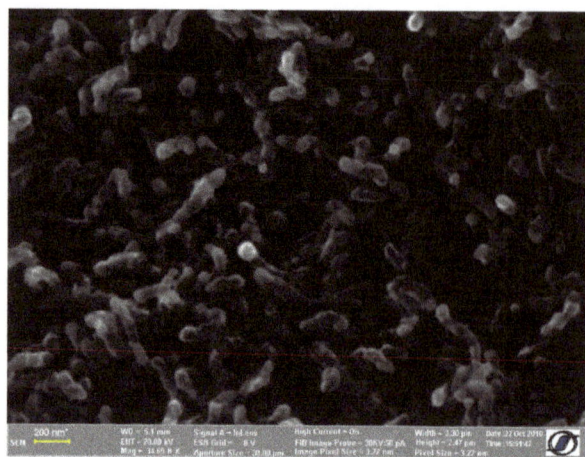

Figure 2. SEM image of new CBN.

To confirm the structure of these new CBN, a SEM-EDS mapping analysis was conducted. As it can be seen from Figure 3, a higher concentration of carbon leads to a more compact material structure. The results on SEM-EDS analysis are presented in Table 1 in terms of chemical constituents for the four investigated spectra.

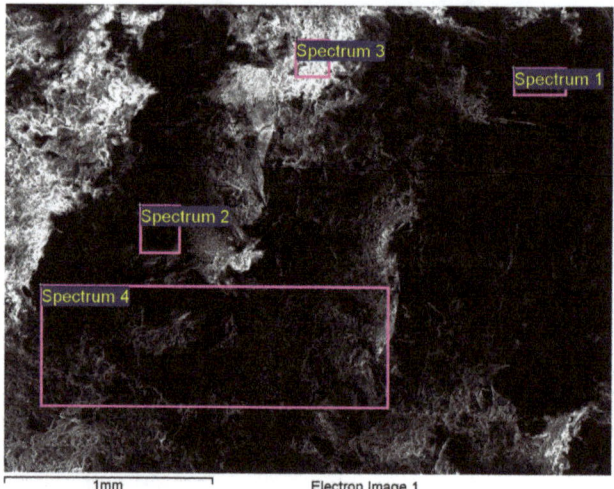

Figure 3. SEM-EDS analysis of new CBN.

Table 1. Composition of the four investigated spectra.

Spectrum	C	O	Na	Mg	Al	Si	S	K	Ca	Ti	Fe
Spectrum 1	7.98	59.33	-	-	21.88	10.57	-	-	0.24	-	-
Spectrum 2	18.13	66.85	-	0.31	0.98	3.94	0.35	0.07	9.20	-	0.16
Spectrum 3	15.37	66.95	-	0.67	1.59	4.97	0.28	0.08	9.57	0.07	0.45
Spectrum 4	-	76.73	0.29	0.54	1.66	7.23	0.45	0.12	12.61	-	0.35

The mix proportions used in this research are shown in Table 2. The quantities stated in Table 2 were enough to cast 3 prisms of 40 mm × 40 mm × 160 mm that were used to determine the mechanical and dynamic elastic properties. A total of three batches were cast for each mix proportion resulting in 9 prisms per mix. This would imply that for each of the mix proportions 9 values for the flexural tensile strength of mortar and 18 values for the compressive strength would be obtained.

Table 2. Mix proportions.

Mix Designation	Cement (g)	Sand (g)	Water/Cement	SDS (mMol)	Tributyl Phosphate/Cement (%)	CNT/Cement (%)	CBN/Cement (%)
Ref	500	2050	0.6	-	-	-	-
CNT *				1	-	0.025	-
CNT2 **				1	0.13	0.025	-
CBN				-	-	-	0.025

* mix with MWCNT and SDS (sodium dodecyl sulfate) only; ** mix with MWCNT, SDS, and anti-foaming agent (tributyl phosphate).

The MWCNTs/cement (CNT mix) and CBN/cement (CBN mix) content was chosen as 0.025%. The scientific literature reports a wide range of CNT content from 0.005% [12] to 1.5% [25]. A 0.025% content of CNTs from cement mass was considered to be satisfactory in terms of expected improvements in the values of the mechanical properties of mortars and to assess whether the new CBNs would be a viable alternative to classical MWCNTs. The choice of such a content percentage of CNT was based on the results reported in the scientific literature. Mohsen et al. [26] concluded in their study that a 0.03% content of CNT would not only be the optimum concrete mix in terms of strength gain but also in terms of cost saving. Xu et al. investigated the influence of MWCNTs on the mechanical properties and microstructure of cement pastes. A 6.25% increase in the value of the flexural strength was obtained for 0.025% MWCNTs by mass of cement [27]. In the study of Morsy et al., it was concluded that a 0.02% MWCNT by mass of cement resulted in the highest increase in the values of the mechanical properties of mortar mixed with nano-clay [28]. It was also suggested that lower percentages of MWCNT would prevent re-agglomeration during the mixing phase with cement and aggregates.

2.2. Methods

The cement and the sand were dry mixed before water (in case of Ref mix) or water with nano-tubes (in case of CNT and CBN mixes) was added. The MWCNTs in the CNT mix were dispersed by ultra-sonication for 1 h into a 1 mM sodium dodecyl sulfate (SDS) solution at room temperature. No other anti-foaming agent was used. For the CNT2 mix, however, tributyl phosphate was used as defoaming agent in order to reduce the volume of voids [20]. The content of defoaming agent was chosen at 0.13% wt. of cement, similar to the data provided in [29]. The new CBN was mixed with tap water without any ultrasonication or surfactants being used.

The resulting mortar was cast into 40 mm × 40 mm × 160 mm prismatic molds and covered by wet cloths and plastic foil to prevent the excessive drying of the mortar surface. After 24 h, the prisms were demolded and place in water for curing until the age of 28 days.

The prisms were measured and weighed after demolding and before being tested. Four measurements were taken for the length of each prism and six values for the cross-sectional dimensions. The height and the width of the cross-section were measured at both ends of the prism and at the middle. This was performed both to check for changes in the dimensions and to determine the density of the hardened mortar. The obtained data were then used to compute the dynamic modulus of elasticity for mortar mixes presented in Table 2.

The dynamic modulus of elasticity, E_d, was determined in accordance with ASTM C215:14 [30] and was based on the first resonant frequency obtained from the impact echo method. The dynamic modulus of elasticity for the prisms was computed as shown in Equation (1):

$$E_d = D \cdot m \cdot f_{ln}^2 \quad (1)$$

where m is the mass of the sample [kg], f_{ln} is the fundamental longitudinal frequency of vibration (Hz) and D is a coefficient that depends on the dimensions of the prism (Equation (2)):

$$D = 4 \cdot \frac{L}{b \cdot t} \quad (2)$$

where L is the length of the prism [m], and b and t are the cross-sectional dimensions [m].

The flexural strength of mortar mixes was determined at the age of 28 days by means of 3-point bending test conducted at a loading rate of 50 N/s in accordance with [31], as shown in Figure 4a. The resulting half prisms were examined for signs of visible cracks and then subjected to uniaxial compression test, Figure 4b, with a loading rate of 2400 N/s [31].

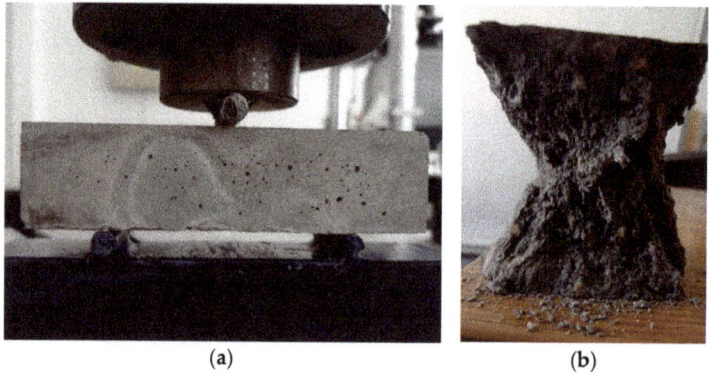

Figure 4. Testing of mortar prisms: (**a**) flexural test; (**b**) uniaxial compression (failed half-prism).

The structure of a single modified MWCNT (CBN) was assessed by means of an ultra-high resolution transmission electron microscope, UHR-TEM LIBRA 200MC. The obtained results are shown in Figure 1b.

The SEM (Carl Zeiss NEON 40EsB, Iasi, Romania) analysis was conducted on a Carl Zeiss NEON 40EsB cross-beam system with thermal Schottky field emission emitter and accelerated Ga ions column. The SEM is connected to an EDS (energy dispersive x-ray spectroscope, Carl Zeiss NEON 40EsB, Iasi, Romania) unit which allows a characteristic X-ray spectrum to be displayed.

The XRD (BRUKER AXS D8-Advance X-ray Diffractometer, Iasi, Romania) measurements were conducted by means of a Powder X-ray diffractometer equipped, BRUKER AXS D8-Advance X-ray Diffractometer, with a Cu X-ray source with a wavelength $\lambda = 1.5406$ Å. The powder diffraction covered the $10° < 2\theta < 90°$ range with $0.02°$ steps.

3. Results

3.1. Density

Figure 5 shows the change in the values of density 1 day from casting to the day of testing. The values of the density represent the mean value of nine determinations. Minor variations, less than 0.6%, were observed between the two sets of values. This would suggest a geometrical stability of the specimens and the fact that the available mixing water was replaced by cement hydration products. Similar trends were reported in the scientific literature for cement mortar incorporating superabsorbent polymers [32]. The influence of the SDS surfactant on the values of density is significant with a 20% decrease compared to the reference mix. This may be due to the increased porosity of the hardened mortar, as reported in the scientific literature [33]. As previously mentioned, no anti-foaming agent was used in the CNT mix. For the CNT2 mix, where the anti-foaming agent was used, the density was similar to the reference mix and CBN mix. The value was 5.43% and 6.26% smaller compared to the reference mix and CBN mix, respectively. The influence of the surfactant on the density of mortar could not be entirely reduced by the use of tributyl phosphate.

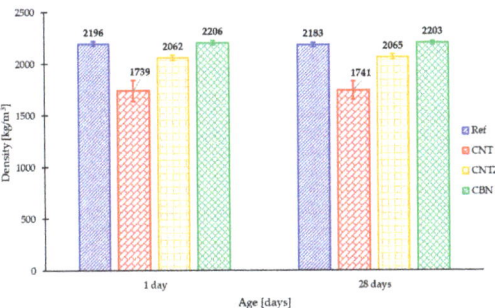

Figure 5. Density of mortar mixes.

The error bars presented in Figure 5, representing the standard deviation from the mean value, show that there was a larger scattering of the values of density for the CNT mix compared to the other three mixes. This would suggest either a non-uniform distribution of the MWCNTs in the mortar mix or a non-uniform pore distribution. Using an anti-foaming agent reduced the scattering and led to an increase in the value of density for the CNT2 mix. In general, the scattering of the results decreased with the increase in the curing age of the mortar.

Considering the error bars for both CNT2 and CBN mixes, it can be concluded that: (a) in the case of the CNT2 mix, the use of surfactant and sonication procedure ensured a uniform distribution of the MWCNTs in the mortar mix. Additionally, the use of an anti-foamer led to an improvement in the material structure by reducing the number of pores in the mix. (b) The use of the new, modified, MWCNTs in the CBN mix, without surfactant and sonication procedure, proved to be effective in obtaining a uniform structure of the mortar mix.

3.2. Dynamic Modulus of Elasticity

The determination of the dynamic modulus of elasticity was based on the first resonant frequency of the prismatic specimen which was determined by means of the impact echo method. It was determined on all nine prisms for each of the four mixes shown in Table 2. For each prism, at least four determinations were conducted to check for the consistency of the recorded data. The values of the dynamic modulus of elasticity may offer insightful information in terms of the quality of the investigated material, be it mortar, concrete or any other material. Low values of the dynamic modulus of elasticity may indicate internal damages of the material or a large number of voids/pores. Considering the shape of Equation (1), this would also be reflected in the values for the density of the material.

The free vibration response of all specimens (nine for each of the mix proportions presented in Table 2) was recorded, as shown in Figure 6a for the Ref mix, using the experimental set-up presented and considerations described in [34]. The fast Fourier transform (FFT) was applied to the recorded signal in order to obtain the response spectrum of the specimens, as shown in Figure 6b. The latter was used to assess the fundamental longitudinal frequency of vibration for each specimen. For each prism, at least four measurements were considered from which the fundamental frequency of vibration was calculated as the average value.

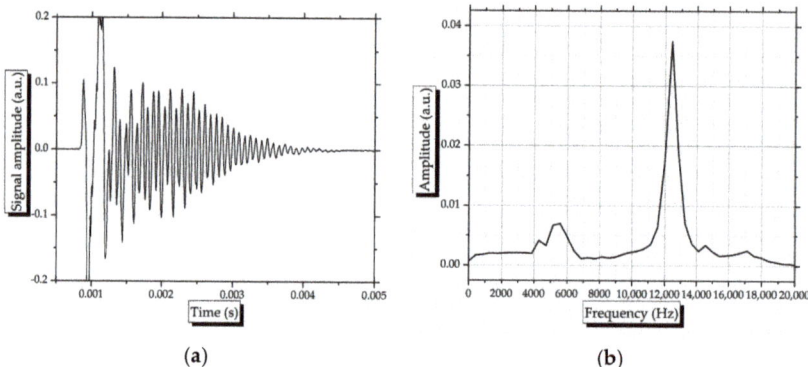

Figure 6. Free damped vibration response of Ref mix: (**a**) recorded signal; (**b**) response spectrum.

The obtained dynamic moduli of elasticity are presented in Figure 7. It can be observed that the influence of the SDS surfactant is very important, and it results in a sharp decrease in the value of E_d, by as much as 55%, compared to the reference mix. The obtained result is consistent with the already observed trend in terms of density, as shown in Figure 5, as well as the values for the fundamental frequency of vibration.

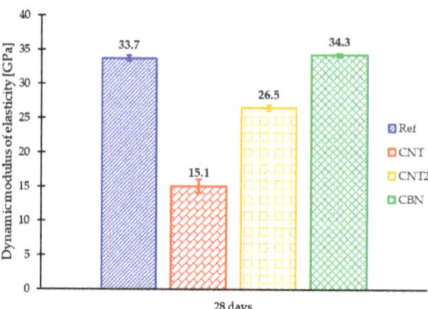

Figure 7. Dynamic modulus of elasticity.

Using an anti-foaming agent led to an increase in the value of the dynamic modulus of elasticity, compared to the mix where only SDS was used, but the obtained value was still lower than that of the reference sample. The decrease, in this case, was 21.36%, still significant considering the benefits one expects from using CNTs. However, taking into account the small percentage of CNTs, the trend observed for the density, and considering the shape of Equation (1), the obtained values of the modulus of elasticity exhibit the same pattern.

On the other hand, a small addition of the new CBN, 0.025% of the cement mass, resulted in a 1.78% increase in the value of the dynamic modulus of elasticity. This is not a very impressive increase but considering the production process of CBN coupled with the fact that there was no need for using surfactant and ultra-sonication procedure, the obtained results are encouraging.

The fact that the value of the modulus of elasticity increased for the CBN mix with respect to the Ref mix suggests that the new MWCNTs were uniformly dispersed in the mortar volume and contributed to achieving a better bond between the sand and the cement paste, as earlier studies suggested [35].

3.3. Flexural Strength

The obtained results from three-point bending tests on nine prisms for each mix proportion are shown in Figure 8. The use of surfactant without an anti-foaming agent results in a significant decrease, 53.83%, in the value of the flexural strength for the CNT mix compared to the Ref one. On the other hand, the use of new CBN results in an 18.76% increase in flexural strength compared to the reference mix. Similar trends, although by a much smaller percentage, were reported in the scientific literature for the same MWCNT dosage in the case of cement pastes [27]. The contribution of both MWCNTs and the new CBN in bridging the micro-cracks in the matrix is quite important, especially in the case of a CNT mix where the pore volume is expected to be large.

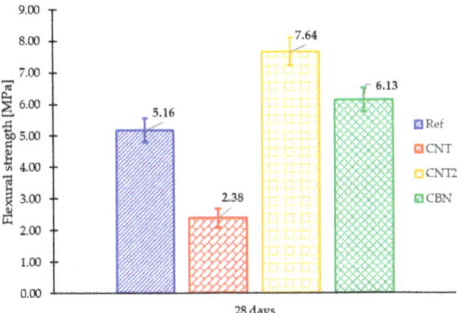

Figure 8. Flexural strength of mortars.

The use of an anti-foaming agent reduced the porosity of the samples which was reflected in the values obtained for the density. Since CNTs are longer than CBNs, they are more effective in bridging the micro-cracks and preventing their further development [11,36,37]. This would explain the higher value obtained for the flexural tensile strength of the CNT2 mix compared to the values of reference and CBN mixes.

3.4. Compressive Strength

The values of the compressive strength were assessed from 18 determinations for each considered mix presented in Table 2. The resulting parts of prisms after the flexural tests were examined for visible cracks and then subjected to uniaxial compression tests. Figure 9 summarizes the obtained results. The trend was similar to what was already observed in the case of dynamic modulus of elasticity and flexural strength.

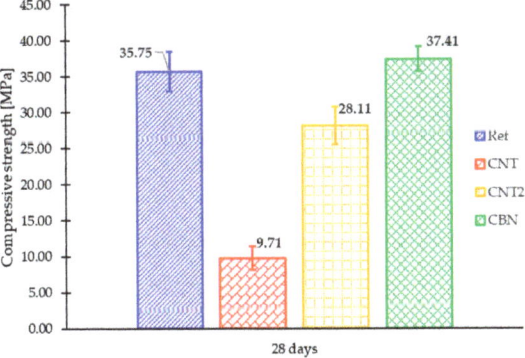

Figure 9. Compressive strength of mortars.

The compressive strength seems to be the most affected material property by the use of SDS surfactant and in the absence of anti-foaming agents. Increased porosity was also reported in the scientific literature when MWCNTs were used, even in the presence of foam reducing admixtures [33,38]. As with previously reported results, the values of the compressive strength for the CNT2 mix was lower than that of both the reference and the CBN mixes. The use of new CBN results in a 4.64% increase in the compressive strength compared to the reference mix.

4. Discussions

Figure 10 presents the magnified view of the CBN mix as compared to the Ref mix. As it can be seen, the CBN mortar shows a more compact structure whereas in the Ref mix a more pronounced pore concentration was observed. Upon a higher magnification factor, 5000×, Figure 11, the CSH gels in both mixes became more evident. The Ref mix exhibited a less organized and more porous arrangement of the ettringite needles in the structure.

Figure 10. Structure of mortars at 300× magnification factor: (**a**) CBN mix; (**b**) Ref mix.

Figure 11. Structure of mortars at 5000× magnification factor: (**a**) CBN mix; (**b**) Ref mix.

This offers a possible explanation of the improved mechanical properties of the CBN mix even for a very low percentage of nanotubes of only 0.025%. Similar observations were found in the scientific literature but for longer MWCNTs used together with surfactants and anti-foaming agents [25]. The CBNs that were used in the present study helped attain the same level of performance of the mortar mix but without the need of using surfactant and ultra-sonication procedures. Their geometry and structure had an impact on the macro-scale properties of mortar [39].

The use of MWCNTs with SDS surfactant resulted both in the increase in porosity as well as promoting the hydration reaction. Figure 12 shows CH (calcium hydroxide) hexagonal crystals as well as ettringite needles formed in a porous structure. Such a porous structure has an impact on the values of both elastic and mechanical properties of the CNT mix for which both SDS surfactant was used as well as ultra-sonication procedure in order to ensure a uniform distribution within the aqueous solution of the MWCNTs.

Figure 12. SEM image of the CNT mortar structure.

The XRD patterns of the three mixes are shown in Figure 13. Comparing the patterns leads to the conclusion that the introduction of nano-tubes promoted the development of additional crystallographic phases, with a different orientation of SiO_2 in the case of CNT mix compared to the Ref mix (at 2 theta = ~22 degrees) as well as in the case of CBN mix (at 2 theta = ~26 degrees).

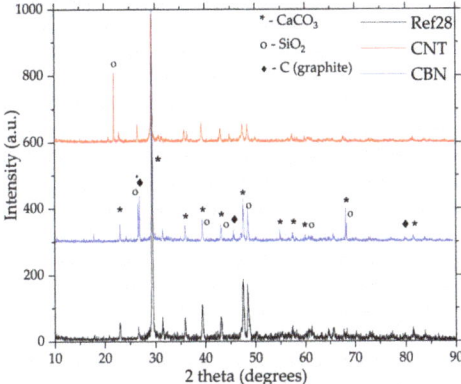

Figure 13. XRD patterns of mortars.

Table 3 presents the summary of the collected data during the preliminary stages of the research. It can be observed that the largest scattering of the results was obtained for the CNT mix with SDS surfactant and without an anti-foaming agent. Taking into account the material structure presented in Figure 12, the large number of pores induced by the SDS solution severely impacted the strength characteristics of the mortar.

On the other hand, the Ref, CNT2, and CBN mixes show a much lower scattering of the results, with respect to the median value owing to a more compact structure and to the use of the anti-foaming agent in the case of the CNT2 mix. The new type of MWCNTs provide a better material structure, are able to bridge the nano-cracks more efficiently [10,11,36,37] and thus resulting in overall better elastic and mechanical properties compared to the reference mix. The use of CNTs with an anti-foaming agent resulted in higher values of flexural tensile strength owing to longer CNTs that were able to bridge the cracks more effectively compared to CBNs.

Table 3. Statistical characterization of the experimental data (mean value ± standard deviation).

Mix	Density [kg/m^3]	Dynamic Modulus of Elasticity [GPa]	Compressive Strength [MPa]	Flexural Strength [MPa]
Ref	2183 ± 20.76 COV = 0.95	33.7 ± 0.35 COV = 1.04	35.75 ± 2.74 COV = 7.66	5.16 ± 0.39 COV = 7.51
CNT	1741 ± 85.62 COV = 4.92	15.1 ± 1.03 COV = 6.82	9.71 ± 1.64 COV = 16.84	2.38 ± 0.31 COV = 12.95
CNT2	2065 ± 24.17 COV = 1.17	26.5 ± 0.44 COV = 1.67	28.11 ± 2.62 COV = 9.34	7.64 ± 0.44 COV = 5.70
CBN	2203 ± 17.55 COV = 0.8	34.3 ± 0.65 COV = 1.89	37.41 ± 1.69 COV = 4.51	6.13 ± 0.38 COV = 6.23

Taking into account that, in the case of the CBN mix, there was no SDS added to the mix and that the ultra-sonication procedure was not applied, the obtained results warrant further investigations on the effect of the new type of MWCNTs on the properties of cement-based materials. Lower values for standard deviations as well as coefficient of variation suggest that a uniform distribution of the carbon-based nano-tubes was achieved without the help of classical, by now, methods.

5. Conclusions

Based on the obtained results from the laboratory investigations, the following conclusions can be drawn:

1. A new type of MWCNT was developed which has a similar structure to the SWCNTs in the sense that it is capped at both ends. The aspect ratio is smaller compared to "traditional" MWCNTs which results in a better bridging effect of the nano-cracks and a denser material structure compared to the reference mix. However, "traditional" MWCNTs are better in that aspect due to their increased lengths.
2. The new type of MWCNT achieved a uniform dispersion within the mortar mix, proven by consistently lower values of standard deviations and coefficients of variation for each of the investigated parameters. The data are even more encouraging and suggest that further research should be conducted in this direction taking into account that they are produced using the same technology applied for MWCNTs but the use of surfactants and, consequently, of anti-foaming agents is no longer necessary. Moreover, the ultra-sonication procedure required to ensure the distribution of nano-tubes within the aqueous solution, which should be carefully applied in case of long MWCNTs, is also not necessary.
3. The data are however limited to a rather small number of specimens and should be completed by large-scale laboratory investigations where significantly larger datasets should be collected before more generally valid conclusions could be drawn.
4. The improvement of mechanical properties of mortar using the new type of MWCNT is small, 18.76% in the case of flexural strength and only 4.64% in the case of compressive strength. However, these improvements were achieved with a very small percentage of CNTs in the mix, 0.025% by mass of cement. Additional percentages should be considered, and observations should be made as to whether or not the highlighted trends at this stage of the research are confirmed. Last but not least, durability studies should be conducted in order to assess the behavior of the mortar mixes to different types of chemical attacks and weathering conditions.

Author Contributions: Conceptualization, I.-O.T., S.-M.A.-S. and I.-A.G.; methodology, I.-O.T., S.-M.A.-S., G.S. and I.-A.G.; validation, I.-O.T., D.C. and A.-M.T.; formal analysis, S.-M.A.-S., G.S. and I.-A.G.; investigation, S.-M.A.-S., G.S., I.-A.G., A.-M.T., D.C. and I.-O.T.; resources, G.S. and I.-A.G.; data curation, I.-O.T.; writing—original draft preparation, S.-M.A.-S., G.S., I.-A.G. and A.-M.T.; writing—review and editing, D.C. and I.-O.T.; supervision, S.-M.A.-S. and I.-O.T.; project administration, S.-M.A.-S. and I.-O.T.; funding acquisition, I.-O.T. All authors have read and agreed to the published version of the manuscript.

Funding: The APC was funded by The "Gheorghe Asachi" Technical University of Iasi.

Institutional Review Board Statement: Not applicable.

Informed Consent Statement: Not applicable.

Data Availability Statement: The data presented in this research are available, upon request, from the corresponding author.

Conflicts of Interest: The authors declare no conflict of interest.

References

1. Irshidat, M.R.; Al-Saleh, M.H. Thermal performance and fire resistance of nanoclay modified cementitious materials. *Constr. Build. Mater.* **2018**, *159*, 213–219. [CrossRef]
2. Moro, C.; Francioso, V.; Velay-Lizancos, M. Nano-TiO$_2$ effects on high temperature resistance of recycled mortars. *J. Clean. Prod.* **2020**, *263*, 121581. [CrossRef]
3. Nuaklong, P.; Boonchoo, N.; Jongvivatsakul, P.; Charinpanitkul, T.; Sukontasukkul, P. Hybrid effect of carbon nanotubes and polypropylene fibers on mechanical properties and fire resistance of cement mortar. *Constr. Build. Mater.* **2021**, *275*, 122189. [CrossRef]
4. Zhan, P.; He, Z.; Ma, Z.; Liang, C.; Zhang, X.; Abreham, A.A.; Shi, J. Utilization of nano-metakaolin in concrete: A review. *J. Build. Eng.* **2020**, *30*, 101259. [CrossRef]
5. Zhang, M.; Zhu, X.; Liu, B.; Shi, J.; Gencel, O.; Ozbakkaloglu, T. Mechanical property and durability of engineered cementitious composites (ECC) with nano-material and superabsorbent polymers. *Powder Technol.* **2022**, *409*, 117839. [CrossRef]
6. Song, H.; Li, X. An Overview on the Rheology, Mechanical Properties, Durability, 3D Printing, and Microstructural Performance of Nanomaterials in Cementitious Composites. *Materials* **2021**, *14*, 2950. [CrossRef]
7. Gdoutos, E.E.; Konsta-Gdoutos, M.S.; Danoglidis, P.A. Portland cement mortar nanocomposites at low carbon nanotube and carbon nanofiber content: A fracture mechanics experimental study. *Cem. Concr. Compos.* **2016**, *70*, 110–118. [CrossRef]
8. Song, C.; Hong, G.; Choi, S. Effect of dispersibility of carbon nanotubes by silica fume on material properties of cement mortars: Hydration, pore structure, mechanical properties, self-desiccation, and autogenous shrinkage. *Constr. Build. Mater.* **2020**, *265*, 120318. [CrossRef]
9. da Silva Andrade Neto, J.; Santos, T.A.; de Andrade Pinto, S.; Dias, C.M.R.; Ribeiro, D.V. Effect of the combined use of carbon nanotubes (CNT) and metakaolin on the properties of cementitious matrices. *Constr. Build. Mater.* **2021**, *271*, 121903. [CrossRef]
10. Lee, H.S.; Balasubramanian, B.; Gopalakrishna, G.V.T.; Kwon, S.-J.; Karthick, S.P.; Saraswathy, V. Durability performance of CNT and nanosilica admixed cement mortar. *Constr. Build. Mater.* **2018**, *159*, 463–472. [CrossRef]
11. Siddique, R.; Mehta, A. Effect of carbon nanotubes on properties of cement mortars. *Constr. Build. Mater.* **2014**, *50*, 116–129. [CrossRef]
12. Jarolim, T.; Labaj, M.; Hela, R.; Michnova, K. Carbon nanotubes in cementitious composites: Dispersion, implementation, and influence on mechanical characteristics. *Adv. Mater. Sci. Eng.* **2016**, *2016*, 7508904. [CrossRef]
13. Madni, I.; Hwang, C.-Y.; Park, S.-D.; Choa, Y.-H.; Kim, H.-T. Mixed surfactant system for stable suspension of multiwalled carbon nanotubes. *Colloids Surf. A Physicochem. Eng. Asp.* **2010**, *358*, 101–107. [CrossRef]
14. Torabian Isfahani, F.; Li, W.; Redaelli, E. Dispersion of multi-walled carbon nanotubes and its effects on the properties of cement composites. *Cem. Concr. Compos.* **2016**, *74*, 154–163. [CrossRef]
15. Chaipanich, A.; Nochaiya, T.; Wongkeo, W.; Torkittikul, P. Compressive strength and microstructure of carbon nanotubes–fly ash cement composites. *Mater. Sci. Eng. A* **2010**, *527*, 1063–1067. [CrossRef]
16. Coppola, L.; Cadoni, E.; Forni, D.; Buoso, A. Mechanical characterization of cement composites reinforced with fiberglass, carbon nanotubes or glass reinforced plastic (GRP) at high strain rates. *Appl. Mech. Mater.* **2011**, *82*, 190–195. [CrossRef]
17. Sobolkina, A.; Mechtcherine, V.; Khavrus, V.; Maier, D.; Mende, M.; Ritschel, M.; Leonhardt, A. Dispersion of carbon nanotubes and its influence on the mechanical properties of the cement matrix. *Cem. Concr. Compos.* **2012**, *34*, 1104–1113. [CrossRef]
18. Alafogianni, P.; Dassios, K.; Tsakiroglou, C.D.; Matikas, T.E.; Barkoula, N.M. Effect of CNT addition and dispersive agents on the transport properties and microstructure of cement mortars. *Constr. Build. Mater.* **2019**, *197*, 251–261. [CrossRef]
19. Tragazikis, I.K.; Dassios, K.G.; Exarchos, D.A.; Dalla, P.T.; Matikas, T.E. Acoustic emission investigation of the mechanical performance of carbon nanotube-modified cement-based mortars. *Constr. Build. Mater.* **2016**, *122*, 518–524. [CrossRef]

20. Abedi, M.; Fangueiro, R.; Correia, A.G. An effective method for hybrid CNT/GNP dispersion and its effects on the mechanical, microstructural, thermal, and electrical properties of multifunctional cementitious composites. *J. Nanomater.* **2020**, *2020*, 6749150. [CrossRef]
21. Parveen, S.; Rana, S.; Fangueiro, R.; Paiva, M.C. Characterizing dispersion and long term stability of concentrated carbon nanotube aqueous suspensions for fabricating ductile cementitious composites. *Powder Technol.* **2017**, *307*, 1–9. [CrossRef]
22. SR EN 197-1; Cement. Part I: Composition, Specifications and Conformity Criteria for Normal Use Cements. ASRO (Romanian Standards Association): București, Romania, 2011.
23. Inoue, Y.; Kakihata, K.; Hirono, Y.; Horie, T.; Ishida, A.; Mimura, H. One-step grown aligned bulk carbon nanotubes by chloride mediated chemical vapor deposition. *Appl. Phys. Lett.* **2008**, *92*, 213113. [CrossRef]
24. Alexa-Stratulat, S.-M.; Covatariu, D.; Toma, A.-M.; Rotaru, A.; Covatariu, G.; Toma, I.-O. Influence of a novel carbon-based nano-material on the thermal conductivity of mortar. *Sustainability* **2022**, *14*, 8189. [CrossRef]
25. Shao, H.; Chen, B.; Li, B.; Tang, S.; Li, Z. Influence of dispersants on the properties of CNTs reinforced cement-based materials. *Constr. Build. Mater.* **2017**, *131*, 186–194. [CrossRef]
26. Mohsen, M.O.; Alansari, M.; Taha, R.; Senouci, A.; Abutaqa, A. Impact of CNTs' treatment, length and weight fraction on ordinary concrete mechanical properties. *Constr. Build. Mater.* **2020**, *264*, 120698. [CrossRef]
27. Xu, S.; Liu, J.; Li, Q. Mechanical properties and microstructure of multi-walled carbon nanotube-reinforced cement paste. *Constr. Build. Mater.* **2015**, *76*, 16–23. [CrossRef]
28. Morsy, M.S.; Alsayed, S.H.; Aqel, M. Hybrid effect of carbon nanotube and nano-clay on physico-mechanical properties of cement mortar. *Constr. Build. Mater.* **2011**, *25*, 145–149. [CrossRef]
29. Wang, B.; Han, Y.; Liu, S. Effect of highly dispersed carbon nanotubes on the flexural toughness of cement-based composites. *Constr. Build. Mater.* **2013**, *46*, 8–12. [CrossRef]
30. ASTM C215-14; Standard Test Method for Fundamental Transverse, Longitudinal, and Torsional Resonant Frequencies of Concrete Specimens. ASTM International: West Conshohocken, PA, USA, 2014.
31. SR EN 196-1; Methods of Testing Cement—Part 1: Determination of Strength. ASRO (Romanian Standards Association): București, Romania, 2016.
32. Jamnam, S.; Sua-iam, G.; Maho, B.; Pianfuengfoo, S.; Sappakittipakorn, M.; Zhang, H.; Limkatanyu, S.; Sukontasukkul, P. Use of cement mortar incorporating superabsorbent polymer as a passive fire-protective layer. *Polymers* **2022**, *14*, 5266. [CrossRef]
33. Hu, S.; Xu, Y.; Wang, J.; Zhang, P.; Guo, J. Modification effects of carbon nanotube dispersion on the mechanical properties, pore structure, and microstructure of cement mortar. *Materials* **2020**, *13*, 1101. [CrossRef]
34. Alexa-Stratulat, S.M.; Covatariu, D.; Toma, I.O.; Covatariu, G. Computation and experimental considerations on dynamic testing of cement mortar. *IOP Conf. Ser. Mater. Sci. Eng.* **2021**, *1141*, 012019. [CrossRef]
35. Hawreen, A.; Bogas, J.A.; Dias, A.P.S. On the mechanical and shrinkage behavior of cement mortars reinforced with carbon nanotubes. *Constr. Build. Mater.* **2018**, *168*, 459–470. [CrossRef]
36. Guo, J.; Yan, Y.; Wang, J.; Xu, Y. Strength analysis of cement mortar with carbon nanotube dispersion based on fractal dimension of pore structure. *Fractal Fract.* **2022**, *6*, 609. [CrossRef]
37. Chen, J.; Akono, A.-T. Influence of multi-walled carbon nanotubes on the hydration products of ordinary Portland cement paste. *Cem. Concr. Res.* **2020**, *137*, 106197. [CrossRef]
38. Li, Q.; Liu, J.; Xu, S. Progress in research on carbon nanotubes reinforced cementitious composites. *Adv. Mater. Sci. Eng.* **2015**, *2015*, 307435. [CrossRef]
39. Mohsen, M.O.; Taha, R.; Abu Taqa, A.; Al-Nuaimi, N.; Al-Rub, R.A.; Bani-Hani, K.A. Effect of nanotube geometry on the strength and dispersion of CNT-cement composites. *J. Nanomater.* **2017**, *2017*, 6927416. [CrossRef]

Disclaimer/Publisher's Note: The statements, opinions and data contained in all publications are solely those of the individual author(s) and contributor(s) and not of MDPI and/or the editor(s). MDPI and/or the editor(s) disclaim responsibility for any injury to people or property resulting from any ideas, methods, instructions or products referred to in the content.

Article

Experimental Study on the Bending Resistance of Hollow Slab Beams Strengthened with Prestressed Steel Strand Polyurethane Cement Composite

Jin Li [1,*], Yongshu Cui [1], Dalu Xiong [2], Zhongmei Lu [2], Xu Dong [1,*], Hongguang Zhang [2], Fengkun Cui [1] and Tiancheng Zhou [1]

1 School of Transportation Civil Engineering, Shandong Jiaotong University, Jinan 250357, China
2 Jinan Kingyue Highway Engineering Company Limited, Jinan 250220, China
* Correspondence: sdzblijin@163.com (J.L.); dongxu512@126.com (X.D.)

Abstract: In order to explore the toughening performance and failure mechanism of hollow slab beams strengthened with prestressed steel strand polyurethane cement composite, three test beams (L1–L3) were strengthened and one test beam (L0) was used as a comparison. The influence of different tensile stresses of steel strand and fiber additions on the flexural bearing capacity of the hollow slab beams, was studied. The cracking characteristics, load deflection relationship, ductility and strain of each test beam were compared and analyzed. The test results showed that the toughened material was well bonded to the hollow slab beam and the steel strand, which effectively inhibited the development of cracks in the test beams. The flexural bearing capacity of the strengthened test beams was significantly improved. The use of prestressed steel strand polyurethane cement composite material effectively improved the flexural bearing capacity of the test beams, and this reinforcement process can be further extended to engineering applications.

Keywords: steel strand polyurethane cement; hollow slab beam; bending test; reinforcement process; material properties

1. Introduction

With the rapid growth of traffic volume in China, the vehicle transport load continuously increases. Due to the effects of rain, snow, weathering and other natural conditions, a large number of bridges built at various times in the past have appeared to gradually deteriorate and erode [1–4]. In particular, the prefabricated concrete hollow slab beams put into operation at the end of the 20th century suffer from these conditions [5,6], affecting the normal passage of the road and causing huge economic losses. Although existing reinforcement methods [7–13] could effectively improve the bearing capacity of old bridges, there are still some limitations in controlling the self-weight of the structure, shortening the reinforcement period and simplifying the reinforcement process. In recent years, polyurethane cement composite, a lightweight and high strength corrosion resistant material [14–17], has become a hot research topic in the field of bridge reinforcement.

In recent years, polyurethane cement composite has gradually been applied to practical projects. Due to characteristics such as simplicity, high strength and corrosion resistance, it has become a new type of reinforcement method. Many experts have studied the field of polyurethane reinforcement to a certain extent: Zhou Yonghong [18] proposed a design and construction technology scheme to improve the lateral load distribution and increase the overall stiffness of bridges by using MPC composite reinforcement under the condition of uninterrupted railroad transportation by relying on the load test before reinforcement of the project. They verified the reliability of the reinforcement scheme by establishing a finite element model and performing load tests after reinforcement. Sun Quansheng et al. [19,20] analyzed the reinforcement effects of polyurethane cement wire rope through tension and compression tests on polyurethane cement and flexural loadbearing damage tests on 3 m ordinary reinforced concrete T-beams with different polyurethane wire rope reinforcement schemes. They verified the practical engineering value of polyurethane wire rope flexural reinforcement, showing that it could effectively improve the flexural loadbearing capacity of the test beams. Wang Jianlin et al. [21], taking the actual project as an example, deduced the theoretical calculation formula of using MPC composite material to reinforce the normal section of a hollow plate girder bridge, and used a super strong, high-toughness MPC composite material to design and construct the hollow plate girder bridge reinforcement. They concluded that their method could effectively improve the ultimate bearing capacity in normal use, and that maintenance and reinforcement could be achieved without interrupting traffic operation.

Based on this, the authors of the present work designed a flexural bearing capacity test for hollow slab beams by using polyurethane cement composite material and steel strand reinforcement. This test was used to explore improvements of the toughness of hollow slab beams using a prestressed steel hinge line and polyurethane cement composite, and the failure mechanism of test beams under load. By reducing the size of the model beam, the cost was reduced during the test. The stated goal was to provide a basis for the subsequent reinforcement of a hollow slab solid bridge.

2. Test Overview

2.1. Test Materials

The design strength grade of concrete was C50, and the average compressive strength of cube specimens was 52.6 MP. Four longitudinal reinforcement bars were arranged in the compression zone and tension zone of the hollow slab beam, and a stirrup was arranged every 15 cm along the longitudinal reinforcement bars. Two support reinforcement bars were arranged at each end of support. The polyurethane cement composite material used for toughening the test beam was composed of isocyanate, modified polyether, cement, defoaming agent, catalyst and other components, as shown in Table 1. The density of the test block formed by this ratio was 1400 kg/m^3.

Table 1. Composition and mass proportions of polyurethane cement composite.

Composition	Proportion (%)
Isocyanate	30.5
Modified polyether	35
Cement	33.2
Defoaming agent	0.5
Catalyst	0.8

Polyurethane toughening test with carbon fiber and glass fiber: see Figure 1. Performance indexes are shown in Table 2.

The type, size, mechanical index and other parameters of the steel strand used for toughening the hollow plate beam, are shown in Table 3.

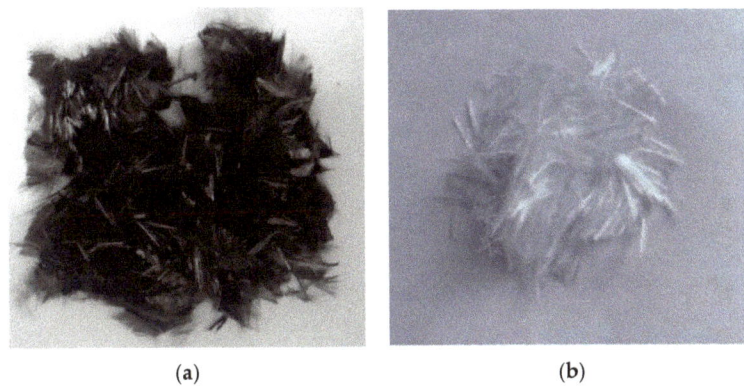

Figure 1. Test fibers: (**a**) carbon fiber, (**b**) glass fiber.

Table 2. Fiber material properties.

Fiber	Fiber Performance Index			
	Diameter (μm)	Length (mm)	Density (g/cm^3)	Elastic Modulus (GPa)
Carbon fiber	4	12	1.8	280
Glass fiber	9	12	2.5	73

Table 3. Specifications and parameters of steel hinge lines.

Category	Nominal Diameter (mm)	Effective Cross Section (mm^2)	Design Value of Tensile Strength (MPa)	Elastic Modulus (GPa)	Poisson's Ratio
1 × 7 standard	15.2	140	1860	0.195	0.3

2.2. Design of Test Beam

A total of 4 test beams, numbered L0–L3, were formed in this test. The control beam L0 was not strengthened, and the test beams L1–L3 were toughened in different ways. The length, width and height of the test beams were 300, 50 and 40 cm, respectively. Circular channels with a diameter of 25 cm were arranged in the cross section of the hollow slab beam. Hollow slab beams at both ends of the pouring support measured a support size length, width, and height of 50, 15, and 10 cm, respectively; in the test beams L0–L3, reserved tension channels were supported for the insertion of steel strands. The specific test beam sizes are shown in Figures 2 and 3.

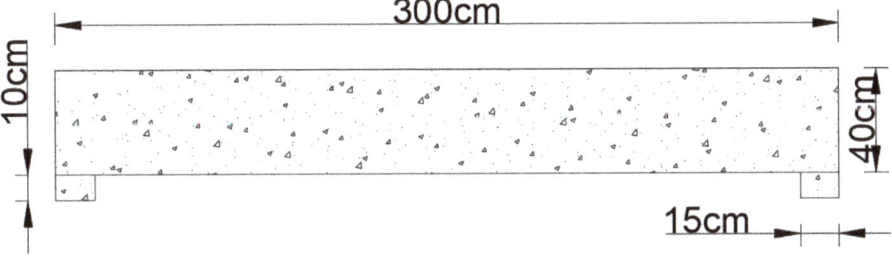

Figure 2. Longitudinal section of the test beam.

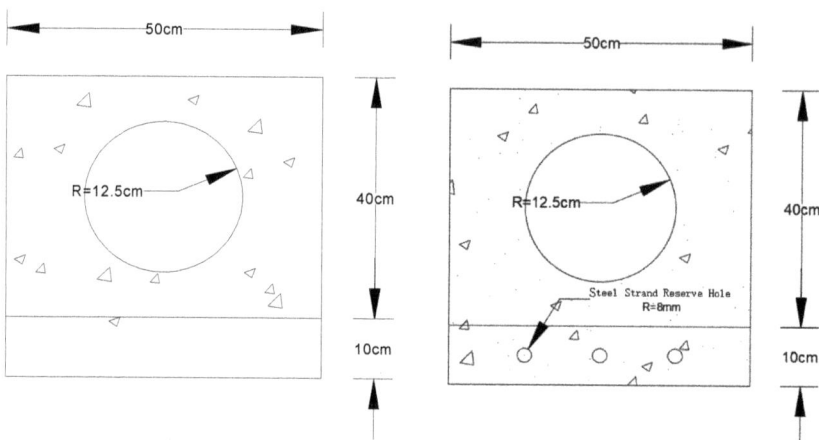

Figure 3. Cross section of the test beam.

2.3. Toughening Scheme

In order to study the failure phenomenon, bearing capacity and failure mechanism of hollow slab beams under load, the pouring temperature and thickness of polyurethane cement composite should be strictly controlled in the reinforcement process. The same is true of the type of steel strand, reinforcement ratio and tensile stress, among other factors. In this experiment, the toughening material was poured in the room at an ambient temperature of 20 °C.

See Table 4 for the reinforcement scheme of test beams L0–L3.

Table 4. Reinforcement scheme of test beams L0–L3.

Test Beam No.	Quantity of Steel Strand (Root)	Tension (MPa)	Reinforcement Material	Fiber Add	Thickness of the Material (cm)
L0	0	0	/	/	/
L1	3	300	Steel strand, polyurethane cement composite material	/	4
L2	3	400	Steel strand, polyurethane cement composite material	/	4
L3	3	300	Steel strand, polyurethane cement composite material	0.04 and 0.04% carbon fiber and glass fiber	4

After the material was cured for 14 days, flexural bearing capacity testing of each hollow slab beam was carried out. Test beam L1 was compared with beam L0 (without toughening treatment), and test beams L2 and L3 were, respectively, compared with test beam L1 in order to observe the failure phenomenon of each test beam under load and evaluate their bearing capacities.

2.4. Measuring Point Arrangement and Loading Scheme

In order to better reflect the strain situation along the beam height in the middle of the test beam span under all levels of load, a strain measuring point was set up on the middle side of the test beam span, with an interval of 6.6 cm between two adjacent measuring points. A total of 6 measuring points were set up on test beam L0, and 7 measuring points were set up on test beams L1–L3. In order to observe the tensile changes of the beam bottom under various loads, two strain measuring points were arranged at the bottom of the mid-span beam.

In order to more clearly reflect the displacement changes of the test beam at the key section under all levels of load, deflection measurement points were placed at the bottom support, at 1/4 of the test beam and at the middle of the span. A total of 5 deflection measurement points were arranged for each test beam. See Figure 4 for the layout of the strain and deflection measurement points.

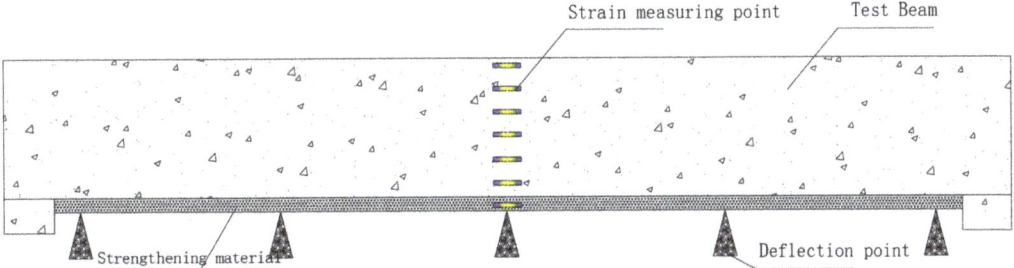

Figure 4. Layout diagram of strain and deflection measurement points.

The test used a reaction frame to load through the jack, and the load was applied to the test beam through the steel plate, the pressure sensor, the distribution beam and so on. The distribution beam interval was 25 cm. Before the test, an initial load of 10 KN was applied to observe whether the strain gauge, displacement sensor and pressure sensor were working normally through the static acquisition system. After the inspection, the jack was unloaded, the data of the static acquisition system was cleared, and the formal loading began, with 15 KN loading per level. During the loading process, the phenomena of the test beam under each load were recorded, and a crack width meter was used to observe and record the crack development on the test beams at all times. Additionally, attention was paid to cracking sounds.

3. On-Site Reinforcement Processes

In total, 12 steel strands of 4 m length were cut. One end of the reserved channel was anchored with a single anchor head, and the other end was laid with an anchor plate and tensioning jack. The tensioning stress was controlled through the tensioning stress table and elongation. See Figure 5 for tension and anchorage of the steel strands.

(a) (b)

Figure 5. Field photograph of steel strands: (**a**) tensioning, (**b**) anchorage.

Two layers of sponge glue were pasted on the upper edge at 20 cm on both sides of the test beam. In order to prevent the reinforcement material from sticking to the mold during casting and ensure a smooth demold, the side of the template used for casting was wrapped with plastic film, and the treated template was pasted to the side of the beam and firmly fixed with steel nails. After fixing, foam glue was applied evenly to the joint of the template to avoid material spillover during pouring. Before fixing the formwork, the debris on the upper part of the beam was cleaned to ensure that the toughening material was in full contact with the bottom surface of the beam and the steel strand.

The toughening material was weighed and mixed according to the ratio and then poured into the mixing bucket for full stirring with a hand-held agitator. After the materials were mixed evenly, the mixture was poured directly onto the beam surface. Due to the fast condensation of the reinforcement material, it was stirred continuously in the process of mixing and pouring to prevent consolidation of the material. After demolding and curing at 20 °C for 14 days, flexural capacity testing of the hollow slab beam was carried out. The site pouring and loading are shown in Figures 6 and 7.

Figure 6. Site pouring.

Figure 7. Field loading.

4. Test Results and Analysis

4.1. Experimental Phenomenon

During loading, cracks appeared in all beams, and multiple cracks appeared with the acceleration of deflection growth, rapid crack extension, and crack spread to the top of the beams. Before cracks appeared in L1–L3 reinforced beams, the deflection and strain increased regularly with the increase in load. The deflection of the midspan section increased weakly, and the strain changes at different heights of beams were not obvious. When the crack spread to the top of the beam, it was accompanied by the fracture of the polyurethane-reinforced material. The concentrated forces, corresponding to the midspan sections of L0 and L1–L3 of the reinforced beam at each state, are shown in Table 5. The cracks and fractures of each beam are shown in Figures 8–11.

Table 5. Loading state load of each test beam.

Test Beam No.	Cracks Appear	Multiple Cracks and Increasing Deflection	Rapid Crack Extension	Fracture of Toughened Material
L0	150 kN	165 kN	195 kN	240 kN (crack propagation beam top)
L1	225 kN	330 kN	370 kN	450 kN
L2	315 kN	330 kN	450 kN	495 kN
L3	240 kN	300 kN	380 kN	465 kN

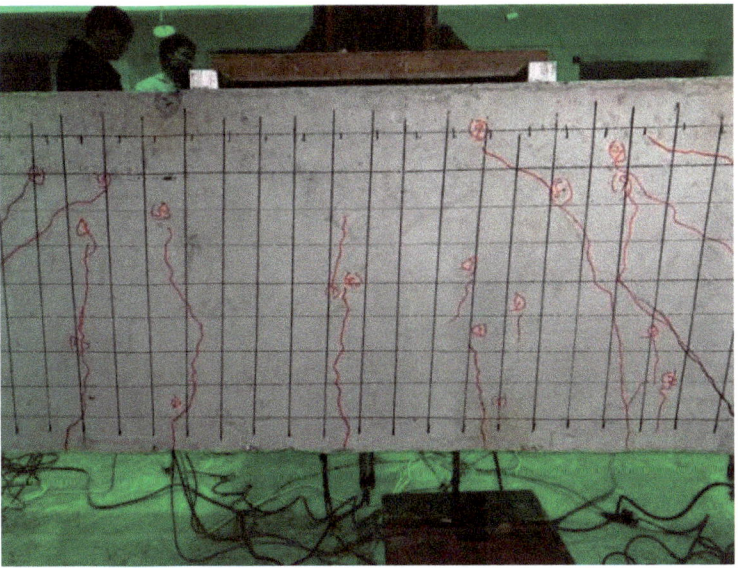

Figure 8. Crack pattern of L0 beam.

Figure 9. Fracture of L1 polyurethane.

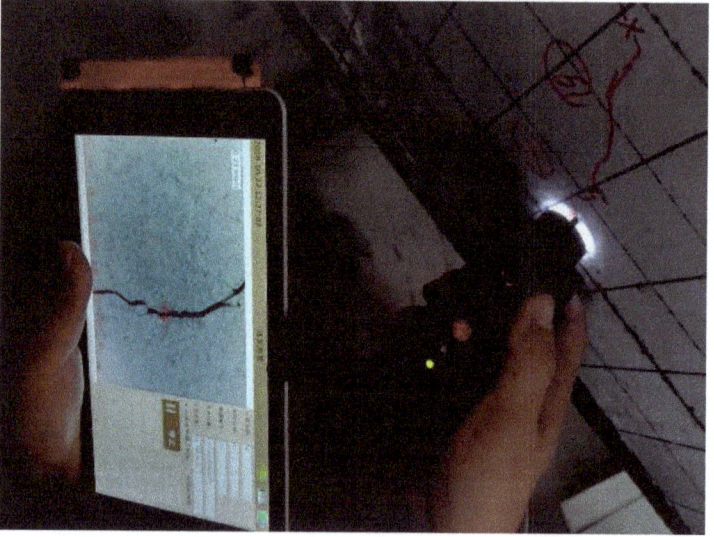

Figure 10. Crack opening assessment for L2 beam body.

Figure 11. L3 crack penetration.

4.2. Load–Deflection

The whole process of stressing and deformation of the test beams L0–L3 under load was experienced as follows: no cracks in the concrete, initial crack appearance and development, and finally, failure. When there was no crack in the test beam, the applied concentrated force was lower than the cracking load, the load–deflection curve was basically linear, and the test beam was in the state of elastic force. After cracking, the slope of the load–deflection curve gradually decreased, showing a nonlinear trend. When the bending moment reached a certain value under the load, the deformation of the beam was limited under the constraint of the steel strand, and the crack width and deflection of the test beam increased slowly with the increase in the load. When the test beam was damaged, the deflection increased rapidly, the cracks developed rapidly and spread above the neutral axis, the concrete compressive strain at the edge of the compression zone reached the ultimate compressive strain, the concrete was crushed, the reinforcement layer broke, and the beam side in the middle of the span appeared through cracks. The load–deflection curve is shown in Figure 12.

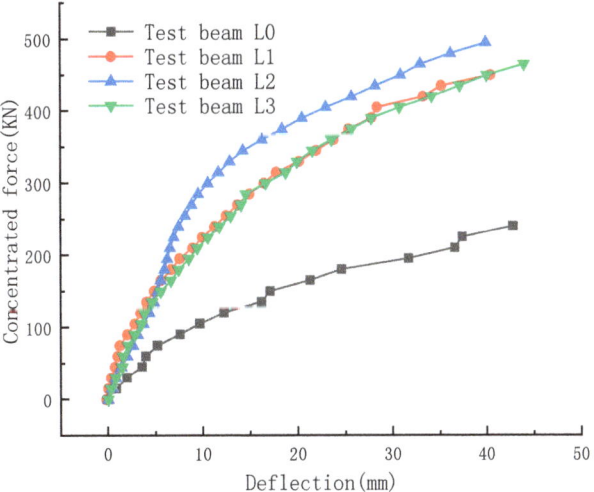

Figure 12. Load–deflection curve (crack development yield limit).

4.3. Ductility Analysis

The ductility coefficient was divided into displacement, curvature and energy ductility coefficients. In this paper, the prestressed steel strands and polyurethane cement composite material were used to reinforce the bottom of the test beam. The prestressed steel strand had an obvious yield point, so the displacement ductility coefficient was used. The displacement ductility coefficient was used to analyze the resistance ability of concrete test beams L0–L3 to inelastic deformation under load, under the condition that the flexural capacity of test beams L0–L3 did not decrease significantly, which mainly indicated the deformation ability of the test beam at the stage from yield load to ultimate load.

The displacement ductility coefficient μ was calculated by the ratio of the mid-span ultimate deflection δ_u of the test beam to the yield deflection δ_y. The results of the ductility of the test beams L0–L3 are shown in Table 6, and the scatter distributions of the ductility of each test beam are shown in Figure 13.

Table 6. Ductility test results of each test beam.

Test Beam	δ_y/mm	δ_u/mm	μ
L0	31.7	42.8	1.35
L1	25.4	40.5	1.59
L2	23	38.6	1.67
L3	27.8	44.7	1.60

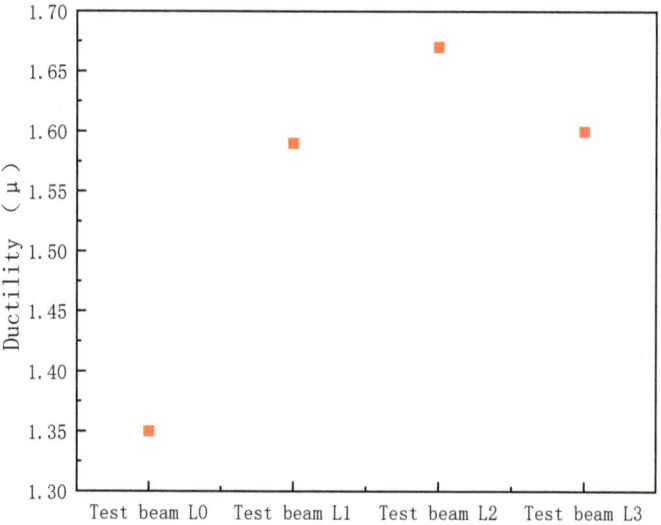

Figure 13. Ductility of each test beam.

According to the distribution of ductility scatter in Figure 13, compared with the test beam L0, the ductility levels of the test beams L1–L3 significantly increased, by 17.8%, 23.7% and 18.5%, respectively. This indicated that the combined reinforcement of prestressed steel hinge lines and polyurethane cement composite could effectively improve the ductility of hollow slab beams and provide a greater safety reserve for structural applications. Compared with test beam L1, test beam L2 improved structural ductility by 5% by increasing the tensile stress of each strand by 100 MPa, and the steel strand tensile stress had some effect on the structural ductility. Meanwhile, test beam L3 improved ductility slightly, by mixing 0.04% carbon fiber and 0.04% glass fiber. Therefore, it was speculated that the toughened fiber and polyurethane materials were not fully mixed evenly during mixing, or the density of the toughened fiber was greater than that of the polyurethane cement toughened material, which sank to the bottom during the mixing process, i.e., part of the fiber stayed at the bottom of the mixing barrel during pouring. By using the combination of prestressed steel strands and polyurethane cement composite reinforcement, the toughness and ductility of hollow slab beams was enhanced, which could provide greater safety to reinforced solid bridges.

4.4. Beam Bottom Strain

The variation trend of the strain at the bottom of the test beam was roughly the same as that of the load–deflection curve, and the three stages of stress and deformation of the test beam under load were also verified. As shown in Figure 14, load–beam bottom strain, the test beam was in the elastic stress stage before cracks appeared, and the load–beam bottom strain at this stage was basically linear—that is, with continuous increase in load, the beam bottom strain did not greatly change. At this stage, the ground strand and polyurethane cement composite materials played a small role, mainly in beam force. When the concentrated force reached the cracking load of the test beam, the test beam immediately entered the elastic–plastic stage. At this stage, the load–beam bottom strain curve showed a nonlinear relationship. At this stage, the strain increased significantly with the incremental increase in load. At this stage, steel strand and polyurethane cement composite materials played a major role. When the load reached the yield load, the test beam entered the failure stage. At this stage, the steel bar reached the yield state, and the concrete of the test beam could be crushed or the polyurethane cement composite could be pulled off. The variation trends of L1 and L3 of the test beams were close to that of the load–deflection, which had similar characteristics. The load–beam bottom strain curve could, thus, effectively verify the variation trend of the load–deflection curve and the performance characteristics of the test beam under load, and more scientifically and accurately describe the mechanical properties of the test beam.

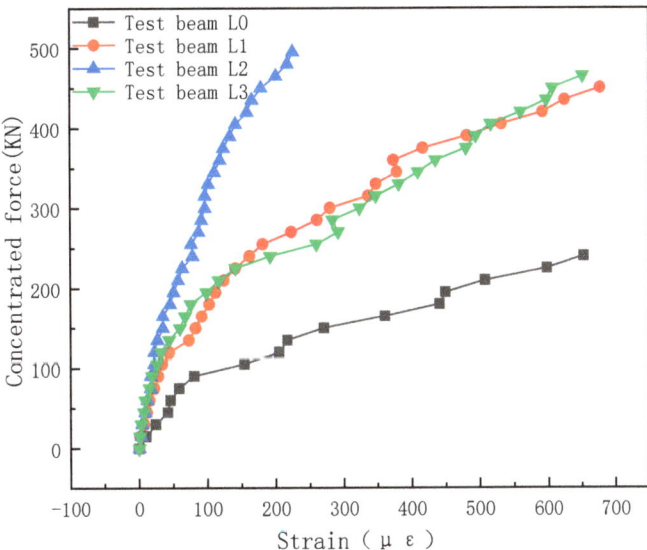

Figure 14. Load and strain at the bottom of the test beams.

4.5. Strain along Beam Height

The assumption of the plane section was verified by analyzing the strain along the high section of the beam. When the concrete beam was purely bent under the action of the external load, any cross section on the beam rotated around the axis of the beam. At this time, the longitudinal material of the concrete beam did not squeeze—that is, the test beam did not show concave and convex warping, and the axes of the deformed beam were still perpendicular to each other.

Through the flexural capacity test of the normal section of the test beam, the concentrated force was applied incrementally in the middle of the span. The concrete strain values of different sections of the beam height under different loads were obtained using the static strain collection instrument, and a distribution map of the measured average strain was drawn. As shown in Figures 15–18, when the test beam was in the elastic stress stage, there was no damage to the concrete beam, and the neutral axis of the section gradually moved to the compression zone as the load increased. At this time, the applied load was small, the concrete below the neutral axis was under tensile stress, and the concrete above the neutral axis was under compressive stress. At this time, the strain was an inclined straight line, and the strain distribution along the high section of the beam under the load was a triangle, with the top and bottom opposite. The concrete test beam at this stage fully conformed to the assumption of the plane section. When the test beams in the elastic–plastic stress test (L1–L3) reached the cracking load, below the neutral axis of the concrete cracks, reinforcement with the surrounding concrete occurred, producing relative displacement. However, at this stage, the prestressed steel hinge line and polyurethane cement composite began to undertake the main external load, offsetting part of the tensile stress. Under the action of load, the strain distribution along the high section of the beam was approximately triangular from top to top, and the average strain distribution of other sections of the beam across the crack was still consistent with the assumption of plane section. When the test beam was in the failure stage, the crack width in the tensile zone of the test beams L0–L3 increased sharply and the steel bar yield in the tensile zone occurred in a finite length range. The concrete in the compression zone was crushed within a certain length range, and the polyurethane cement composite did not peel or fracture. The section deformation at this stage was still approximately consistent with the assumption of the plane section.

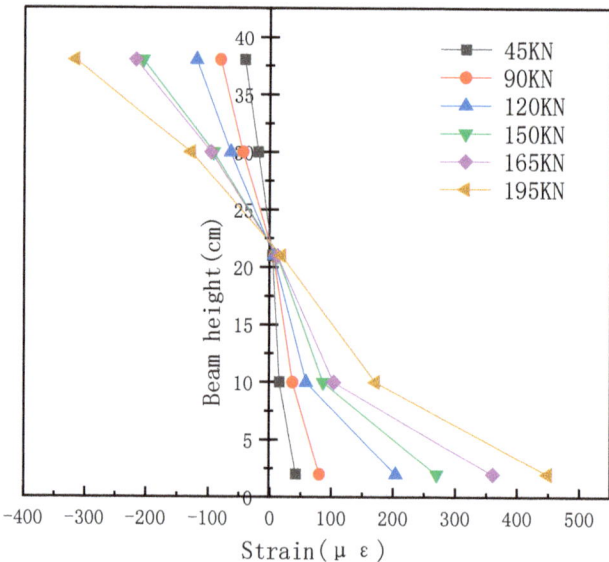

Figure 15. Strain of test beam L0 along beam height.

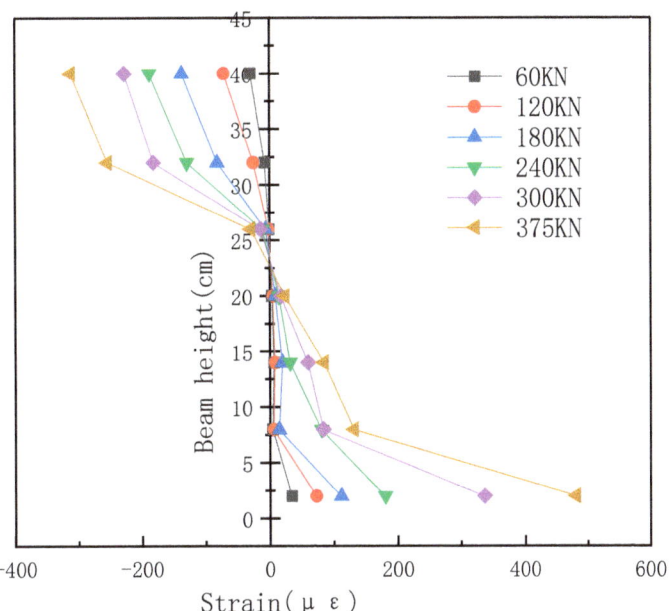

Figure 16. Strain of test beam L1 along beam height.

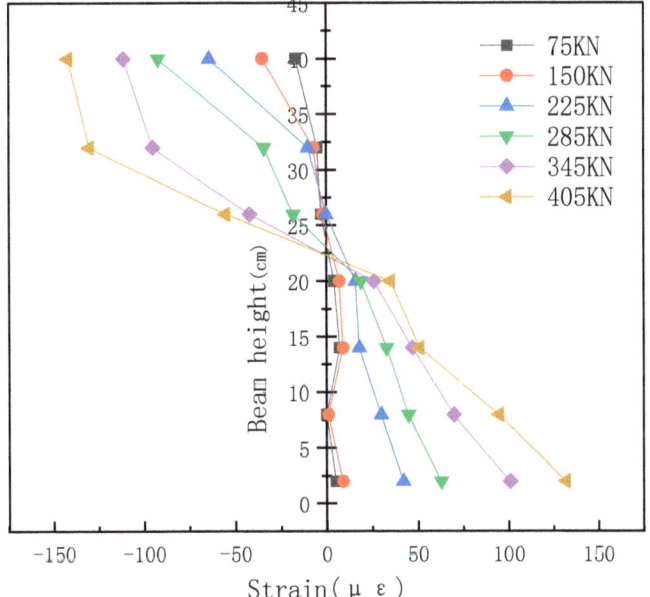

Figure 17. Strain of test beam L2 along beam height.

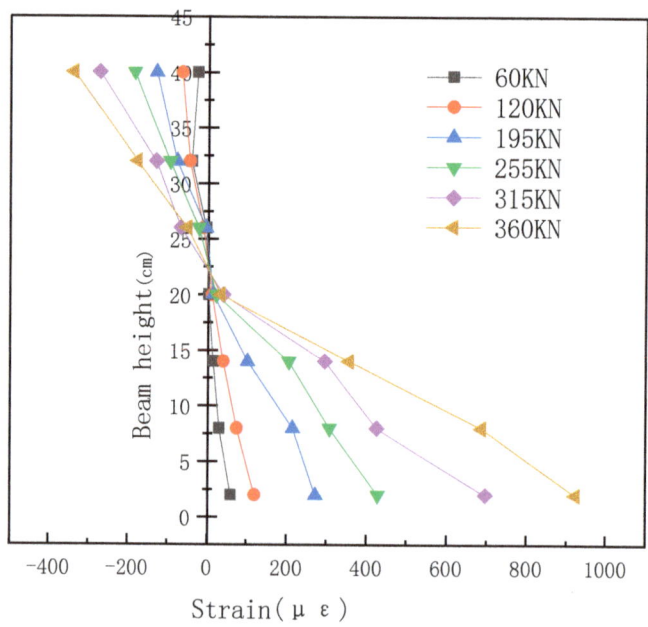

Figure 18. Strain of test beam L3 along beam height.

4.6. SEM Observation

The good adhesion of polyurethane cement composite could be clearly observed by SEM. There was no gap or crack between the fiber and the polyurethane cement composite, as shown in the figures below. As shown in Figures 19 and 20. It can be seen that the fiber played a positive reinforcement role [14].

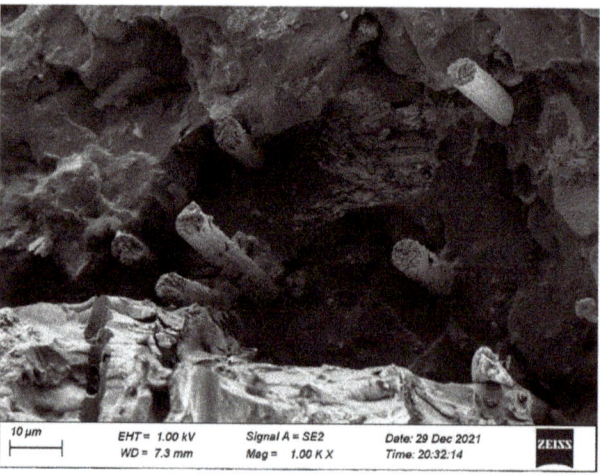

Figure 19. A 10 μm SEM micrograph.

Figure 20. A 5 μm SEM micrograph.

5. Conclusions

In this study, the failure phenomena, load–deflection curve, ductility, strain at the bottom of the beam and strain along the beam height in the middle of the span were analyzed through the flexural loading of the test beam. The following conclusions were drawn:

(1) Prestressed steel strand polyurethane cement composite material was used for reinforcement, which significantly improved the crack resistance of the test beam. The crack resistance was as follows: test beam L2 > test beam L3 > test beam L1 > test beam L0;

(2) The slope of the load–deflection curve of the test beam without cracks reflected the rigidity of the test beam at this stage. The rigidity of the test beam was as follows: test beam L2 > test beam L3 ≈ test beam L1 > test beam L0. The combined reinforcement of prestressed steel strand and polyurethane cement composite material reduced the deflection of the test beam under load and effectively improved the rigidity of the test beam;

(3) The ductility of test beams L0, L1, L2 and L3 under load was 1.35, 1.59, 1.60 and 1.67, respectively. The capacity of the test beam to resist inelastic deformation was as follows: test beam L2 > test beam L3 > test beam L1 > test beam L0. The combined toughening of the prestressed steel strand and polyurethane cement composite material effectively improved the ground ductility of hollow slab beams, providing greater safety reserves for the subsequent application of bridge reinforcement;

(4) According to the high strain diagram along the beam of each test beam, it was noted that the high strain along the beam of test beams L0, L1, L2 and L3 under load conformed to the assumption of the plane section when the test beam was under elastic stress, elastic–plastic stress and failure stage;

(5) Use of prestressed steel strand and polyurethane cement composite material to strengthen hollow slab beams effectively improved the flexural bearing capacity of the test beam, and this strengthening technology can be further extended to engineering applications.

Author Contributions: Conceptualization, Y.C. and J.L.; methodology, D.X.; software, Y.C. and T.Z.; validation, F.C., Y.C. and J.L.; formal analysis, J.L.; investigation, J.L.; resources, J.L.; data curation, Z.L.; writing—original draft preparation, Y.C.; writing—review and editing, J.L. and T.Z.; visualization, X.D.; supervision, H.Z.; project administration, F.C.; funding acquisition, D.X. All authors have read and agreed to the published version of the manuscript.

Funding: This study was supported by the Science and Technology Program of Shandong Provincial Department of Transportation (No. 2017B97) and the Key R&D Program of Science and Technology Department of Shandong Province (No. 2019GGX102041).

Institutional Review Board Statement: Not applicable.

Informed Consent Statement: Not applicable.

Data Availability Statement: Not applicable.

Conflicts of Interest: The authors declare no conflict of interest.

References

1. Yang, G.-L.; Li, R.-G. Diseases Analysis on an Old Arch Bridge and Reinforcement Design and Construction. *China Concr. Cem. Prod.* **2013**, *33*, 71–74.
2. Qi-Lin, H. Disease analysis and reinforcement of Yongan Daxi Bridge. *Highway* **2006**, *51*, 90–92.
3. Luo, Y.; Wu, Z.-L.; Tang, G.; Mao, J.-P. Analysis and Treatment of Crack Disease in Existing Highway Bridge with Hollow Slab. *J. China Foreign Highw.* **2020**, *40*, 151–153.
4. Bin-Jie, Z.; Ping, Y. Analysis of horizontal displacement disease and reinforcement technology of pile bearing bridge abutment during operation. *Highway* **2021**, *66*, 385–389.
5. Bo, S.; Deng-Yan, Z.; Ming, Y. Hollow plate disease analysis and improved design. *Highway* **2016**, *61*, 105–108.
6. Pu, G.-N.; Zhong, R.; Meng, T.-L.; Ding, X.-L. Cause analysis and countermeasure of 30 m prestressed concrete hollow slab. *Highway* **2008**, *53*, 175–179.
7. Zhang, K.-P.; Jiang, Y.-L.; Zeng, X.-F. Development and prospect of bridge reinforcement. *Highway* **2005**, *50*, 299–301.
8. Wei, J.-G.; Huang, L.; Li, Y.-P.; Wu, Q.-X. Research on Continuous Transformation and Reinforcement for Old Simply Supported Hollow Slab Bridge. *J. Archit. Civ. Eng.* **2014**, *31*, 103–109.
9. Guo, R.; Wang, R.-X.; Zhao, S.-W. Experiment of Bending Performance of Hollow Slab Beam Strengthened by Planting Bars and External Prestressing. *J. Highw. Transp. Res. Dev.* **2013**, *30*, 40–45.
10. Fu, X.-R.; Hu, C.-Z.; Gao, H.-B. Research on Anchoring External Prestressing Tendons with Wedges in Long-Span T Beam Strengthening. *World Bridges* **2021**, *49*, 101–106.
11. Zhou, J.-G.; Xu, Z.-M. Repair of fatigue cracks in steel box girders by adhesive steel plate technology. *Highway* **2020**, *65*, 224–230.
12. Xu, X.-Y.; Jia, Y.; Shi, G.-S. Application Analysis of Ultra-high Performance Fiber Reinforced Concrete in Highway Bridge Reinforcement. *Highw. Eng.* **2020**, *45*, 92–95+135.
13. Yang, X.-F. Comparative Analysis for Reinforcement Methods of Prestressed Concrete Bridges with Hollow Slabs. *J. China Foreign Highw.* **2019**, *39*, 94–96.
14. Yang, N.; Sun, Q.-S. Experimental Study on Mechanical Properties of Carbon Fiber Polyurethane Cement. *Polyurethane Ind.* **2019**, *34*, 35–38.
15. Zhang, K.X.; Sun, Q.S. Research on mechanical properties of the high-toughness polyurethane-cement composite (PUC) materials. *New Build. Mater.* **2018**, *45*, 126–128.
16. Liu, J.-C.; Xu, C.-C.; Jiang, K.; Long, G.-X.; Wang, Y.-C.; Li, Z.-H. Influencing Factors on the Properties of Ultra-High Toughness Polyurethane Composite. *Eng. Plast. Appl.* **2021**, *49*, 104–108.
17. Yang, N.; Sun, Q.-S.; Li, C.-W. Experimental Study on Pressure Sensitive Characteristics of Carbon Fiber Reinforced Polyurethane Cement Composites. *Sci. Technol. Eng.* **2018**, *18*, 27–32.
18. Zhou, Y.-H. Basic study on rapid reinforcement of railway Bridges with MPC composite materials. *Highway* **2017**, *62*, 271–275.
19. Zhang, L.-D.; Sun, Q.-S. Application of Simply Supported T-Beam Strengthened by MPC High Toughness Polymer. *Highway* **2016**, *61*, 90–94.
20. Gu, D.-D.; Sun, Q.-S. Study of Static Tests for Using MPC Composite Material to Strengthen Void Plate Girder Bridge. *World Bridges* **2015**, *43*, 88–92.
21. Wang, J.-L.; Liu, G.-W.; Ye, L.S. Research on application technology of reinforced hollow slab bridge with MPC composite material. *Highway* **2013**, *58*, 39–43.

Disclaimer/Publisher's Note: The statements, opinions and data contained in all publications are solely those of the individual author(s) and contributor(s) and not of MDPI and/or the editor(s). MDPI and/or the editor(s) disclaim responsibility for any injury to people or property resulting from any ideas, methods, instructions or products referred to in the content.

Article

The Influence of Substitution of Fly Ash with Marble Dust or Blast Furnace Slag on the Properties of the Alkali-Activated Geopolymer Paste

Brăduț Alexandru Ionescu [1], Alexandra-Marina Barbu [2,3,*], Adrian-Victor Lăzărescu [1,*], Simona Rada [4], Timea Gabor [4] and Carmen Florean [1,4]

[1] NIRD URBAN-INCERC Cluj-Napoca Branch, 117 Calea Floresti, 400524 Cluj-Napoca, Romania
[2] NIRD URBAN-INCERC, 266 Sos. Pantelimon, 021652 Bucharest, Romania
[3] Doctoral School, Technical University of Civil Engineering Bucharest, 122-124 Lacul Tei Bvd., 020396 Bucharest, Romania
[4] Faculty of Materials and Environmental Engineering, Technical University of Cluj-Napoca, 103-105 Muncii Boulevard, 400641 Cluj-Napoca, Romania
* Correspondence: alexandra.barbu@incd.ro (A.-M.B.); adrian.lazarescu@incerc-cluj.ro (A.-V.L.)

Abstract: Worldwide, it is now known that industrial by-products rich in silicon (Si) and aluminum (Al) can be transformed by alkaline activation into so-called "green concrete", an efficient and sustainable material in the field of construction; the geopolymer material. In this work, geopolymer materials produced using fly ash and marble dust or blast furnace slag were studied to assess the influence of these substitutions on the performances of the final product. Geopolymer materials have been characterized by physico-mechanical methods, FTIR spectroscopy and microscopically. The analysis of the results indicates the reduction of the mechanical strength performance by substituting the fly ash as the raw material.

Keywords: fly ash; alkali-activation; blast furnace slag; geopolymerization

1. Introduction

Fly ash (FA) is a by-product resulting from the coal burning in thermal power plants. Millions of tons of fly ash are produced globally. Fly ash is partially eliminated in the dumps and waste storage, but also in the atmosphere, when stored unproperly. Current policies at the global level require, on the one hand, the reduction of pollutant emissions, and on the other hand, the identification of new possibilities for sustainable use of non-renewable raw materials, with special emphasis on the implementation of the main concept regarding sustainable development of construction Materials Circular Economy [1–5]. The use of fly ash as a raw material in the production of new, "ecological concrete", by exploiting the alkaline geopolymerization mechanism, actively and efficiently responds to the need for sustainable development and the implementation of the Circular Economy principles. Moreover, when compared to the ordinary Portland cement concrete, one of the advantages of heat-treated geopolymers are a shorter gain of mechanical properties' time (generally seven days), while for cementitious composites, the hydrolysis-hydrolysis reactions of cement are continuous [1,2]. The resulting geopolymer product has a three-dimensional structure, similar to that of natural silico-aluminous materials [1–3].

The ecological concrete concept includes a whole series of new techniques, the common feature of which is to minimize the impact on the environment, either directly in the manufacturing process, or indirectly, contributing through its use to the reduction of other types of errors, as well as reducing costs. Another common element is the nature of the performance of these techniques, which must ensure similar mechanical behavior to the technique they replace to be validated, as this will affect durability and subsequent conservation costs.

There is no doubt that recently, and increasingly in the future, the recycling of construction products is promoted by public administrations, both for economic and environmental reasons [1–5].

Ecological concrete plays an important role in the field of civil engineering. This technology can effectively improve the internal structure and improve the mechanical properties and durability of the material. Among the many advantages of using it in construction are the following: reduced costs for maintenance and repair, increased life of structures with lower costs, cement economy by not justifying such a large number of new constructions, reduction of green-house gas emissions, use of cheap and environmentally friendly materials, local (fly ash, blast furnace slag), as well as reducing the consumption of non-renewable resources. The reduction of costs will be made based on increasing the period between two repairs, without the safety of the structures suffering. Moreover, it will reduce the environmental problems related to the intervention works and the waste resulting from this activity [1–5].

The geopolymerization process is a heterogeneous chemical reaction between a silico-aluminum solid material and a hydroxydic solution, strongly alkaline [2] and is based, according to Duxon et al., [3] on the following stages: dissolution, oligomerization and geopolymerization. The alkaline elements involved in the geopolymerization process (Na and K) have the role of generating a pH high enough to activate the solid material in the reaction and balancing the chemical species formed in the mold of the silico-aluminous gel [4]. Dissolution is the stage at which the Si-O-Si, Al-O-Al and Al-O-Si bonds are broken. The above-mentioned chemical bonds exist in the source material of solid aluminosilicate (fly ash (FA), used as a raw material). The release of silicate and aluminium elements occurs into the liquid phase, most likely in the form of monomers [3,5]. The oligomerization process consists in the formation of oligomers, which are molecules that form the 3D networks of the geopolymer binders. These elements are considered the main binding units that produce the geopolymerization process [3,6]. When polycondensation occurs, during the stage of oligomerization, the coagulated structures of the geopolymer are formed by the dissolution of the monomers. During this process, several different structures are formed: tetramers, dimers, trimmers, and larger molecules of the polymeric covalent bond [3], called oligomers. Several oligomers that form the geopolymer binder are Poly (siloxane) Si-O-Si-O, Si-O-Al-O, and poly (silate-disiloxo) Si-O-Al-O-Si-O-Si-O [5]. According to Koleżyński et al., 2018 [7], 2 or more basic (primary) oligomers built from atoms of Si, Al, O, Na and H, can connect by the means of 2, 3 or more oxygen bridges, forming complex oligomers; the geopolymer structural models made can reach about 200 atoms, having different Si:Al ratios: 4, 5, 6 and 10 [7]. Geopolymerization is the stage at which a rearrangement and binding of oligomers takes place, leading to the formation of a three-dimensional aluminosilicate network which further generates the formation of the geopolymer binder [3]. A structural geopolymer model may include more than 800 atoms with a Si:Al ratio of 2.81 [7]. In general, the chemical formula of a geopolymer is $Mn[-(SiO_2)Z-AlO_2]nwH_2O$, where M is a cation of Na, K, Ca, or Li; n is the degree of polycondensation; w represents the number of water molecules in the system and z is 1, 2, 3, or a number much greater than 3 [8]. Therefore, it is currently known that a parallel cannot be drawn between the mechanism for strengthening the grout that underlies the understanding of the structure of cement stone and concrete and the mechanism of geopolymer formation that underlies the understanding of the structure of the so-called "green concrete". If in the case of cement paste the hardening occurs through hydration—hydrolysis reactions of calcium-based compounds, silicates, and aluminates, in the case of geopolymer paste, the calcium oxide component does not have a significant contribution, the main components of the raw material really valuable in this case being the suppliers of Si and Al.

Literature has reported numerous studies in which the geopolymerization mechanism process was assessed using different analytical and instrumental techniques [5,9,10]. FTIR analysis of the geopolymer binder, at different time intervals and different curing methods, is considered an optimal tool to identify the gelling mechanism of the geopolymer binder.

Several researchers [5,9–11] have studied the geopolymerization process using different raw materials in the alkali-activation process and different curing stages. The formation of the geopolymer gel in two stages was initially suggested by Fernandez-Jiménez et al. [12]. Results have shown that the geopolymerization process began with the formation of an aluminum-rich gel because of the dissolution of the weaker Al-O bond, when compared to the Si-O bonds. This mainly happened because the concentration of Si was higher than Al and its leaching began later due to the more stable Si-O connection. The Si-O bonds present in the second phase were substantially higher, which led to the geopolymerization. In terms of FTIR analysis, studies have shown that the geopolymerization process begins with the initial occurrence of a peak at 1003 cm^{-1}, which represents the formation of the aluminum-rich gel (gel A) [10]. Afterwards, the gel is converted into a gel rich in Si (gel B) represented by the appearance of a peak at 1018 cm^{-1}. The current mechanism was found consistent with the modelling study of the geopolymerization process assessed in the current study. Similar results were provided by other studies in the literature, using more precisely techniques such as ATR-FTIR (attenuated total reflection) and FTIR techniques [9,10].

To study the chemical bonds of Si, Al, O, H and alkaline cations, as important parameters of geopolymers, Fourier Transformation Infrared Spectroscopy (FTIR) were used to assess the formation and microstructure of geopolymer gels. FTIR is a non-destructive method used for the characterization of materials' microstructure by measuring the infrared spectrum of absorption of a solid, gas or liquid in the range of 4000–400 cm^{-1} [13]. Generally, most of the bonds occur in FITR models in wide bands or in the form of hangers [13]. FTIR bands regarding Si-O-T bonds (T is Si or Al) show an asymmetrical vibration band in the range 900–1300 cm^{-1} [14–19]. This is usually attributed to the main strip of geopolymer gels. The main band can be attributed to the raw materials used in the production of the geopolymer binder, since they are also rich in Si and Al, with wave numbers close to 1000 cm^{-1}. Analyzing the adjustment of the peaks and the deconvolution of the main peaks provided valuable information about the nature and extent of the link in geopolymers [14–17]. Studies have shown that the wave number of the main band in geopolymers is lower than the one characterizing the raw materials [9,10,14–17]. Therefore, the main bands of raw materials and geopolymers are always compared with each other in terms of this parameter, in order to assess the formation of geopolymer gels [14,15]. The main band moves to a lower wave number when the geopolymer is formed and this change can be seen in the FTIR analysis with result changes from 1054, 1080 or 1100 cm^{-1} in raw materials at a wave number less than 1000 cm^{-1} [14,20,21]. In terms of geopolymer binder production, another band can be formed around 840–900 cm^{-1} and it can be assigned to the formation of the Si-OH link. This confirms the formation of the geopolymer gel by its appearance, being observed only in geopolymers, but not in raw materials [14,22–24]. The FTIR peaks commonly seen in geopolymers include analyzed bands with a range between 1440 cm^{-1} (which represent sodium bicarbonate), around 3500 and 1600 cm^{-1} (for water) and O-H bonds. During the geopolymerization process, the sodium bicarbonate is formed when Na$^+$ and atmospheric CO$_2$ react. This occurs when the geopolymer samples are exposed to air [12,24,25]. Subsequently, water is removed through the pores or absorbed by the surface. The decrease in water strips phenomenon can be used to observe the hardening mechanism of geopolymer samples [12,24]. Results obtained in the literature regarding the geopolymerization process, analyzed using FTIR analysis, are presented in Table 1. Although results in the literature are consistent, there are several contradictions regarding the interpretation of FTIR bands. Symmetrical Al-O bonds were assigned to several values [14,23,24]. Reports from the literature, in agreement with the fact that the appearance of the bands centered around 870 cm^{-1}, can be used to indicate the formation of geopolymer gels [23,24].

Table 1. Assignment of IR bands from the geopolymer structure according to the specialized literature.

Bond	Wave Number (cm^{-1})	Ref.
	~3500 and ~1600	[12,24,25]
Symmetric and antisymmetric O-H bond elongation vibrations in water	3445	[26]
H-O-H deformation vibrations in water	3400–3650	[27]
Antisymmetric O-H bond elongation vibrations in water	1650	[28]
	1640	[9]
Elongation vibrations of the C-O bond in the carbonate ion	~1440–1453	[12,25]
Antisymmetric elongation vibrations of the Al-O bond	~1400	[19]
Vibrations of antisymmetric deformation of angles Si-O-Al	1180	[17]
	990	[29]
Antisymmetric deformation vibrations of the Angles Si-O-Si (from quartz and mulit)	1100	[11]
Si-O-T angle deformation vibrations (T=Si or Al from silicate alumino gel)	1025–1091	[30]
Antisymmetric deformation vibrations of angles Si-O-T (T=Si or Al)	900–1300	[14–19]
Symmetrical elongation vibrations T-O-Si (T=Si or Al)	1020	[31]
Si-O/Al-O of the aluminosilicate network reflecting the formation of amorphous aluminosilicate gel in binary systems	1015	[32]
Si-OH bond elongation vibrations	840–900	[14,22–24]
Elongation vibrations of the Si-O link	~800–810	[10,24]
Al-O bond elongation vibrations in AlO$_4$	750–900	[10,19,25,29]
	680	[14]
Vibration deformation Si-O-Al	700	[25,33]
Al-O bond elongation vibrations	667	[34]
Si-O bond elongation vibrations	575	[28]
	530	[26]
Si-O-Al angle deformation vibrations	569	[25,33]
Si-O-T deformation vibrations (T=Si or Al) bending	540–555	[33]
Elongation vibrations of the T-O bond (T=Si or Al)	475	[24,29]
O-Si-O deformation vibrations	454	[12]

The Circular Economy, an integral part of the concept of sustainable development, is based on a series of principles that can be summarized in the 4Rs (Reduce, Reuse, Recycle, Recover) [35–38]. However, it should be noted that during the evolution of this concept there has been a correlation with several other principles that underpin European good practice, one of which is found in international environmental law, the Precautionary Principle (PP) [35–39]. At present, it is not possible to talk about the successful implementation of the Circular Economy without considering contextual links with guiding principles of international environmental law [40,41]. Moreover, the implementation of the Circular Economy cannot be achieved in a single sector of activity and, moreover, this approach would not be successful because if in one sector of activity the product is waste or an industrial by-product, in another sector the same product may be a valuable and under-exploited raw material. An example of this is blast furnace slag, a useful raw material in the development of geopolymer mixtures, but currently seen as a huge polluting waste dump. In the case of Romania, a specific case that supports this example is that of the SOMETRA non-ferrous metal plant, Copșa Mică, Sibiu County, which, at the time of sale, had a clause in the contract requiring the buyer to "green the slag waste dump". This contractual obligation concerned "the greening of the current landfill by waterproofing the slopes, covering it with a layer of clay, covering it with a layer of fertile soil and fixing it with a vegetable carpet" [42–46], followed by "the construction of a new ecological landfill on a new site with waterproofing of the substrate, a drainage and collection system for the rainwater percolating down the ramp, a neutralization and dewatering station for the resulting water and its discharge into a surface watercourse" [42–46]. According to Mangau et al. [35], respecting the principle of prevention would have required a risk analysis of the complex situation in this case, since, in addition to the main waste, blast furnace slag, there are several other atmospheric, soil and/or water pollutants, the entire ecosystem being affected, and there is also a major risk factor in terms of the health of the population [42–46]. When considering slag properties, it is of interest to analyze its chemical composition to assess the possibility of using this waste as a raw material in the production of alkali-activated geopolymer materials, namely the so-called "green concrete".

Similarly, fly ash, waste from thermal power plants, with its specific chemical composition, induces several environmental problems and storage costs [47–55]. Its contamination with various elements such as arsenic, barium, beryllium, boron, cadmium, chromium, thallium, selenium, molybdenum and mercury or traces of heavy metals, make it difficult to store it safely to eliminate the risk of contamination of soil or groundwater. Research in recent years has also focused on evaluating the possibilities of recycling this waste by using it in geopolymer composites [47–55]. In Romania there are several thermal power plants, most of which are currently equipped with electrostatic precipitators that have the capacity to filter the fly ash removed with the flue gases (e.g., Govora Power Plant, Vâlcea County, Romania) [47]. This, once again, supports the possibility and the need to identify more and more possibilities to consider this product as a raw material and not as waste.

Therefore, also considering the specifics of the local problem and the chemical characteristics of the Romanian raw materials, the aim of this research is to analyze the importance of the chemical composition of the raw materials used in the preparation of alkali-activated geopolymers, on the polymerization reaction products, on the microstructure and on the physical—mechanical performance of the binder.

2. Materials and Methods

2.1. Materials

Raw materials from Romania were used to produce the alkali-activated geopolymer binder. The raw material used as the main geopolymerization material was a low-calcium fly ash, obtained from the Rovinari Power Plant, Gorj County, Romania.

In order to assess to possibility of using other waste/by-products in the production of geopolymer materials, fly ash was partially substituted with marble dust (MD), obtained from the Ruschița marble deposit, Caraș-Severin County, Romania and blast furnace slag (BFS), obtained from the ArcelorMittal Steel Factory, Galați County, Romania. The waste samples were sieved to obtain a maximum particle size of 0.063 mm. The chemical composition of the raw materials used in the mixtures was established by X-ray fluorescence analysis (XRF analysis) (Table 2).

Table 2. Raw materials' chemical composition.

Oxides	Fly Ash (FA) (wt.%)	Marble Dust (MD) (wt.%)	Blast Furnace Slag (BFS) (wt.%)
SiO_2	46.94	0.28	30.20
Al_2O_3	23.83	1.37	10.05
Fe_2O_3	10.08	0.17	14.70
CaO	10.72	54.63	37.40
MgO	2.63	0.43	4.05
SO_3	0.45	-	-
Na_2O	0.62	-	0.20
K_2O	1.65	-	0.38
P_2O_5	0.25	-	-
TiO_2	0.92	-	<0.52
Cr_2O_3	0.02	-	<0.05
Mn_2O_3	0.06	-	2.15
ZnO	0.02	-	-
SrO	0.03	-	-
CO_2	-	42.65	-
L.O.I. *	2.11	0.37	-
$SiO_2 + Al_2O_3$	70.77	1.65	40.25

* Loss on Ignition.

2.2. Methods

The alkaline activator used in the production of the geopolymer samples was prepared using 8M sodium hydroxide solution (NaOH) and aqueous solution of sodium silicate (Na_2SiO_3) with a concentration of 35%–40%. The mass ratio between the two solutions was $Na_2SiO_3/NaOH = 2.5$. The NaOH solution was prepared in the laboratory by dissolving 99% purity NaOH pearls in distilled water until the desired molar concentration was obtained (e.g., 320 g of NaOH pearls were dissolved in water, for one liter of solution to obtain 8M NaOH solution) and the Na_2SiO_3 was conditioned at 23 ± 2 °C. The alkaline activator solution was prepared 24 h prior mixing. The mix-design ratio for the alkali-activated geopolymer binder samples is presented in Table 3.

Table 3. Alkali-activated geopolymer binder mix design.

Material	Mixture	NaOH (M)	FA (wt.%)	MD or BFS (wt.%)	$Na_2SiO_3/NaOH$	AA/ (FA + MD/BFS)
FA	C	8	100	-		
Samples subjected to heat treatment (70 °C for 24 h)						
FA + MD	P1	8	90	10		
	P2		75	25		
	P3		50	50		
FA + BFS	P4	8	90	10	2.5	0.78
	P5		75	25		
	P6		50	50		
Samples subjected to laboratory conditions						
FA + MD	P7	8	90	10		
	P8		75	25		
	P9		50	50		
FA + BFS	P10	8	90	10		
	P11		75	25		
	P12		50	50		

After casting into 40 × 40 × 160 mm molds, with a corresponding 10 min vibration, control sample C and samples P1–P6 were subjected to heat treatment (holding at 70 °C for 24 h), in order to study the effect of the fly ash substitution on the mechanical properties of the material.

Subsequently, samples with the partial substitution of fly ash with marble dust and blast furnace slag were also produced (P7–P12) but were not subjected to heat treatment. The samples were stored in laboratory conditions to study the effect of heat treatment on the mechanical properties of the material. The demolding of these samples was achieved only 48 h after casting because the lack of heat treatment caused a delay in the geopolymerization process.

After demolding, the geopolymer samples were stored in laboratory conditions at the temperature T = 23 ± 2 °C and relative humidity RH = 60 ± 5% until mechanical strength tests were conducted (7 days).

The density of the geopolymer samples was measured by weighing the samples and relating them to their volume. Initially, the density of the samples was measured at the end of the mixing, according to EN 12350-6:2019 [56]. After demolding (24 h for samples subjected to heat treatment and 48 h for samples stored in laboratory conditions), density was measured by weighing the samples and relating them to their volume (according to EN 12390-7:2019 [57]. Before conducting the mechanical strength tests (7 days), apparent density was also measured for each sample.

To obtain results regarding the mechanical performances of the geopolymer binder, a minimum of three samples were tested to determinate the average value of the assessed parameters. Tests were performed at the age of 7 days. The flexural strength of the alkali-activated geopolymer paste samples was determined by adopting the three-point bending (3PB) test, according to EN 196-1:2016 [58], the standard method for evaluation of mechanical performances of OPC paste and standard type mortars. Using the half prismatic test specimens resulting from the three-point bending test (3PB), compressive strength of the samples was determined according to the same standard on 40 × 40 mm samples [58].

Subsequently, geopolymer samples were subjected to infrared analysis using the Fourier Transformed Method (FTIR) to assess their microstructure. Laboratory tests by the FTIR method were performed using a Jasco FT/IR-6200 Fourier Transform Infrared Spectrometer apparatus (JASCO, Tokyo, Japan). Subsequently, a microscopic analysis was performed using a Leica DNC2900 optical stereomicroscope (Leica, Wetzlar, Germany), to assess possible changes in terms of pore size and distribution in the geopolymer matrix, as well as its homogeneity.

The analysis and interpretation of the results were performed both from the point of view of the values recorded for each parameter separately, as well as the increase/decrease compared to the control sample for alkali-activated geopolymer samples subjected to heat treatment. Discussions were also made based on comparing the results obtained on samples subjected to heat treatment and the ones conditioned in laboratory conditions to assess the influence of this parameters on the mechanical properties of the material.

3. Results and Discussions

3.1. Influence of Fly Ash Substitution with Marble Dust on Alkali-Activated Geopolymer Samples Subjected to Heat Treatment

The effect of fly ash substitution on alkali-activated geopolymer samples subjected to heat treatment was evaluated on samples P1–P3. The amount of fly ash substituted by marble dust was 10%, 25% and 50%. The results in terms of apparent density, flexural strength and compressive strength are graphically presented in Figure 1. All results obtained are compared to the control sample, produced using only fly ash as a raw material in the production of the alkali-activated geopolymer binder.

When analyzing the geopolymer samples subjected to heat treatment, as the quantity of fly ash (FA) was substituted with marble dust (MD), the following are observed:

- the fresh-state density increases by 6.3% to 15.8%.
- the density of the geopolymer paste immediately after completing the 24 h heat treatment increases by 6.8% to 18.4%.
- the mechanical properties of the geopolymer paste at 7 days are reduced by up to 27.8% in the case of flexural strength values, and by a minimum 11.3% and a maximum 66.7% in the case of compressive strength.

3.2. Influence of Fly Ash Substitution with Blast Furnace Slag on Alkali-Activated Geopolymer Samples Subjected to Heat Treatment

The effect of fly ash substitution on alkali-activated geopolymer samples subjected to heat treatment was evaluated on samples P4–P6. The amount of fly ash substituted by blast furnace slag was 10, 25 and 50%. The results in terms of apparent density, flexural strength and compressive strength are graphically presented in Figure 2. All results obtained are compared to the control sample, produced using only fly ash as a raw material in the production of the alkali-activated geopolymer binder.

When analyzing the geopolymer samples subjected to heat treatment, as the quantity of fly ash (FA) was substituted with blast furnace slag (BFS), the following are observed:

- casting density increases by 2.5% to 16.5%.
- the density of the geopolymer paste immediately after going through the 24 h of heat treatment, increases by 5.4% to 16.3%.
- the density of the geopolymer paste at 7 days after mixing increases by 7.4% to 33.8%.
- the mechanical properties of the geopolymer paste. at 7 days after casting, are reduced by minimum 4% and maximum 78.3% in terms of flexural strength and by minimum 21% and maximum 64.3% in case of compressive strength.

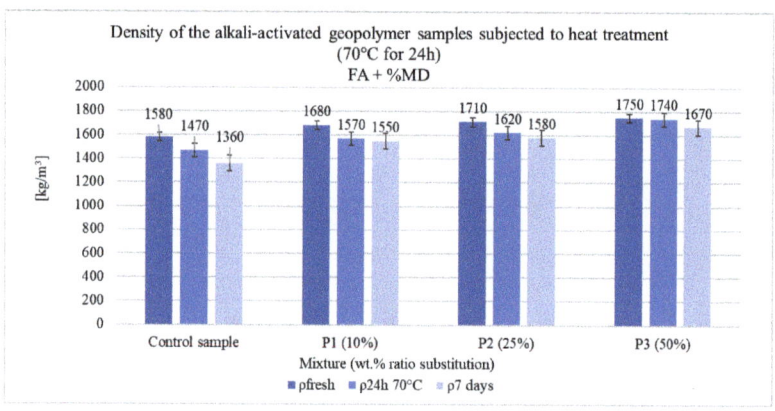

(a)

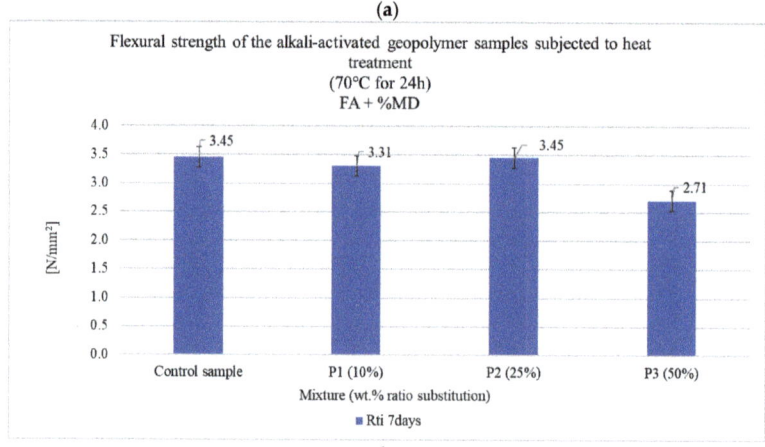

(b)

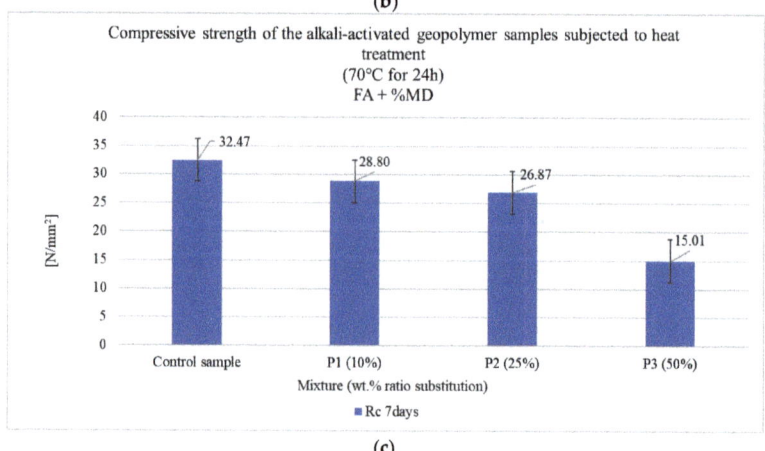

(c)

Figure 1. Physico-mechanical performances of the alkali-activated geopolymer binder using MD as FA substitution subjected to heat treatment: (**a**) Apparent density; (**b**) Flexural strength and (**c**) Compressive strength.

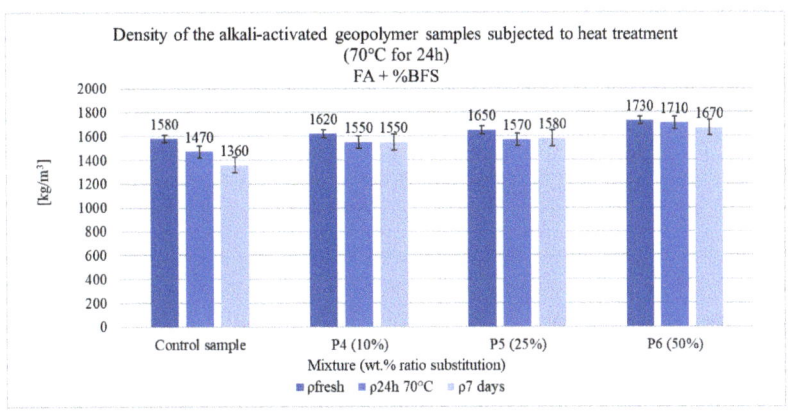

(a)

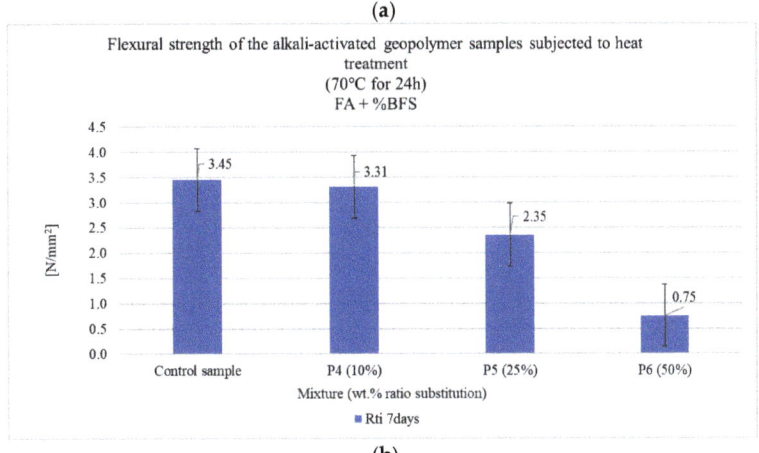

(b)

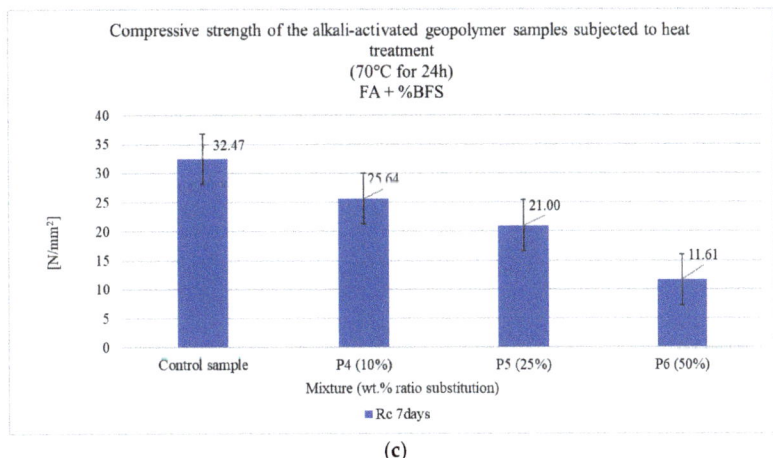

(c)

Figure 2. Physico-mechanical performances of the alkali-activated geopolymer binder using BFS as FA substitution subjected to heat treatment: (**a**) Apparent density; (**b**) Flexural strength and (**c**) Compressive strength.

By analyzing Figures 1 and 2, it can be stated that by increasing the amount of the fly ash substitution with 10%, 25% and 50% of the total amount of dry material (fly ash + substitution), the physico-mechanical properties of the material do not vary proportionally with the percentage of the substitution. Thus, when analyzing the apparent density of the mixtures, they increase, as the amount of marble dust (MD) or blast furnace slag (BFS) increases in the mixture, by a maximum of 11%.

Compared to the same situation (mixtures P1 and P4), the mechanical properties are strongly influenced by the increase in the amount of the raw activating material, less if the substitution is the marble dust and more if the substitute is the blast furnace slag. When assessing marble dust as substitution, a decrease of a maximum 18% is observed in terms of flexural strength and 48% in terms of compressive strength. When using blast furnace slag as substitution, a decrease of a maximum 77% is observed in terms of flexural strength and 55% in terms of compressive strength.

3.3. Influence of Fly Ash Substitution with Marble Dust on Alkali-Activated Geopolymer Samples Subjected to Laboratory Conditions

The effect of fly ash substitution on alkali-activated geopolymer samples subjected laboratory conditions was evaluated on samples P7–P9. The amount of fly ash substituted by marble dust was 10, 25 and 50%. The results in terms of apparent density, flexural strength and compressive strength are graphically presented in Figure 3.

When analyzing the geopolymer samples subjected to laboratory conditions, as the quantity of fly ash (FA) was substituted with marble dust (MD), the following are observed:
- in terms of density, it can be seen that it remains constant within the analysis limits for each sample.
- in terms of the percentage increase in substitution, the density increases as the amount of thermal power plant ash is substituted with marble dust.
- flexural strength increases from 2.49 N/mm^2 for samples produced with a 10% fly ash substitution to 3.25 N/mm^2 for samples with a 50% substitution.
- compressive strength decreases from 23.40 N/mm^2 for samples produced with a 10% fly ash substitution to 10.80 N/mm^2 for samples with a 50% substitution.

3.4. Influence of Fly Ash Substitution with Blast Furnace Slag on Alkali-Activated Geopolymer Samples Subjected to Laboratory Conditions

The effect of fly ash substitution on alkali-activated geopolymer samples subjected laboratory conditions was evaluated on samples P10–P12. The amount of fly ash substituted by blast furnace was 10, 25 and 50%. The results in terms of apparent density, flexural strength and compressive strength are graphically presented in Figure 4.

When analyzing the geopolymer samples subjected to laboratory conditions, as the quantity of fly ash (FA) was substituted with blast furnace slag (BFS), the following are observed:
- in terms of density, it remains constant within the analysis limits for each sample.
- in terms of the percentage increase in substitution, the density increases as the amount of thermal power plant ash is substituted with marble dust.
- flexural strength decreases from 3.73 N/mm^2 for samples produced with 10% fly ash substitution to 1.39 N/mm^2 for samples with 50% substitution.
- compressive strength decreases as the amount of fly ash is substituted. Particularly it can be seen that for a 25% substitution of FA with BFS a slight increase in compressive strength was obtained.

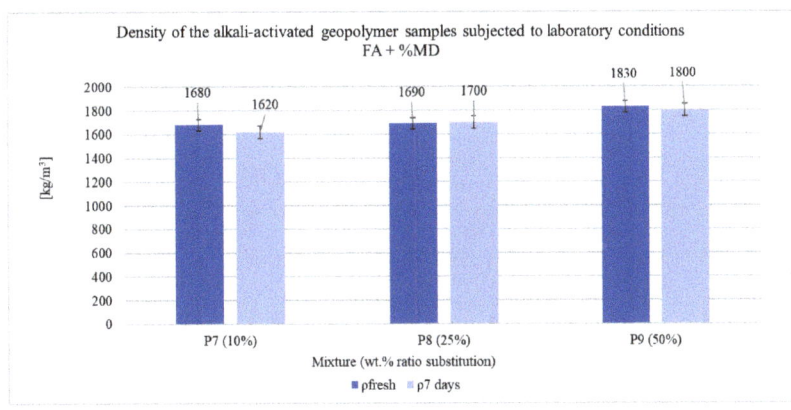

(a)

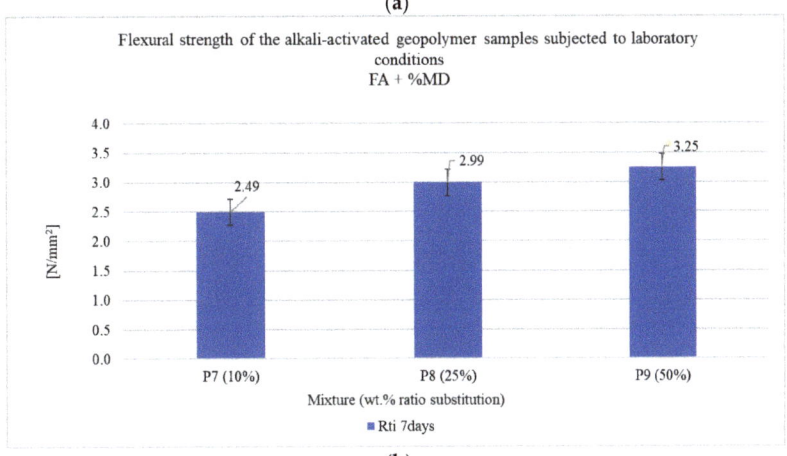

(b)

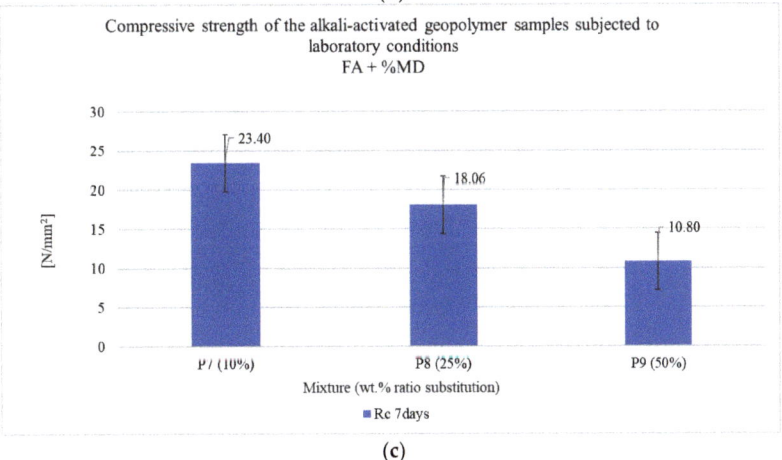

(c)

Figure 3. Physico-mechanical performances of the alkali-activated geopolymer binder using MD as FA substitution subjected laboratory conditions: (**a**) Apparent density; (**b**) Flexural strength and (**c**) Compressive strength.

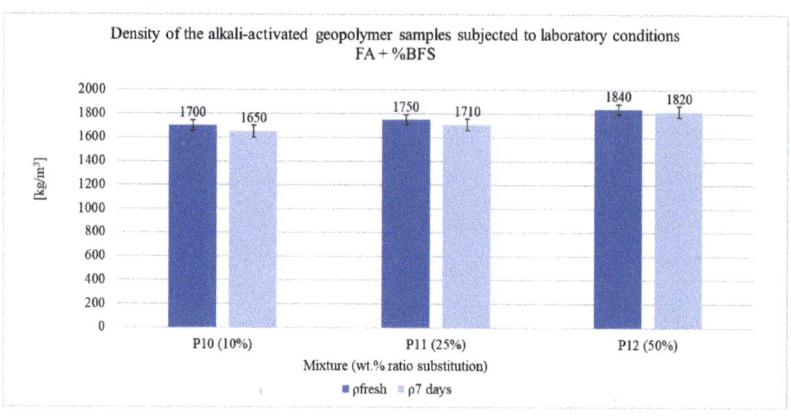

(a)

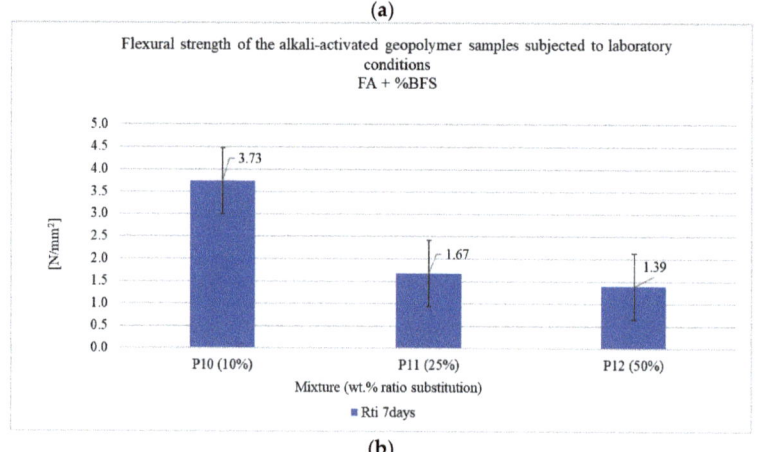

(b)

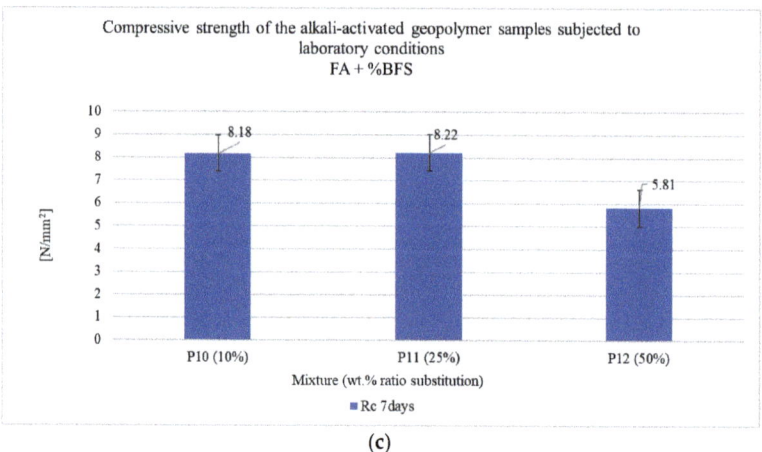

(c)

Figure 4. Physico-mechanical performances of the alkali-activated geopolymer binder using BFS as FA substitution subjected laboratory conditions: (**a**) Apparent density; (**b**) Flexural strength and (**c**) Compressive strength.

3.5. Influence of Heat Treatment on the Mechanical Properties of Alkali Activated Geopolymer Samples Produces Using Marble Dust as Fly Ash Substitution

The effect heat treatment on the mechanical properties of the samples using marble dust as substitution and was studied to assess the differences in mechanical strength. The results in terms of flexural strength and compressive strength are graphically presented in Figure 5.

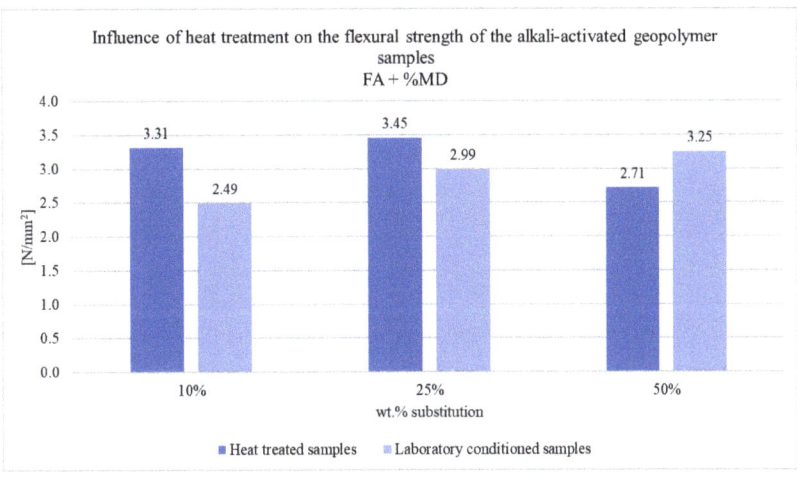

(a)

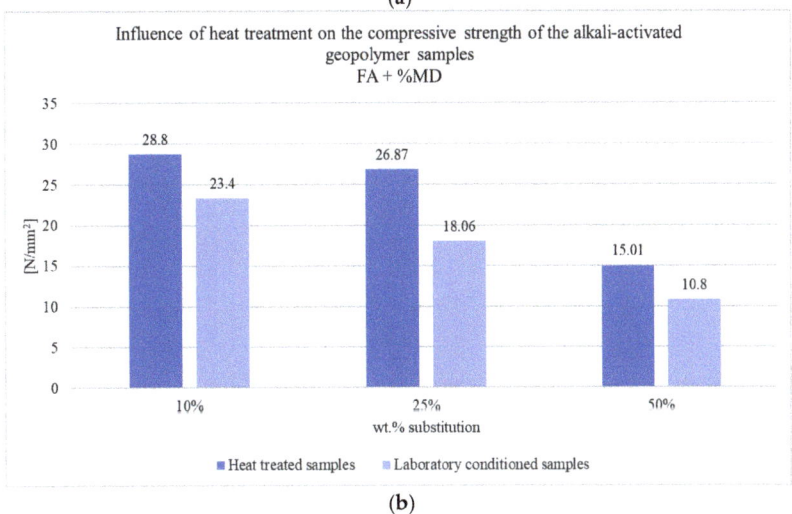

(b)

Figure 5. Influence of heat treatment on the mechanical properties of alkali activated geopolymer samples produces using marble dust as fly ash substitution: (**a**) Flexural strength and (**b**) Compressive strength

By analyzing Figure 5a, it can be seen that the flexural strength of the samples subjected to laboratory condition decreases when the amount of marble dust in the mixture increases from 10% to 25%. In the case of using marble dust as fly substitution up to 50% the flexural strength increases for the samples which were subjected to laboratory conditions.

In terms of compressive strength (Figure 5b), it can be seen that the values obtained for samples subjected to laboratory conditions are lower than the ones subjected to heat-treatment. These results are in accordance with the literature stating that heat treatment not only decreases the demolding time, but also generates a more powerful geopolymerization process, thus, resulting in better mechanical properties of the material [47–54].

3.6. Influence of Heat Treatment on the Mechanical Properties of Alkali Activated Geopolymer Samples Produces Using Blast Furnace Slag as Fly Ash Substitution

By analyzing Figure 6a, it can be observed that, as in the case of the samples produced using marble dust as fly ash substitution, in terms of flexural strength, samples present different variations. For samples with 10 and 50% blast furnace slag an increase in flexural strength is observed for samples subjected to laboratory conditions. For samples with 25% blast furnace slag substitution, a decrease in flexural strength was observed.

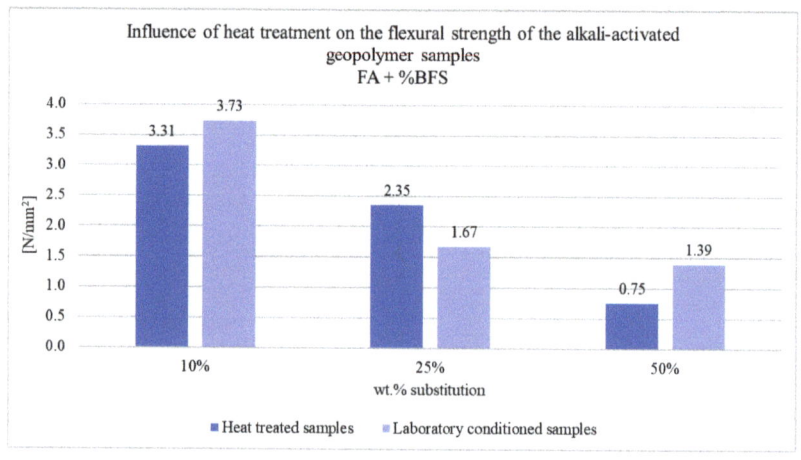

(a)

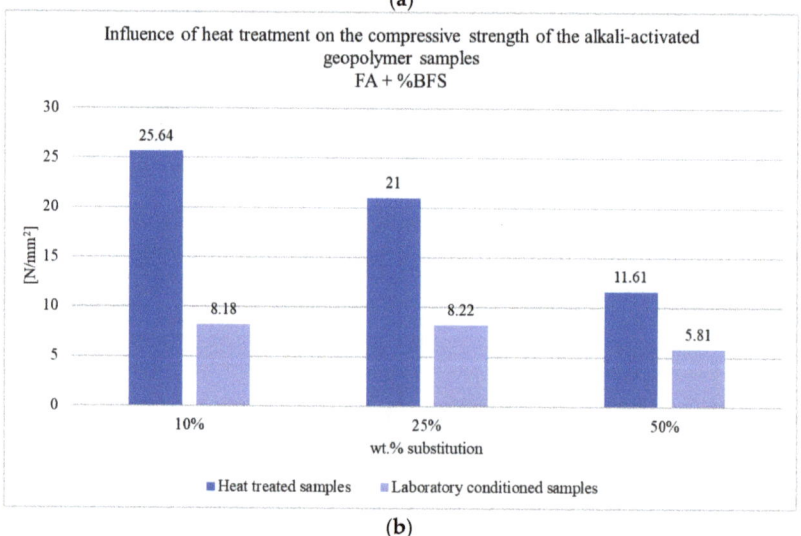

(b)

Figure 6. Influence of heat treatment on the mechanical properties of alkali activated geopolymer samples produces using blast furnace slag as fly ash substitution: (**a**) Flexural strength and (**b**) Compressive strength.

The compressive strength results obtained for samples produced using blast furnace slag as fly ash substitution show the same behavior: the compressive strength of samples subjected to heat-treatment are higher than the ones subjected to laboratory conditions (Figure 6b).

3.7. Influence of Geopolymer Binder Chemical Composition on the Mechanical Properties

The influence of the geopolymer binder chemical composition on the mechanical properties was established by reporting the above experimental results to the percentage content of SiO_2, Al_2O_3, Fe_2O_3 and CaO. The mass percentage of the elements was calculated cumulatively for the dry mixture of raw materials. By analyzing Figure 7, it can be said that the high CaO content is harmful to the physico-mechanical performances of the binder. Unlike the strengthening mechanism of the cement paste in which the CaO content is particularly important, in the case of the strengthening mechanism of the alkali-activated geopolymer a high percentage of CaO is harmful for the geopolymerization process. It is possible that a high CaO content will contribute to the reduction of the curing time of the alkali-activated geopolymer paste in the absence of heat treatment, but in terms of mechanical properties, the damaging effect far exceeds the benefit of reducing the curing time. Therefore, to obtain flexural and compressive strength as high as possible, the chemical analysis of the raw materials is particularly important, being able to identify even the possibilities of substituting fly ash with other materials, if they contribute efficiently through their own intake of SiO_2, Al_2O_3 and Fe_2O_3, the main participants in the geopolymerization mechanism.

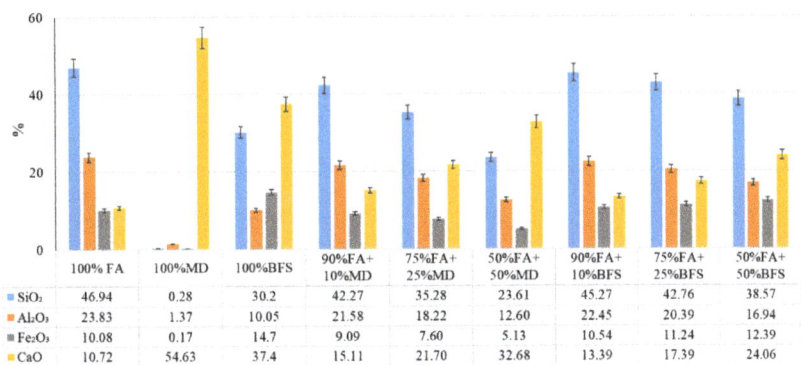

Figure 7. Geopolymer binder chemical composition for different mixtures (wt.%).

3.8. FTIR Analysis of the Alkali-Activated Geopolymer Samples

The FTIR analysis of the alkali-activated geopolymer samples produced using marble dust or blast furnace slag as fly substitutions are shown in Figure 8. The IR band of increased intensity, centered at 1026 cm^{-1} corresponds to deformation vibrations of the Si-O-Si or Si-O-Al angles in the alumino-silicate gel and is responsible for the formation of amorphous aluminosilicate gel in binary systems [32]. The IR band centered at 1443 cm^{-1} comes from elongation vibrations of the C-O bond in the carbonate ion. The IR band located at 3451 cm^{-1} is attributed to symmetrical and antisymmetric elongation vibrations of the H-O bond in water molecules.

Figure 8a shows the 10% and 50% substitutions of fly ash with marble dust (samples P7 and P8). In both cases the position of the IR band centered at 1026 cm^{-1} moves towards smaller numbers, reaching 1014 cm^{-1}. This displacement reflects the Si-O/Al-O bond formation of the aluminosilicate network characteristic of an aluminosilicate gel from binary systems with amorphous structure. The intensity of the IR bands decreases over the entire range between 400 and 4000 cm^{-1} by partially substituting fly ash with marble dust up to 50%. The intensity of the IR bands centered at 475 and 1100 cm^{-1} decreases in intensity, which suggests that the content of Al-O or Si-O bonds from different SiO_2 crystalline phases in quartz and Al_2O_3-SiO_2 in mullite decreases.

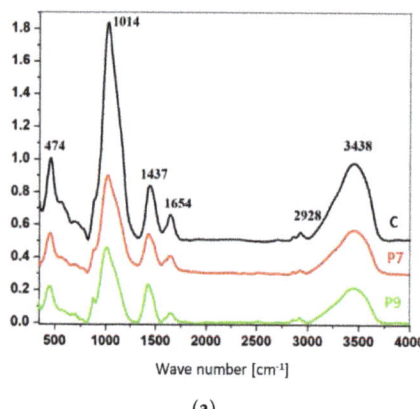
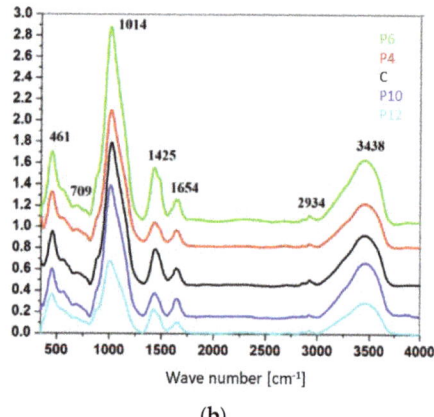

Figure 8. FTIR spectra of geopolymer pastes made with: (**a**) 10% and 50% fly ash substitution with marble dust; (**b**) 10% and 50% fly ash substitution with blast furnace slag.

For sample P7 (10% marble dust, subjected to laboratory conditions), the intensity of the centered band at 1400 cm^{-1} decreases which indicates a decrease in the number of Al-O bonds. The higher the marble dust content (up to 50%), a trend of displacement of the band located at 1014 cm^{-1} towards smaller wave numbers and a decrease in water content can be observed. This evolution highlights the formation of an aluminosilicate gel with a disordered structure, and by substituting the flying ash with marble dust, materials with a lower intake of crystalline phases SiO_2 or SiO_2-Al_2O_3 are obtained.

Figure 8b shows the IR spectra of the samples produced using 10% and 50% blast furnace slag as fly ash substitutions. By subjecting the samples to heat treatment (samples P4 and P6), the intensity of the IR bands increases. The IR bands centered at 1014 and 1400 cm^{-1} move towards higher wavelengths (1022 and 1425 cm^{-1}). The intensity of the IR band centered at 1100 cm^{-1} also increases. These structural developments indicate that the number of deformation vibrations of the Si-O-Al, Si-O-Si bonds and al-O elongation vibrations increases and as a result the degree of crystallinity of the gel increases. For samples subjected to laboratory conditions, the geopolymerization degree increases and the P12 sample indicates a lower absorption of water.

The vibrations of the bound water molecules, recorded in 3438, 2934/2928 and 1654 cm^{-1} are attributed to stretching (–OH) and bending (H–O–H), respectively [40–42]. These results can be seen for both types of geopolymer binder in Figure 8a (samples produced using marble dust as fly ash substitution) and in Figure 8b (samples produced using blast furnace slag as fly ash substitution).

The IR bands due to asymmetrical elongation vibrations of the Al-O (1400 cm^{-1}) and O-H (1654 and 3438 cm^{-1}) bonds are shifted to higher wavelengths with the partial substitution of the fly ash. According to the literature, when analyzing the samples produced only using fly ash, an indicator strip of sodium bicarbonate formation is recorded at 1443 cm^{-1}. The bond analyzed in correlation with the movement trend of the samples prepared with marble dust or blast furnace slag (Figure 8a,b), could be moved below the limit indicated in the literature (~1440–1453 cm^{-1}) [12,39], at wavelengths 1437 and 1425 cm^{-1}. In the case of the P4 sample, the sodium bicarbonate content is lower than for the P6 sample.

3.9. Microscopic Analysis of the Alkali-Activated Geopolymer Samples

The microscopic analysis (Figure 9) was performed using a Leica DNC2900 optical stereomicroscope, at 1× magnification, aiming at possible changes in terms of pore size and distribution in the composite matrix, as well as its homogeneity.

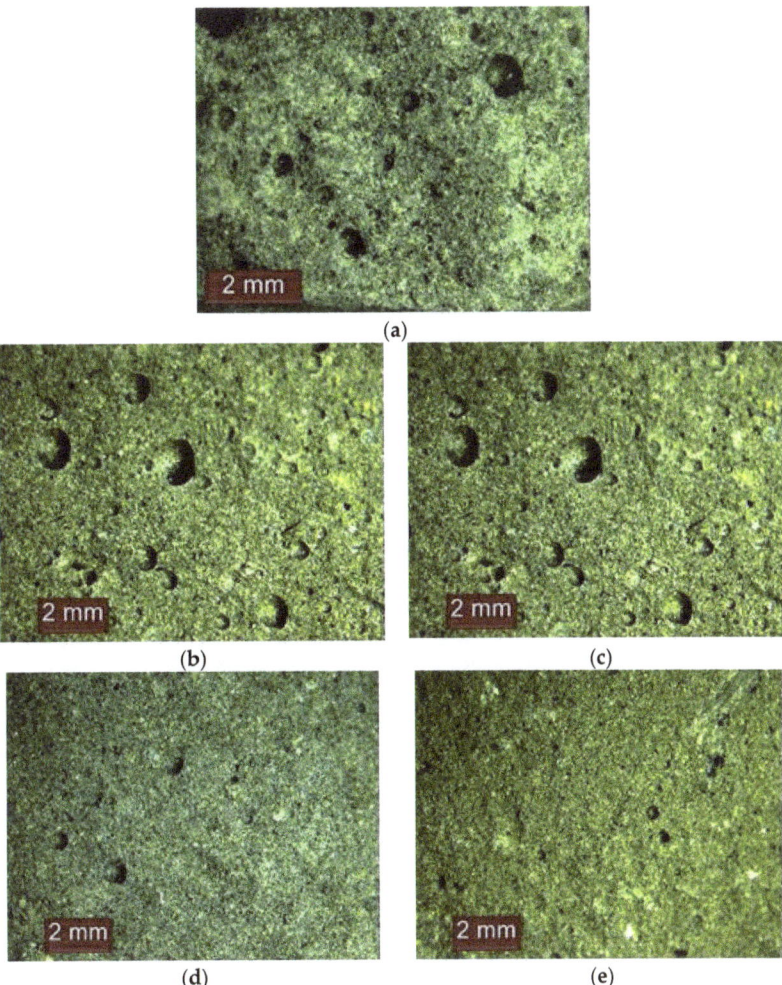

Figure 9. Microscopic images of geopolymer pastes with different substitutions, made at 1× magnification for mixtures: (**a**) C; (**b**) P1; (**c**) P7; (**d**) P6 and (**e**) P12.

As the substitution of fly ash with marble dust (sample P1) and with blast furnace slag (sample P4) increases, the dimensions and/or number of pores in the matrix increases. This can be attributed to the fact that the samples were subjected to heat-treatment.

Comparing the two microscopic figures (P1 vs. P7, and P6 vs. P12) in the case samples P1 and P6 (subjected to heat treatment), the size and number of pores are larger than in the case of sample P7 and P12 subjected to laboratory conditions. These results are in correlation with the lower bulk densities at 7 days of age. It is estimated that the heat treatment facilitates the formation of the geopolymer gel through a more uniform and homogeneous process, hence a better compressive strength, but influences the structural microscopy.

In the case of geopolymer samples subjected to laboratory conditions, their structure is much denser and the pores are smaller and fewer.

4. Conclusions

The aim of the work was to analyze the influence of the raw material on the properties of mechanical strength and on the reaction of geopolymerization by microstructural analysis.

As fly ash (FA) is replaced by marble dust (MD) or blast furnace slag (BFS) compared to the control sample, it is noticed that, with the increase in the quantity of the raw material, the fresh state density of the geopolymer paste and the hardened state density after the heat treatment and at the age of 7 days increases. In terms of mechanical performances results obtained on geopolymer samples showed that the parameters decrease as the quantity of fly ash in the mixture is substituted from 10 to 50%.

Increasing the amount of the substitution by 10%, 25% and 50% of the total amount of dry material (fly ash + substitute), it is observed that it is not possible to identify a proportionality function among the measurable indicators for physico-mechanical properties modifying the preparation mixture. Thus, in the case of densities, they increase, as the amount of marble dust (MD) or blast furnace slag (BFS) increases in the mixture, by a maximum of 11% compared to the values recorded for the situation with the lowest substitution of the fly ash (substitution 10%).

Comparing the results obtained for the same mixtures, in the case of substitution of fly ash (FA) with marble dust (MD), with and without heat treatment, it is noticed that the lack of heat treatment will cause an increase in the apparent density of the matured material 7 days after casting, compared to the density of the material subjected to heat treatment, as well as a reduction of mechanical resistances, both for flexural and compressive strength by a maximum of 48%, variable, depending on the amount of the substituted fly ash.

Comparing the results obtained for the same mixtures, in the case of substitution of fly ash (FA) with blast furnace slag (BFS), with and without heat treatment, it is noticed that the lack of heat treatment will cause an increase in the apparent density of the matured material 7 days after casting, compared to the density of the material subjected to heat treatment and a reduction of the compressive strength, by over 50%–77%, depending on the amount of the substituted fly ash.

After analyzing the microscopic images taken at $1\times$ magnification, it was found that geopolymer pastes contain less pores with the substitution of 10% fly ash with marble dust and blast furnace slag. Geopolymer samples produced without heat treatment (23 °C) compared to those made with heat treatment (70 °C) are denser. Although geopolymer pastes are much more compact, their mechanical strengths decrease with the substitution of fly ash and are smaller compared to those produced using heat treatment. Thus, it can be said that the dimensions and number of pores do not influence their mechanical performances in terms of flexural and compressive strength.

In conclusion, it can be said that heat treatment helps the geopolymerization process both from the point of view of the formation of reaction products and the homogeneity of the samples. On the other hand, although there is a possibility of substituting fly ash with various other Si and Al supplying material, it is particularly important that these raw materials are analyzed in terms of their intake in Si, Al, and Ca, so that the geopolymer product is characterized by an optimal Si/Al ratio for the possibility of obtaining satisfactory mechanical performances.

Further studies are in progress in order to evaluate the possible direction for the development of alkali-activated geopolymer concrete, by adding aggregates to the mixtures and finding possible applications of the material based on specific mechanical requirements.

Author Contributions: Conceptualization, B.A.I. and A.-V.L.; methodology, B.A.I., A.-M.B., A.-V.L., S.R., T.G. and C.F.; validation, B.A.I., S.R. and T.G.; formal analysis, B.A.I., A.-M.B. and A.-V.L.; investigation, B.A.I., A.-M.B., A.-V.L., S.R., T.G. and C.F.; data curation, B.A.I., A.-V.L., S.R. and C.F.; writing—original draft preparation, B.A.I. and A.-V.L.; writing—review and editing, A.-M.B. and A.-V.L.; visualization, B.A.I., A.-M.B., A.-V.L., S.R., T.G. and C.F. All authors have read and agreed to the published version of the manuscript.

Funding: This research received no external funding.

Institutional Review Board Statement: Not applicable.

Informed Consent Statement: Not applicable.

Data Availability Statement: Not applicable.

Acknowledgments: This paper was supported by the Program Advanced research on the development of eco-innovative solutions, composite materials, technologies and services, in the concept of a circular economy and increased quality of life, for a sustainable digitized infrastructure in a built and urban environment resilient to climate change and disasters, "ECODIGICONS", Program code: PN 23 35 05 01: "Innovative sustainable solutions to implement emerging technologies with cross-cutting impact on local industries and the environment and to facilitate technology transfer through the development of advanced, eco-smart composite materials in the context of sustainable development of the built environment", financed by the Romanian Government.

Conflicts of Interest: The authors declare no conflict of interest.

References

1. Komnitsas, K.; Zaharaki, D. Geopolymerisation: A review and prospects for the minerals industry. *Miner. Eng.* **2007**, *20*, 1261–1277. [CrossRef]
2. Rashad, M.; Khalil, A.; Mervat, H. A preliminary study of alkali-activated slag blended with silica fume under the effect of thermal loads and thermal shock cycles. *Constr. Build. Mater.* **2013**, *40*, 522–532. [CrossRef]
3. Duxson, P.; Fernández-Jiménez, A.; Provis, J.L.; Lukey, G.C.; Palomo, A.; van Deventer, J.S.J. Geopolymer technology: The current state of the art. *J. Mater. Sci.* **2007**, *42*, 2917–2933. [CrossRef]
4. Skorina, T. Ion exchange in amorphous alkali-activated aluminosilicates: Potassium based geopolymers. *Appl. Clay Sci.* **2014**, *87*, 205–211. [CrossRef]
5. Komnitsas, K.; Zaharaki, D. *Structure, Processing, Properties and Industrial Applications PART II: Manufacture and Properties of Geopolymers*; Provis, J., van Deventer, J.S.J., Eds.; CRC Press: Boca Raton, FL, USA; Woodhead Publishing Ltd.: Oxford, UK, 2009.
6. Davidovits, J. *Geopolymer Chemistry and Applications*, 3rd ed.; Institute Geopolymer: Saint-Quentin, France, 2011.
7. Koleżyński, A.; Król, M.; Żychowicz, M. The structure of geopolymers—Theoretical studies. *J. Mol. Struct.* **2018**, *1163*, 465–471. [CrossRef]
8. Wang, W.C.; Wang, H.Y.; Lo, M.H. The fresh and engineering properties of alkali activated slag as a function of fly ash replacement and alkali concentration. *Constr. Build. Mater.* **2015**, *84*, 224–229. [CrossRef]
9. Hajimohammadi, A.; Provis, J.L.; van Deventer, J.S.J. Time-resolved and spatially resolved infrared spectroscopic observation of seeded nucleation controlling geopolymer gel formation. *J. Colloid Interface Sci.* **2011**, *357*, 384–392. [CrossRef]
10. Rees, C.A.; Provis, J.L.; Lukey, G.C.; van Deventer, J.S.J. Attenuated total reflectance Fourier transform infrared analysis of fly ash geopolymer gel aging. *Langmuir* **2007**, *23*, 8170–8179. [CrossRef]
11. Ruscher, C.H.; Mielcarek, E.; Lutz, W.; Ritzmann, A.; Kriven, W.M. Weakening of alkali-activated metakaolin during aging investigated by the molybdate method and infrared absorption spectroscopy. *J. Am. Ceram. Soc.* **2010**, *93*, 2585–2590. [CrossRef]
12. Fernández-Jiménez, A.; Palomo, A. Mid-infrared spectroscopic studies of alkali-activated fly ash structure. *Micropor. Mesopor. Mater.* **2005**, *86*, 207–214. [CrossRef]
13. Jaggi, N.; Vij, D.R. Fourier transform infrared spectroscopy. In *Handbook of Applied Solid State Spectroscopy*; Springer: New York, NY, USA, 2006; pp. 411–450.
14. Lyu, S.J.; Wang, T.T.; Cheng, T.W.; Ueng, T.H. Main factors affecting mechanical characteristics of geopolymer revealed by experimental design and associated statistical analysis. *Constr. Build. Mater.* **2013**, *43*, 589–597. [CrossRef]
15. García-Lodeiro, I.; Fernandez-Jimenez, A.; Palomo, A.; Macphee, D.E. Effect of Calcium Additions on N-A-S-H Cementitious Gels. *J. Am. Ceram. Soc.* **2010**, *93*, 1934–1940. [CrossRef]
16. Criado, M.; Fernández-Jiménez, A.; Palomo, A. Alkali activation of fly ash: Effect of the SiO_2/Na_2O ratio. *Micropor. Mesopor. Muter.* **2007**, *106*, 180–191. [CrossRef]
17. Ismail, I.; Bernal, S.A.; Provis, J.L.; Hamdam, S. Drying-induced changes in the structure of alkali-activated pastes. *J. Mater. Sci.* **2013**, *48*, 3566–3577. [CrossRef]
18. Guo, X.; Shi, H.; Dick, W.A. Compressive strength and microstructural characteristics of class C fly ash geopolymer. *Cem. Concr. Compos.* **2010**, *32*, 142–147. [CrossRef]
19. Liew, Y.M.; Kamarudin, H.; Al Bakri, A.M.M.; Bnhussain, M.; Luqman, M.; Khairul, I.; Ruzaidi, C.M.; Heah, C.V. Optimization of solids-to-liquid and alkali activator ratios of calcined kaolin geopolymeric powder. *Constr. Build. Mater.* **2012**, *37*, 440–451. [CrossRef]
20. Nasab, G.M.; Golestanifard, F.; MacKenzie, K.J.D. The Effect of the SiO_2/Na_2O Ratio in the Structural Modification of Metakaolin-Based Geopolymers Studied by XRD, FTIR and MAS-NMR. *J. Ceram. Sci. Technol.* **2014**, *5*, 185–192.
21. Zhang, Z.; Provis, J.L.; Wang, H.; Bullen, F. Quantitative kinetic and structural analysis of geopolymers. Part 1. The activation of metakaolin with sodium hydroxide. *Thermochim. Acta* **2012**, *539*, 23–33. [CrossRef]
22. Lyu, S.J.; Hsiaso, Y.H.; Wang, T.T.; Cheng, T.W. Microstructure of geopolymer accounting for associated mechanical characteristics under various stress states. *Cem. Concr. Res.* **2013**, *54*, 199–207. [CrossRef]
23. Juenger, M.C.G.; Winnefeld, F.; Provis, J.L.; Ideker, J.H. Advances in alternative cementitious binders. *Cem. Concr. Res.* **2011**, *41*, 1232–1243. [CrossRef]

24. Rashad, A.M. A comprehensive overview about the influence of different admixtures and additives on the properties of alkali-activated fly ash. *Mater. Des.* **2014**, *53*, 1005–1025. [CrossRef]
25. Yunsheng, Z.; Wei, S.; Zongjin, L. Composition design and microstructural characterization of calcined kaolin-based geopolymer cement. *App. Clay Sci.* **2010**, *47*, 271–275. [CrossRef]
26. Tchadjié, I.N.; Djobo, J.N.Y.; Ranjbar, N.; Tchakouté, H.K.; Kenne, B.B.D.; Elimbi, A.; Njopwouo, D. Potential of using granite waste as raw material forgeopolymer synthesis. *Ceram. Int.* **2016**, *42*, 3046–3055. [CrossRef]
27. Krol, M.; Minkiewicz, J.; Mozgawa, W. IR Spectroscopy Studies of Zeolites in Geopolymeric Materials derived from Kaolinite. *J. Mol. Struct.* **2016**, *1126*, 200–206. [CrossRef]
28. Djobo, J.N.Y.; Elimbi, A.; Tchakouté, H.K.; Kumar, S. Reactivity of volcanic ash in alkaline medium, microstructural and strength characteristics of resulting geopolymers under different synthesis conditions. *J. Mater. Sci.* **2016**, *51*, 10301–10317. [CrossRef]
29. Rees, C.A. Mechanisms and Kinetics of Gel Formation in Geopolymers. Ph.D Thesis, Department of Chemical and Biomolecular Engineering, The University of Melbourne, Melbourne, Australia, 2007.
30. Spătaru, I. Contributions Concerning the Obtaining of Geopolymers by Using Production Residues. PhD Theis, Dunărea de Jos from Galați University, Galați, Romania, 2016.
31. Phair, J.W.; van Deventer, J.S.J. Effect of the silicate activator pH on the microstructural characteristics of waste-based geopolymers. *Int. J. Miner. Process.* **2002**, *66*, 121–143. [CrossRef]
32. Robayo-Salazar, R.A.; Mejía, R.; Gutiérrez, D.; Puertas, F. Effect of metakaolin on natural volcanic pozzolan-based geopolymer cement. *Appl. Clay Sci.* **2016**, *132–133*, 491–497. [CrossRef]
33. Gao, X.; Yu, Q.L.; Brouwers, H.J.H. Reaction kinetics, gel character and strength of ambient temperature cured alkali activated slag–fly ash blends. *Constr Build Mater.* **2015**, *80*, 105–115. [CrossRef]
34. Kym, S. Mechanical properties of sodium and potassium activated metakaolin-based geopolymers. Master's Thesis, Texas A&M University, College Station, TX, USA, 2010.
35. Mangau, A.; Vermesan, H.; Paduretu, S.; Hegyi, A. An Incursion into Actuality: Addressing the Precautionary Principle in the Context of the Circular Economy. *Sustainability* **2022**, *14*, 10090. [CrossRef]
36. Merli, R.; Preziosi, M.; Acampora, A. How do scholars approach the circular economy? A systematic literature review. *J. Clean Prod.* **2018**, *178*, 703–722. [CrossRef]
37. Vermesan, H.; Mangau, A.; Tiuc, A.-E. Perspectives of Circular Economy in Romanian Space. *Sustainability* **2020**, *12*, 6819. [CrossRef]
38. Velenturf, A.P.M.; Purnell, P. Principles for a sustainable circular economy. *Sustain. Prod. Consum.* **2021**, *27*, 1437–1457. [CrossRef]
39. Stefansson, H.O. On the Limits of the Precautionary Principle. *Risk Anal.* **2019**, *39*, 6. [CrossRef] [PubMed]
40. Politica de Mediu. Available online: https://www.mae.ro/node/35846 (accessed on 23 October 2022).
41. Politica de Mediu: Principii Generale si Cadrul de Baza. Available online: https://www.europarl.europa.eu/factsheets/ro/sheet/71/politica-de-mediu-principii-generale-si-cadrul-de-baza (accessed on 23 October 2022).
42. Muntean, E.; Muntean, N.; Mihăiescu, T. Cadmium and lead soil pollution in Copsa Mica area in relation with the food chain. *Res. J. Agric. Sci.* **2010**, *42*, 731–734.
43. Ungureanu, A. Aspects of Soil Pollution by Heavy Metals in Cop,sa Mica and Media,s, Sibiu County. Buletinul Institutului Politehnic din Ia,si. 2010. Available online: https://hgim.tuiasi.ro/documente/buletin/2010/HIDRO2din2010.pdf#page=9 (accessed on 9 August 2022).
44. Damian, F.; Damian, G.; Lăcătusu, R.; Iepure, G. Heavy metals concentration of the soils around Zlatna and Cop,sa Mică smelters Romania. Carpth. *J. Earth Environ. Sci.* **2008**, *3*, 65–82.
45. Muntean, O.L.; Drăgut, L.; Baciu, N.; Man, T.; Buzilă, L.; Ferencik, I. *Environmental Impact Assessment as a Tool for Environmental Restoration: The Case Study of Cop,sa-Mică Area, Romania*; Springer: Dordrecht, The Netherlands, 2008; pp. 461–474.
46. Szanto, M.; Micle, V.; Prodan, C.V. Study of Soil Quality in Copsa Mica Areawith the Aim of their Remediation. *ProEnvironment* **2011**, *4*, 251–255.
47. Lăzărescu, A.; Szilagyi, H.; Baeră, C.; Hegyi, A. Alternative Concrete—Geopolymer Concrete. Emerging Research and Oportunities. In *Materials Research Foundations*; Materials Research Forum LLC.: Millersville, PA, USA, 2021; Volume 109, 138p.
48. Sandu, A.V. Obtaining and Characterization of New Materials. *Materials* **2021**, *14*, 6606. [CrossRef]
49. Jamil, N.H.; Abdullah, M.M.A.B.; Ibrahim, W.M.A.W.; Rahim, R.; Sandu, A.V.; Vizureanu, P.; Castro-Gomes, J.; Gómez-Soberón, J.M. Effect of Sintering Parameters on Microstructural Evolution of Low Sintered Geopolymer Based on Kaolin and Ground-Granulated Blast-Furnace Slag. *Crystals* **2022**, *12*, 1553. [CrossRef]
50. Sofri, L.A.; Abdullah, M.M.A.B.; Sandu, A.V.; Imjai, T.; Vizureanu, P.; Hasan, M.R.M.; Almadani, M.; Aziz, I.H.A.; Rahman, F.A. Mechanical Performance of Fly Ash Based Geopolymer (FAG) as Road Base Stabilizer. *Materials* **2022**, *15*, 7242. [CrossRef]
51. Ibrahim, W.M.W.; Abdullah, M.M.A.B.; Ahmad, R.; Sandu, A.V.; Vizureanu, P.; Benjeddou, O.; Rahim, A.; Ibrahim, M.; Sauffi, A.S. Chemical Distributions of Different Sodium Hydroxide Molarities on Fly Ash/Dolomite-Based Geopolymer. *Materials* **2022**, *15*, 6163. [CrossRef]
52. Jamaludin, L.; Razak, R.A.; Abdullah, M.M.A.B.; Vizureanu, P.; Bras, A.; Imjai, T.; Sandu, A.V.; Abd Rahim, S.Z.; Yong, H.C. The Suitability of Photocatalyst Precursor Materials in Geopolymer Coating Applications: A Review. *Coatings* **2022**, *12*, 1348. [CrossRef]

53. Tahir, M.F.M.; Abdullah, M.M.A.B.; Rahim, S.Z.A.; Mohd Hasan, M.R.; Sandu, A.V.; Vizureanu, P.; Ghazali, C.M.R.; Kadir, A.A. Mechanical and Durability Analysis of Fly Ash Based Geopolymer with Various Compositions for Rigid Pavement Applications. *Materials* **2022**, *15*, 3458. [CrossRef] [PubMed]
54. Zailan, S.N.; Mahmed, N.; Abdullah, M.M.A.B.; Rahim, S.Z.A.; Halin, D.S.C.; Sandu, A.V.; Vizureanu, P.; Yahya, Z. Potential Applications of Geopolymer Cement-Based Composite as Self-Cleaning Coating: A Review. *Coatings* **2022**, *12*, 133. [CrossRef]
55. Luhar, I.; Luhar, S.; Abdullah, M.M.A.B.; Razak, R.A.; Vizureanu, P.; Sandu, A.V.; Matasaru, P.-D. A State-of-the-Art Review on Innovative Geopolymer Composites Designed for Water and Wastewater Treatment. *Materials* **2021**, *14*, 7456. [CrossRef]
56. *EN 12350-6*; Testing Fresh Concrete—Part 6: Density. National Standardisation Body—ASRO: Bucharest, Romania, 2019.
57. *EN 12390-7*; Testing Hardened Concrete—Part 7: Density of Hardened Concrete. National Standardisation Body—ASRO: Bucharest, Romania, 2019.
58. *EN 196-1*; Methods of Testing Cement—Part 1: Determination of Strength. National Standardisation Body—ASRO: Bucharest, Romania, 2016.

Disclaimer/Publisher's Note: The statements, opinions and data contained in all publications are solely those of the individual author(s) and contributor(s) and not of MDPI and/or the editor(s). MDPI and/or the editor(s) disclaim responsibility for any injury to people or property resulting from any ideas, methods, instructions or products referred to in the content.

Article

Identification Fluidity Method to Determine Suitability of Weathered and River Sand for Constructions Purposes

Haoyu Zuo [1,*], Jin Li [1,*], Li Zhu [2], Degang Cheng [2] and De Chang [2]

[1] School of Transportation Civil Engineering, Shandong Jiaotong University, Jinan 250357, China
[2] Jinan Kingyue Highway Engineering Company Ltd., Jinan 250409, China
* Correspondence: 21107008@stu.sdjtu.edu.cn (H.Z.); sdzblijin@163.com (J.L.); Tel.: +86-18678092667 (H.Z.); +86-13678824225 (J.L.)

Abstract: At present, in order to comply with the development of the "the Belt and Road Initiatives", the country is accelerating the pace of construction and increasing the demand for construction river sand. However, the quality of construction river sand is uncontrollable, and its shape is very similar to that of weathered sand. Therefore, using inferior weathered sand and mixed sand as inferior substitute sand in the market is prohibited, resulting in an increase in the difficulty coefficient of quality control of concrete fine aggregate in actual projects. This lays hidden dangers for the construction quality of the project. It is urgent to improve the quality control, testing, and detection process of river sand. Due to the long-term weathering of weathered sand, its density is small, and there are many pores, which leads to the material's water absorption rate is higher than that of standard sand and river sand during fluidity tests. This paper takes this as a breakthrough point, reveals the variation law of fluidity loss under different variables, and explores a method for effectively screening low-quality sand and gravel. Through the silt content test (screening and washing method), the low-quality sand is preliminarily screened out, the mortar ratio is designed, and the fluidity test is carried out to compare the difference in fluidity loss of different types of mortar; determine the loss threshold range (mobility loss ≤ 15 mm) according to the mobility test results of the control group, and determine the qualification standard by comparing the measured mobility loss of the unknown sample with the loss threshold range. When the mobility loss is within the loss threshold range, the sample is qualified river sand. Otherwise, it is weathered sand or chowder sand. This method establishes a complete detection scheme for distinguishing weathered sand and river sand through mud content tests and mobility loss tests, solves the difficult problem of river sand quality control in engineering applications, and effectively eliminates the phenomenon of using low-quality weathered sand as river sand in the sand and gravel material market, thus avoiding congenital defects in concrete homogeneity.

Keywords: river sand; weathered sand; mud content; fluidity preface

Citation: Zuo, H.; Li, J.; Zhu, L.; Cheng, D.; Chang, D. Identification Fluidity Method to Determine Suitability of Weathered and River Sand for Constructions Purposes. *Coatings* **2023**, *13*, 327. https://doi.org/10.3390/coatings13020327

Academic Editors: Ofelia-Cornelia Corbu, Ionut Ovidiu Toma and Valeria Vignali

Received: 26 December 2022
Revised: 23 January 2023
Accepted: 28 January 2023
Published: 1 February 2023

Copyright: © 2023 by the authors. Licensee MDPI, Basel, Switzerland. This article is an open access article distributed under the terms and conditions of the Creative Commons Attribution (CC BY) license (https://creativecommons.org/licenses/by/4.0/).

1. Introduction

In evaluating the quality of engineering construction in civil engineering, similarly to other types of construction, it is necessary to apply a systemic approach. The quality in the previous international standard ISO 8402 was defined as the sum of characteristics of the product or service that reflect their ability to meet the stated and implied needs of the customers. The product or the construction product of a building process is the engineering construction representative of the highly expensive product from a majority of the range of works [1]. The standard EN ISO 9000 [2] defines quality as the degree to which a set of inherent characteristics fulfills requirements. Pavement roads are required to be designed, built, maintained, and disposed of at a reasonable price, with reasonable quality, respecting the relevant requirements of users and their surrounding residents and the principles of sustainable development during the life cycle.

Construction river sand is an indispensable part of the fine aggregates in pavement road engineering construction. It refers to the construction materials with certain quality standards formed by the action of natural forces, such as the impact and erosion of river water. After drying and screening, it can be widely used in various dry mortars and plays an irreplaceable role in the construction industry. Weathered sand is a kind of material that is broken and loose after a long exposure to solar radiation, the atmosphere, and water. Its durability is poorer than ordinary soil materials, its strength is weaker, its physical and mechanical properties are unstable, and it contains a certain amount of fine soil particles.

Compared with river sand, weathered sand is cheap and easy to obtain, but its various property indicators cannot meet the requirements of construction materials. If it is used for concrete mixing, it will have a huge impact on the quality of concrete, causing rapid loss of concrete slump and affecting its strength, durability, and workability. Due to improper use of weathered sand, the concrete strength is seriously lower than the design strength grade, and the phenomenon of concrete cracking frequently occurs, which brings some difficulties to construction quality control.

Generally, the surface of river sand is smooth, the particles are smooth and relatively clean, and the strength is high, so it is not easy to twist with fingers. However, the weathered sand particles are angular, with a prickly feeling when rubbed by hand. The size distribution is uneven and contains a large number of fine soil particles. At present, the common discrimination method in the industry is to distinguish by hand and eye. If only judged by experience, the distinction between the two is vulnerable to subjective factors, and the discrimination error is large, leading to the increase in the difficulty coefficient of raw material quality control in engineering applications. As the quality of inland river sand in Shandong Province is relatively poor and is mixed with weathered sand, the durability of weathered sand is poorer than that of ordinary soil materials, and its physical and mechanical properties are more unstable, it is easy to lay hidden dangers for project construction quality.

Yan Zhenqiang [3], from the Shandong Jianzhu University, systematically studied the road performance and microscopic characteristics of cement-stabilized weathered sand. First, based on mastering the basic physical and mechanical properties of weathered sand, a series of indoor tests were carried out to determine the influence of cement content and curing age on the mechanical properties of weathered sand. Through the dry shrinkage test, dry wet cycle test, and freeze-thaw cycle test, the influence of environmental changes on the durability of cement-stabilized weathered sand is analyzed. In addition, the microstructure characteristics of cement-stabilized weathered sand under different working conditions are analyzed through a scanning electron microscope test. Based on this, the strong growth mechanism of cement-stabilized weathered sand and the deterioration mechanism under dry, wet, and freeze-thaw cycles are discussed.

Lin Yunken [4] of the Yongjia Rongchang Concrete Co., Ltd. (Wenzhou, China) studied the influence of river sand silt content on the performance of machine-made sand concrete and tested the amount of a mixture, working performance and compressive strength of concrete mixed with river sand with different silt content. It is found that river sand containing mud will increase the absorption of admixtures, reduce the working performance of concrete, reduce the slump retaining capacity of concrete, and thus reduce the engineering properties of concrete.

Zhao Wenkun [5] of the China Communications Construction Company First Harbor Engineering Co., Ltd. (Beijing, China) studied the difference in the microstructure of manufactured sand in Shiling Quarry and river sand in Qingping Quarry through Nikon SMZ800N body microscope and computer graphics processing technology and then analyzed its impact on the mechanical properties of concrete from the difference in roundness coefficient and particle morphology.

Zhang Xiao [6] of the Liaoning Provincial Communications Planning and Design Institute Co., Ltd. (Shenyang, China) found that with the increase in river sand content, the mechanical properties of UHPC first increased and then decreased, and the working performance gradually improved. When river sand content was 30%, the compressive strength, and flexural strength reached the maximum, the tensile strength of UHPC continued to decrease, and the durability of UHPC was better due to the improvement of chloride ion penetration resistance. Domestic and foreign scholars have conducted a lot of research on the microscopic characteristics of weathered sand and river sand and the improvement and application of machine-made sand and river sand in concrete [7–26].

Through a series of tests, this paper compares the differences in physical properties between weathered sand and river sand and then puts forward a method to judge the quality of river sand. It uses scientific and reasonable means to evaluate the quality of sand and gravel, providing scientific standards and norms for practical engineering applications. The river sand and unknown sand samples required for the test in this paper are taken from the beach along the Yellow River in Qihe River, Dezhou, Shandong Province, and the weathered sand samples are taken from the weathered rock in Wulian, Rizhao, Shandong Province. In the application process, this method can reflect the problem of water absorption of materials to reflect the workability and fluidity of the tested materials used for mixing concrete. The test equipment and methods used are the existing mud content and fluidity test equipment and methods. The process is simple, convenient, fast, saves manpower and material resources, and is suitable for standardized management with high accuracy. It avoids the subjective factors affecting the results only based on subjective experience. The identification of defects with large errors reduces the difficulty of raw material quality control in engineering applications, ensures the safety of engineering construction, and can create more economic value and social benefits.

2. Basic Physical Properties Test of River Sand and Weathered Sand

2.1. Natural Water Content Test

Sample the soil samples and place them in a cool place for sealed storage and drying. Take a clean and dry aluminum box and say its mass is m_1. Put the soil sample in the sealed bag into the aluminum box, close the lid and call its mass m_2. Set the temperature of the oven to 105 °C; when the oven reaches the set temperature, take off the lid of the aluminum box in the oven for drying. After drying, take out the aluminum box and quickly cover the box cover. After the sample cools, weigh the aluminum box and the dried sample mass m_3. When two consecutive weighing differences are unchanged, drying ends. The natural water content test results are shown in Table 1.

Table 1. Natural water content test.

Aluminum Box Number	1	2	3	4
Aluminum box mass/g	88.6	102.6	88.6	102.6
Aluminum box + wet soil total mass/g	160.3	155.9	123.6	143.4
Aluminum box + dry soil total mass/g	163.4	159.1	127.5	147.5
Water content/%	3.1	3.2	3.9	4.1
Average moisture content/%		3.6		

Calculate the natural moisture content of weathered sand:

$$\omega = \frac{m_2 - m_3}{m_3 - m_1} \times 100\% \tag{1}$$

ω—moisture content (%); m_1—Mass of aluminum box (g); m_2—Total mass of the aluminum box and wet soil (g); m_3—Total mass of the aluminum box and dry soil (g)

Test results:

According to the above method, the average water content of weathered sand measured is 3.6%. Similarly, the average water content of the river sand sample is 5.8%. The average water content of river sand is greater than that of weathered sand. However, the water content will produce large errors due to environmental impact during the transportation and test of sand and gravel materials, so it cannot be used as a basis for distinguishing weathered sand from river sand.

2.2. Particle Analysis Test

(1) Test method

Representative samples of dry soil were taken out by the quartering method, and 2 mm samples were screened in batches. Samples larger than 2 mm are passed through each layer of the coarse sieve in turn, and the remaining soil samples on the sieve are weighed separately. Shake the soil sample under the 2 mm sieve through the vibrating sieve machine for 10 min, then start with the screen with the largest aperture and gently pat and shake at the bottom of the sieve where the white paper is placed until the mass under the sieve does not exceed 1% of the remaining mass of the sieve level per minute. Place all leaking soil particles in the next level sieve, brush the soil sample on the sieve with a soft brush and weigh separately.

(2) Test results and analysis

The test results of granular analysis of weathered sand and river sand are shown in Table 2, and the particle distribution curve is shown in Figure 1.

Table 2. Particle analysis results.

Screen Size/mm	40	20	10	5	2	1	0.5	0.25	0.075
Aeolian sand passing mass percentage/%	100	96.1	90.2	85.2	73.8	63.7	37.6	18.6	2.7
River sand passing mass percentage/%	100	100	96.4	89.7	79.4	49.3	19.1	5.5	2.2

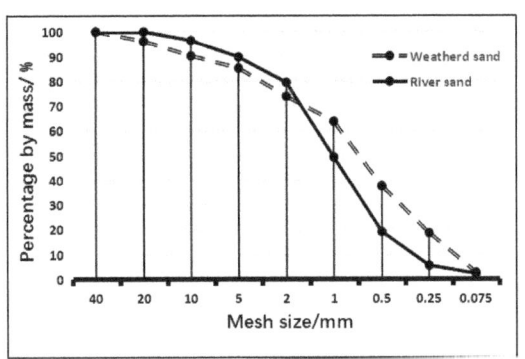

Figure 1. Particle analysis result plot.

From Table 2 and Figure 1, it can be concluded that the content of the weathered gravel grain group of this group of samples is 26.2%, the content of the sand grain group is 73.8%, the content of the fine grain group is less than 5%, the non-uniformity coefficient Cu is 5.6, and the curvature coefficient Cc is 1.1, which are well-graded sands. The content of the river gravel group is 21.6%, the content of the sand group is 78.4%, the content of the fine group is less than 5%, the non-uniformity coefficient Cu is 10.7, and the curvature coefficient Cc is 3.7. Therefore, the size distribution of river sand is more concentrated, and the size distribution of weathered sand is more dispersed.

3. Apparent Density and Bulk Density Test of Sand

3.1. Test Method for Apparent Density of Sand

(1) Weigh 300 g (G_0) of dried river sand and weathered sand samples, respectively;

(2) Fill the volumetric flask with water to the bottleneck scale line, wipe off the water outside the flask, and weigh its mass (G_2);

(3) Pour out the water from the inside of the volumetric flask and the rest to about 1/3 of the height of the ball. Add 300 g of sand into the volumetric flask, tilt the volumetric flask at an angle of about 45 degrees, and make the sample stir fully in the water (remove bubbles). After standing for a period of time, add water to the bottleneck scale line with a dropper, wipe off the water outside the bottle, and weigh its mass (G_1).

The apparent density test of sand is shown in Figure 2.

Figure 2. Sand apparent density test.

3.2. Data Processing

The apparent density of sand is calculated as follows: weathered sand is ρ_1, river sand is ρ_2, and the result is accurate to 10 kg/m³. The correction coefficient of the effect of different water temperatures on the apparent density of sand is shown in Table 3.

$$\rho_0 = \left(\frac{G_0}{G_0 + G_2 - G_1} - \alpha_t \right) \times \rho_{water} \qquad (2)$$

Table 3. Correction coefficient of the influence of different water temperatures.

Water Temperature/°C	15	16	17	18	19	20	21	22	23	24	25
α_t	0.002	0.003	0.003	0.004	0.004	0.005	0.005	0.006	0.006	0.007	0.008

α_t: Correction coefficient of the influence of different water temperatures on the apparent density of sand.

ρ_{water} = 997 Kg/m³, Water temperature = 18 °C.

$$\rho_1 = \left(\frac{300}{300 + 400 - 584.6} - 0.04 \right) \times 997 = 2551.97 \qquad (3)$$

$$\rho_2 = \left(\frac{300}{300 + 400 - 595.3} - 0.04 \right) \times 997 = 2816.85 \qquad (4)$$

Take the arithmetic mean of the two parallel sample test results as the final result. The difference between the measurement results should be less than 20 Kg/m³. Otherwise, it should be reperformed. According to the formula, the apparent density of weathered sand is 2573, and that of river sand is 2810. The apparent density of river sand is slightly higher than that of weathered sand.

3.3. Test Method for Loose Bulk Density of Sand

(1) Weigh the mass (G_1) of the standard container and measure the volume (V) of the standard container. Place the standard container under the blanking hopper to make the hopper align with the center;
(2) Load the sample into the blanking funnel, open the movable door, and let the sample slowly fall into the standard container until it is full and exceeds the opening of the standard container, then remove the funnel;

The loose bulk density test of sand is shown in Figures 3 and 4.

Figure 3. Feeding funnel charging.

Figure 4. Scrape flat.

(3) Use a ruler to scrape the excess sample in the opposite direction along the center line of the barrel mouth and weigh its mass (G_2).

Data processing:

(1) Calculate the bulk density of the sample according to the following formula, accurate to 10 Kg/m^3.

$$\rho_0 = \frac{G_1 - G_2}{V} \tag{5}$$

(2) Take the arithmetic mean of the test results of two parallel samples as the final result, accurate to 10 Kg/m³.

$$\rho_1 = \frac{G_1 - G_2}{V} = 1576.3 \tag{6}$$

$$\rho_2 = \frac{G_1 - G_2}{V} = 1449.5 \tag{7}$$

According to the formula, the loose bulk density of river sand is 1560, and that of weathered sand is 1450.

To sum up, the apparent density and bulk density of river sand is slightly higher than that of weathered sand. It is speculated that the shape of weathered sand particles is irregular, and the void ratio is higher than that of river sand due to long-term weathering and erosion. However, since the difference in data comparison is not obvious, it cannot be used as a basis for distinguishing weathered sand from river sand.

4. Comparative Test on Fluidity Loss of Different Sand and Gravel Materials

By selecting standard sand, qualified river sand, qualified weathered sand with known silt content, and river sand materials with high silt content, the fluidity test is conducted under the condition of a standard temperature of the laboratory 20 ± 2 °C and relative humidity > 50%. The test results are shown in Table 4. Several groups of repeated tests on fluidity loss of river sand and weathered sand under different mud content conditions are carried out, and the results are shown in Figures 5 and 6:

Table 4. Fluidity loss table.

Group	Standard Sand	Qualified River Sand 1	Qualified River Sand 2	Weathered Sand 1	Weathered Sand 2	Silt 3.6% River Sand	Mud-3.2% River Sand
First diameter value (mm)	300	300	297	281	274	262	270
Stand for 30 min diameter value (mm)	300	300	295	212	230	196	211
Fluidity loss s (mm)	0	0	2	69	44	66	59

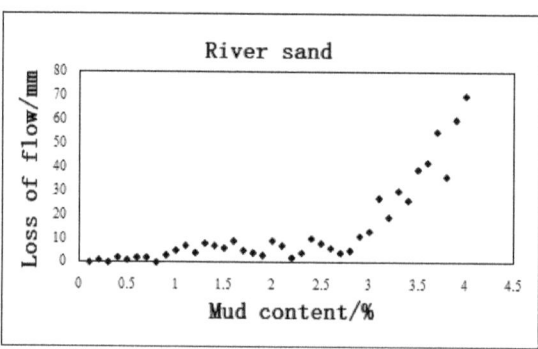

Figure 5. Flow loss of river sand with different silt content.

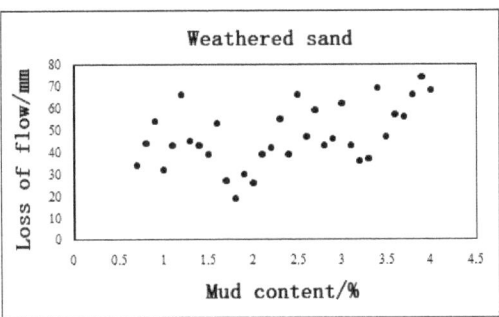

Figure 6. Flowability loss of weathered sand with different mud content.

According to test data, the results show that:

(1) The loss value of fluidity of river sand with unqualified silt content is large. In combination with Figure 5, the relationship between the loss of fluidity of river sand and silt content, the loss value of fluidity of river sand with silt content of more than 3% changes greatly. When the silt content is less than 3%, the loss of fluidity of river sand is less than 15 mm. Therefore, it can be seen that the silt content has a great impact on mobility. First, it is necessary to exclude the impact of the silt content on the material fluidity loss test.

(2) The fluidity loss value of weathered sand is significantly larger than that of standard sand and qualified river sand. In combination with Figure 6, the relationship between fluidity loss and mud content of weathered sand, if the mud content of mobilized materials meets the standard, and the fluidity loss value is >15 mm, it can be judged as weathered sand, that is, the river sand is unqualified.

(3) The fluidity loss of qualified river sand is less than 15 mm, and the fluidity loss of weathered sand or unqualified sand is more than 15 mm, so the threshold range of qualified material fluidity loss is determined: under the condition of qualified mud content when the fluidity loss is more than 15 mm, it can be judged as weathered sand or mixed sand. The smaller the fluidity loss, the better the river sand quality.

5. Quality Inspection Test of Newly Arrived Sand and Gravel

Then, the quality of unknown sand and gravel is detected through two embodiments of mud content and a fluidity test.

5.1. Implementation I

(1) Mud content test

The new materials shall be tested in accordance with the requirements of T 0333-2000 fine aggregate mud content test (sieve washing method) in JTG E42-2005 Test Rules for Aggregates of Highway Engineering [27]. The mud content of the samples shall be calculated to 0.1% according to the formula.

$$Q_n = \frac{m' - m}{m} \times 100 \tag{8}$$

where: Q_n is the mud content of the sample (%); m is the weight of the dried sample before the test (g); m' is the weight of the dried sample after the test (g).

The arithmetic mean of the test results of the two samples was used as the measurement value. When the difference between the two results exceeds 0.5%, a new sample should be taken for testing.

During the mud content test, samples of about 400 g (m) per mass are weighed according to the test procedures. The final mud content shall meet the following requirements:

Q_n 3.0 or less. If the mud content of the incoming material meets the requirements of this standard, proceed to the next step of detection; Otherwise, if the batch of material is unqualified, the yard will not receive it.

For the three new batches of materials, 400 g is taken, respectively, for the mud content detection. The test results are shown in Table 5:

Table 5. Mud content test results of Embodiment 1.

The Sample	Sample 1	Sample 2	Sample 3
Mud content Q_n (%)	3.8	1.9	3.3

Under the same test condition, the mud contents of sample 1 and sample 3 are 3.8% and 3.3%, respectively, which is not satisfied Q_n (mud content) ≤ 3.0 requirements. Sample 2 meets the mud content standard and can be tested for fluidity in the next step after preliminary screening. Sample 2 is selected as the next step of the test material, hereafter referred to as the unknown material.

(2) Fluidity test

The screened sample mortar is prepared from the sample whose mud content is not greater than the set threshold of mud content. The material ratio and mixing of the sample mortar, standard sand mortar, and river sand mortar are based on the requirements specified in GBT 17671-1999 Test Method for Strength of Cement Mortar (ISO Method) [28] and JC/T681-2005 Planetary Cement Mortar Mixer [29].

According to the requirements of GBT 17671-1999 Test Method for Strength of Cement Mortar (ISO Method), the quality mix proportion of cement mortar shall be one part of cement, three parts of sand, and half water, that is, 1:3:0.5 for cement: standard sand: water. In order to better simulate the actual effect of concrete, a 1% water-reducing agent shall be added. The final design mix ratio of cement: sand: water is 2.5:5:1. The specific preparation method is as follows: take water according to the set mass ratio scale and add cement into the pot. The water reducer makes its state select the appropriate adding time. If it is a powder water reducer, add it into the pot together with cement and other powders, and the liquid water reducer is added into the pot together with water. The cement used is ordinary Portland cement.

After using the cement mortar mixer to mix at a low speed for the set time, evenly add the sample into the mixture. The speed of mixing at low speed is: the rotation of the mixer shaft is 140 ± 5 r/min, and the revolution is 62 ± 5 r/min. According to the specification requirements, completely mix the mixture evenly and complete the preparation of the sample mortar. The setting time is 30 s.

The preparation method of standard sand mortar and river sand mortar is the same as that of sample mortar, except that the standard sand mortar is obtained by replacing the sample with ISO standard sand of the same quality, and the river sand mortar is obtained by replacing the sample with river sand of the same quality.

The standard sand group is set as control group 1, river sand 1, river sand 2, and river sand 3 corresponding to control group 2, control group 3, and control group 4, respectively, and the samples to be judged are test groups.

Control group 1: weigh 400 g cement, 800 g standard sand, 160 g water, and 4 g water reducer, respectively, and add water and cement into the pot in turn. The water reducer shall be added at the appropriate time according to its state: powder water reducer shall be added together with cement and other powders, and liquid water reducer shall be added together with water. After 30 s of low-speed mixing with a cement mortar mixer, add standard sand evenly at the beginning of the second 30 s. Mix the materials completely and evenly according to the specification requirements to complete the preparation of standard sand mortar.

In the same way as the control group 1, the control groups 2, 3, 4, and the test group (the standard sand is replaced with the corresponding river sand or sample of the same gram) are mixed with cement mortar to complete the preparation of two groups of river sand mortar and sample mortar.

According to the test procedures of GBT-2419-2005 Method for Determining the Fluidity of Cement Mortar [30], under the condition that the standard temperature of the laboratory is 20 ± 2 °C and the relative humidity is >50%, take the materials separately, and carry out the fluidity test on five groups of newly mixed materials to determine their fluidity values. At this time, their fluidity values are recorded as d. At the same time, stand the mixed mortar for 30 min and measure its fluidity value, which is recorded as d'.

It shall be completed within 6 min from the time of adding water to the mortar to the time of measuring the diffusion diameter. After jumping the table, use a caliper to measure the diameters of the two directions perpendicular to each other on the bottom surface of the mortar, calculate the average value, and take an integer (mm). The average value is the cement mortar fluidity of the water volume.

Calculate the fluidity loss s of each group, and according to the test results of the four control groups, define the fluidity range of qualified sand materials, that is, the loss threshold range: if the fluidity value of the experimental group conforms to the range, the material is available river sand. Otherwise, the material does not meet the material requirements of the stockyard and is unqualified, which is weathered sand or chowder sand (river sand mixed with weathered sand).

Loss of fluidity $s = d - d'$

The fluidity test results are shown in Table 6.

Table 6. Fluidity test results of Embodiment 1.

Group	Control Group 1	Control Group 2	Control Group 3	Control Group 4	Test Group
First diameter value (mm)	300	288	295	300	275
Stand for 30 min diameter value (mm)	300	286	291	300	199
Fluidity loss s (mm)	0	2	4	0	76

Under this test scheme, the performance of control group 1 was good. When it jumped the table 23 times, it exceeded the 300 mm range of the disc table, and the first fluidity value was recorded as 300 mm. In control groups 2, 3, and 4, the initial mobility values were all between 280 mm–300 mm, and there was almost no change in the mobility values after standing for 30 min.

After standing for 30 min, compared with the control test, the material fluidity loss of the experimental group was serious, up to 76 mm, which could be determined that the material of the experimental group was blown sand or mixed sand.

According to the test results, the fluidity loss of problem materials is obviously larger, and the fluidity loss of qualified materials is less than 10 mm. The results of multiple tests can be combined to determine the range of qualified material mobility.

5.2. Implementation II

For the four new batches of materials, 400 g is taken, respectively, for the mud content detection. The test results are shown in Table 7:

Table 7. Mud content test results of Embodiment 2.

The Sample	Sample 1	Sample 2	Sample 3
Mud content Q_n (%)	1.7	3.6	2.1

Among them, the mud content of sample 2 is 3.6%, which does not meet $Qn \leq 3$. The clay content of samples 1 and 3 met the requirements, and they were selected for the next step of the fluidity test.

Standard sand was taken and set as control group 1, river sand 1 and river sand 2 corresponding to control group 2 and control group 3, and sample 1 and sample 3 were set as unknown material 1 and unknown material 2 corresponding to test group 1 and test group 2.

Weigh 400 g cement, 800 g standard sand/river sand 1/river sand 2/unknown material 1/unknown material 2, 160 g water, and 4 g water reducing agent, respectively, and add water and cement into the pot in turn. The appropriate adding time of water reducing agent is selected according to its state: a powder water-reducing agent is added together with cement and other powder, and a liquid water-reducing agent is added together with water. After mixing with the cement mortar mixer at a low speed for 30 s, add the sand evenly at the beginning of the second 30 s, and mix the material completely and evenly according to the time required by the specification.

Five groups of fresh mix materials were tested for fluidity, and the fluidity values of fresh mix and material standing for 30 min were measured, respectively, to calculate the fluidity loss. The fluidity test results are shown in Table 8.

Table 8. Fluidity test results of Embodiment 2.

Group	Control Group 1	Control Group 2	Control Group 3	Test Group 1	Test Group 2
First diameter value (mm)	300	287	291	300	285
Stand for 30 min diameter value (mm)	300	282	288	300	233
Fluidity loss s (mm)	0	5	3	0	52

Under the same test conditions, the fluidity of test group 1 is good, and the fluidity loss is 0, which is qualified river sand. The fluidity loss of test group 2 is serious, and its value is up to 52 mm. It can be judged as weathered sand or mixed sand, and the material is unqualified, so it will not be accepted.

The flow of qualified river sand can be determined from the above tests as shown in Figure 7.

This flow chart provides a method to judge the quality of river sand, including the following steps:

(1) Determine the mud content of sand and gravel with a fineness modulus of 2.3–3.0. If the measured mud content is greater than the set threshold of mud content (mud content ≤ 3.0), the sample is unqualified. When measuring the mud content, carry out at least two mud content measurement tests, and take the arithmetic mean value as the mud content of the sample to be judged.

(2) The sample mortar is prepared from the sample whose mud content is not greater than the set threshold of mud content, and the fluidity of the sample mortar is tested to obtain the fluidity loss of the sample mortar between two test moments. The sample mortar is a mixture of cement, sample, and water with a set mass ratio, and the preparation method of the sample mortar is as follows: weigh the cement and water with a set mass ratio, add the sample after mixing for a set time, continue mixing for a set time, and complete the preparation of the sample mortar. When preparing the sample mortar, add a water reducer, which is mixed with cement and water. Water reducing agent is added in the preparation process of the standard sand mortar and river sand mortar.

(3) Set up multiple groups of control experiments to determine the range of loss threshold, and compare the obtained mobility loss with the range of loss threshold (mobility loss

value > 15 mm). When the mobility loss is within the range of the loss threshold, the sample is qualified as river sand. Otherwise, it is unqualified as weathered sand or mixed sand.

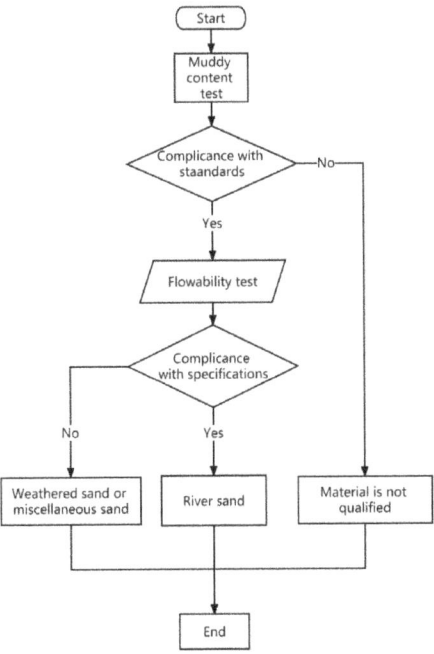

Figure 7. Flow chart of river sand inspection.

The mortar used for the fluidity test in the control experiment is a standard sand mortar and several groups of river sand mortar. The standard sand mortar is a mixture of cement, standard sand, and water with a set proportion, and the river sand mortar is a mixture of cement, river sand, and water with a set proportion.

The fluidity of sample mortar, standard sand mortar, and several groups of river sand mortar are tested by the mortar fluidity tester. The time interval between two test moments is 25 min–35 min.

In addition, the fluidity test is conducted at a temperature of $20 \pm 2\ °C$ and relative humidity of >50%.

6. Conclusions

(1) The average water content of river sand is slightly higher than that of weathered sand, and the water content varies greatly due to environmental factors in different regions, so it cannot be used as a basis for distinguishing weathered sand from river sand;

(2) In the particle analysis test, the non-uniformity coefficient and curvature coefficient show that the size distribution of river sand is more concentrated, and the size distribution of weathered sand is more dispersed. In the actual project, the mixing of different kinds of sand and gravel will affect the results of this parameter, so it cannot be used as a method to identify qualified sand;

(3) The apparent density and bulk density of river sand is slightly higher than that of weathered sand, but because the differences between data are not obvious, it cannot be used as the basis for distinguishing weathered sand from river sand;

(4) When the silt content is ≥3%, the fluidity loss value of river sand is relatively large. Therefore, the influence of silt content on the fluidity loss test of materials should be excluded before the fluidity test of mortar;

(5) The fluidity loss value of weathered sand is significantly larger than that of standard sand and qualified river sand. On the premise that the mud content of the mobilized material meets the standard, the fluidity loss value is >15 mm, which can be judged as weathered sand; that is, the river sand is unqualified;

(6) The fluidity loss of qualified river sand is less than 15 mm, and the fluidity loss of weathered sand or unqualified sand is more than 15 mm, so the threshold range of qualified material fluidity loss is determined: under the condition of qualified mud content when the fluidity loss is more than 15 mm, it can be judged as weathered sand or mixed sand. The smaller the fluidity loss, the better the river sand quality.

In order to improve the soil quality of the country and reduce the occurrence of natural disasters, Koki Nakao [31] of Japan conducted a visual and measurable assessment of the quality and performance of ground improvement through MPS-CAE analysis and worked through computer simulation. A series of operations performed by DRT and common (NT) RS-DMM are extracted using 3D models. Then, the internal condition of the ground and the displacement reduction performance are evaluated during each construction period. In the future, the distinction between weathered sand and river sand can be accurately calculated by similar numerical simulation methods to reduce judgment error. However, due to technical limitations, this method still has certain limitations. The method provided in this paper has a high reference value for the current construction field.

To summarize the above conclusions, this paper explores the method of reasonably distinguishing the weathered sand of river sand through several tests. Whether it is the test comparison of water content, particle size distribution, apparent density, and bulk density, the final difference between the two is almost insignificant and also confirms the similarity between river sand and weathered sand from a scientific point of view. Finally, two parameters, mud content, and fluidity, are selected to judge the quality of sand and stone samples. This method can control the quality of incoming river sand more accurately and make the inspection method standardized, which is convenient for strengthening laboratory management and suitable for popularization and application. In addition, during the application of this method, the water absorption of materials can be reflected, on the other hand, to reflect the workability and fluidity of the tested materials used for mixing concrete. It is suitable for standardized management and can be judged by the objective data obtained from the test with high accuracy. It avoids the defects that the results are affected by subjective factors, and the identification error is large only by the subjective experience judgment, which reduces the difficulty of raw material quality control in the engineering application. It has an important reference significance for the mobilization acceptance of sand and stone materials in the actual project.

Author Contributions: Conceptualization, H.Z.; methodology, H.Z. and J.L.; software, H.Z. and J.L.; resources, H.Z.; writing—original draft preparation, H.Z. and D.C. (Degang Cheng); writing—review and editing, H.Z. and D.C. (Degang Cheng); supervision, H.Z.; project administration, H.Z., L.Z. and D.C. (De Chang); funding acquisition, H.Z. and L.Z. All authors have read and agreed to the published version of the manuscript.

Funding: National Key Research and Development Program of China (2022YFB2601900).

Institutional Review Board Statement: Not applicable.

Informed Consent Statement: Not applicable.

Data Availability Statement: All data that support the findings of this study are included within the article.

Conflicts of Interest: The authors declare no conflict of interest.

References

1. Remišová, E.; Decký, M.; Mikolaš, M.; Hájek, M.; Kovalčík, L.; Mečár, M. Design of Road Pavement Using Recycled Aggregate. In *IOP Conference Series: Earth and Environmental Science*; IOP Publishing: Philadelphia, PA, USA, 2016; Volume 44.
2. *ISO 9000:2015*; Quality Management Systems—Fundamentals and Vocabulary. International Organization for Standardization: Geneva, Switzerland, 2021.
3. Yan, Z. Analysis of Road Performance and Microstructure of Cement Stabilized Weathered Sand. Master's Thesis, Shandong Jianzhu University, Jinan, China, 2021.
4. Lin, Y. The influence of river sand containing mud on the performance of machine-made sand concrete. *Commer. Concr.* **2019**, *27*, 67–70.
5. Zhao, W.; Yang, W.; Zhang, Y. Micro comparative study of machine-made sand and river sand in western Guangdong. *Eng. Qual.* **2020**, *38*, 54–57.
6. Zhang, X. Research on the influence of river sand and aggregate on the performance of UHPC. *Low Temp. Build. Technol.* **2021**, *43*, 22–25.
7. Wang, J. Research on the application and quality control of machine-made sand and river sand in concrete. *Sci. Technol. Innov. Her.* **2018**, *13*, 13,36–37.
8. Fu, X. Study on Weathering Microscopic Characteristics of Sandy Conglomerate in Mogao Grottoes. Master's Thesis, Lanzhou University, Lanzhou, China, 2013.
9. Cai, J. Effect of Stone Powder on the Performance of Machine-Made Sand Concrete and Its Mechanism. Doctor's Thesis, Wuhan University of Technology, Wuhan, China, 2006.
10. Liu, Z. Research on the Influence and Mechanism of Fine Powder in Machine-Made Sand on Concrete Performance. Doctor's Thesis, Wuhan University of Technology, Wuhan, China, 2016.
11. Li, C.-Z.; Qi, Y.-J.; He, G.-M.; Zhang, Q.-G. Experimental study on adaptability of machine-made sand aggregate and water-reducing agent. *J. Build. Mater.* **2008**, *6*, 642–646.
12. Yang, J.; Xu, W.; Zhang, G.; Tang, Y. Unconfined compressive strength of cement stabilized waste weathered sand Strength Test Research. *Environ. Sci. Technol.* **2014**, *37*, 52–56.
13. You, Y.; Xiang, J. Freeze-thaw damage law of strength and microstructure of cement stabilized soil. *Portland Bull.* **2020**, *39*, 453–458.
14. Sathiparan, N.; Anburuvel, A.; Maduwanthi, K.A.P.N.; Dasanayake, S.R.A.C.B. Effect of moisture condition on cement masonry blocks with different fine aggregates: River sand, lateritic soil and manufactured sand. *Sādhanā* **2022**, *47*, 270. [CrossRef]
15. Xie, Q.; Xiao, J.; Zong, Z. Strength and microstructure of seawater and sea sand mortar after exposure to elevated temperatures. *Constr. Build. Mater.* **2022**, *322*, 126451. [CrossRef]
16. Arulmoly, B.; Konthesingha, C.; Nanayakkara, A. Effects of microfine aggregate in manufactured sand on bleeding and plastic shrinkage cracking of concrete. *Front. Struct. Civ. Eng.* **2022**, *16*, 1453–1473. [CrossRef]
17. Arulmoly, B.; Konthesingha, C.; Nanayakkara, A. Effects of blending manufactured sand and offshore sand on rheological, mechanical and durability characterization of lime-cement masonry mortar. *Eur. J. Environ. Civ. Eng.* **2022**, *26*, 7400–7426. [CrossRef]
18. Arulmoly, B.; Konthesingha, C. Pertinence of alternative fine aggregates for concrete and mortar: A brief review on river sand substitutions. *Aust. J. Civ. Eng.* **2022**, *20*, 272–307. [CrossRef]
19. Deng, Y.; Yilmaz, Y.; Gokce, A.; Chang, C.S. Influence of particle size on the drained shear behavior of a dense fluvial sand. *Acta Geotech.* **2021**, *16*, 2071–2088. [CrossRef]
20. Thankam, L.; Neelakantan, T.R.; Christopher Gnanaraj, S. Potential of Fly Ash Polymerized Sand as an Alternative for River Sand in Concrete—A State of the Art Report. *IOP Conf. Ser. Mater. Sci. Eng.* **2020**, *1006*, 012039. [CrossRef]
21. Jiang, Z.; Mei, S. *Machine-Made Sand High-Performance Concrete*; Chemical Industry Press: Beijing, China, 2007.
22. Wang, S.; Yan, Z.; Wang, L.; Wang, H.; Zhang, X.; Hu, W. Study on road performance of weathered sand as roadbed filler. *Eng. Constr. Des.* **2021**, *3*, 61–63.
23. Guo, Y.; Shen, A.; Gao, T.; Li, W. Road performance test and weathering degree evaluation of weathered rock subgrade filler. *J. Transp. Eng.* **2014**, *14*, 15–23.
24. Huo, W. Vigorously Promoting Machine made Sand to Replace River Sand. *Constr. Enterp. Manag.* **2019**, *61*.
25. Ran, R.; Zhu, Y. Research on the preparation of high-quality sand by using tailings sand and fine river sand in a project in West Africa. *Sichuan Water Conserv. Power Gen.* **2021**, *40*, 47–51+73.
26. Ma, Y. Experimental study on the effect of mud content on unconfined compressive strength of cement stabilized weathered sand. *China Water Transport.* **2019**, *19*, 223–224+227.
27. *JTG E42-2005*; Test Rules for Aggregates of Highway Engineering. China Communications Press: Beijing, China, 2005.
28. *GBT 17671-1999*; Test Method for Strength of Cement Mortar (ISO Method). China Standards Press: Bejing, China, 1999.
29. *JC/T 681-2005*; Planetary Cement Mortar Mixer. China Building Materials Press: Bejing, China, 2005.
30. *GBT-2419-2005*; Method for Determining the Fluidity of Cement Mortar. China Standards Press: Beijing, China, 2005.
31. Nakao, K.; Inazumi, S.; Takaue, T.; Tanaka, S.; Shinoi, T. Evaluation of Discharging Surplus Soils for Relative Stirred Deep Mixing Methods by MPS-CAE Analysis. *Sustainability* **2021**, *14*, 58. [CrossRef]

Disclaimer/Publisher's Note: The statements, opinions and data contained in all publications are solely those of the individual author(s) and contributor(s) and not of MDPI and/or the editor(s). MDPI and/or the editor(s) disclaim responsibility for any injury to people or property resulting from any ideas, methods, instructions or products referred to in the content.

Article

Compressive Strength Estimation of Waste Marble Powder Incorporated Concrete Using Regression Modelling

Manpreet Singh [1,†], Priyankar Choudhary [2,†], Anterpreet Kaur Bedi [3,*,†], Saurav Yadav [4,†] and Rishi Singh Chhabra [5,*,†]

1. Department of Civil Engineering, Thapar Institute of Engineering and Technology, Patiala 140412, India
2. Department of Computer Science and Engineering, Indian Institute of Technology, Roopnagar 335073, India
3. Department of Electrical and Instrumentation Engineering, Thapar Institute of Engineering and Technology, Patiala 140412, India
4. Birla Institute of Technology, Pilani 333031, India
5. Indian Institute of Technology, Roorkee 247665, India
* Correspondence: anterpreet.bedi@thapar.edu (A.K.B.); rsinghchhbara@ce.iitr.ac.in (R.S.C.)
† These authors contributed equally to this work.

Abstract: A tremendous volumetric increase in waste marble powder as industrial waste has recently resulted in high environmental concerns of water, soil and air pollution. In this paper, we exploit the capabilities of machine learning to compressive strength prediction of concrete incorporating waste marble powder for future use. Experimentation has been carried out using different compositions of waste marble powder in concrete and varying water binder ratios of 0.35, 0.40 and 0.45 for the analysis. Effect of different dosages of superplasticizer has also been considered. In this paper, different regression algorithms to analyse the effect of waste marble powder on concrete, viz., multiple linear regression, K-nearest neighbour, support vector regression, decision tree, random forest, extra trees and gradient boosting, have been exploited and their efficacies have been compared using various statistical metrics. Experiments reveal random forest as the best model for compressive strength prediction with an R2 value of 0.926 and mean absolute error of 1.608. Further, shapley additive explanations and variance inflation factor analysis showcase the capabilities of the best achieved regression model in optimizing the use of marble powder as partial replacement of cement in concrete.

Keywords: compressive strength; waste marble powder; concrete; machine learning; regression

Citation: Singh, M.; Choudhary, P.; Bedi, A.K.; Yadav, S.; Chhabra, R.S. Compressive Strength Estimation of Waste Marble Powder Incorporated Concrete Using Regression Modelling. *Coatings* **2023**, *13*, 66. https://doi.org/10.3390/coatings13010066

Academic Editors: Ofelia-Cornelia Corbu and Ionut Ovidiu Toma

Received: 25 November 2022
Revised: 19 December 2022
Accepted: 26 December 2022
Published: 30 December 2022

Copyright: © 2022 by the authors. Licensee MDPI, Basel, Switzerland. This article is an open access article distributed under the terms and conditions of the Creative Commons Attribution (CC BY) license (https://creativecommons.org/licenses/by/4.0/).

1. Introduction

Concrete is considered as the second most used material on earth, with cement comprising the primary source of its binder material. Cement production is the source of about 8% of the world's carbon dioxide production. It is also the most expensive concrete component. This forces engineers to choose carefully between high strength and affordability. Numerous studies have been performed in order to introduce newer materials as a replacement for cement. However, the conventional approach of relying solely on laboratory test data is quite costly and inefficient. One requires impractically huge number of controlled testing to reach a reasonable conclusion and thus roll out innovations in the construction industry. In times of computing advancements, it becomes imperative to introduce newer technologies in concrete testing. Marble is another extensively used construction material, representing the most used natural stone in the world [1]. About 500 million metric tons of the material is mined annually [2] out of which, approximately 10% originates from India [3]. The Rajasthan state alone accounts for 85–90% of Indian marble production. The marble industry produces marble dust every year which bears essentially no utility, resulting in:

1. Damage to soil due to dumping of waste;

2. Degradation of groundwater

There is a lack of codal provisions to use marble dust in concrete. Consequently, this has prevented any large-scale commercial use. Well-documented research on the use of a certain amount of marble dust in concrete can significantly reduce costs. Its importance becomes quite evident given the losses that the industry has suffered from the COVID-19 pandemic, damage that will take several years to recover from. Cost reductions can help to bridge the gap. Further, large quantities of solid waste generated from the marble industry also need to be recycled to boost environment protection and economy.

With the advancement of various soft computing techniques, data handling capabilities of researchers have increased and are more efficient now over conventional ways. As a result, many algorithms have gained popularity in the due course of time. However, a detailed comparison between these algorithms still remains less explored and need further study. Use of machine learning (ML) algorithms can provide a reliable mix design for industry. In the long run, the Indian Standard codes can also be updated with marble dust parameters. ML is being increasingly used in civil engineering for the purpose of strength prediction. Elyas Asadi Shamsabadi et al. [4] studied marble-dust-incorporated concrete for strength predictions. Extreme Gradient Boosting (XGB) and ANN were found to be appropriate models, while XGB had fewer errors in prediction, on the other hand, Artificial Neural Network ANN was deemed to be more sensitive to marble dust content. The study also confirmed the non-pozzolanic nature of marble dust incorporated in concrete. Further, Karimipour et al. [5] conducted a soft-computing-based study involving marble dust in steel-fiber-reinforced self-consolidation concrete. In addition, other ingredients such as granite, red mud and limestone were also used. ANN, GMDH-NN and GMDH-Combi models were exploited to predict split tensile strength as well as compressive correction factors. It was observed that ANN using 4 neurons and 1 hidden layer gave better performance than other 2 models. Of the other two, GMDH-NN (neural networks group method of data handling) performed better than GMDH-Combi (combinatorial algorithm group method of data handling). Further, Hong-Hu Chu et al. [6] explored gene-expression programming (GEP) and multi-expression programming (MEP) to predict the compressive strength of geopolymer concrete. Various parameters such as curing regime, silica and superplasticizer content, curing period and age of the sample were related to compressive strength. However, GEP resulted in higher correlation coefficient, minimal statistical error and simplicity. In addition, it covered the impact of each independent parameter, as it was utilized for parametric study and sensitivity as well. Swaidani et al. [7] discussed the use of scoria as a partial replacement for cement in making environmentally friendly concrete. Concrete strength and durability were studied for the same purposes. It was inferred that ANN model was well suited for concrete strength prediction at different curing times for different mix ingredients. Further, it was also observed that compressive strength prediction for concrete comprising ground granulated blast furnace slag could also be achieved using ANN. A Multiple Regression Analysis (MRA) and an ANN model were constructed for comparing the predicted compressive strength of high-performance concrete using nano silica and copper slag as partial replacement and fine aggregate replacement, respectively, [8], by collecting data from laboratory experiments. Levenberg–Marquardt (LM) algorithm was used for generating the ANN model. Models for predicting compressive strength on 22 mixes were generated using MRA analysis and ANN, with ANN resulting in higher accuracy and correlation values. Further, Naddaf et al. [9] proposed an ANN and GEP model to train and study 640 different mix designs and predict various properties by partially replacing cement with nano silica and micro silica by weight. Kazemi et al. [9] presented an ANN model for compressive strength prediction of mortar mixes containing cement of different strengths. The study predicted good accuracy and higher R value in predicting the compressive strength of the mortar. Later, Naderpour et al. [10] exploited ANN to predict the compressive strength of environmentally friendly concrete, comprising recycled aggregate material. The data used for developing the ANN Model were prepared from the literature. Back propagation network was used in the study, resulting in efficient predictions. An Adaptive Neuro Fuzzy Inference System (ANFIS) model was provided by Nejadi et al. [11] that established a relationship between the compressive strength of self-compacting concrete (SCC) and slump flow and mix proportion. In past studies, SCC has proved advantageous in achieving sustainable characteristics, reduction in the overall structural costs, increase in construction rate, quality of casted structure and increase in construction productivity. Poon et al. [12] aimed to predict the compressive strength of concrete comprising recycled aggregate, using ANN. The model constituted 14 different properties of the constituents to predict the 28-day compressive strength of the concrete. Soft computing has been found to have applications in recycled concrete, where deep-learning-based techniques have been found to outperform traditional neural networks in terms of precision, generalization and efficiency [13]. A deep neural network was designed by Ly et al. [14] for predicting compressive strength of concrete with rubber content, resulting in high accuracy. Further, Nunez et al. [15] studied and analysed different ML models predicting the compressive strength of concrete. It was observed that ANN was the best-suited method for prediction, but was accompanied by a lack of clarity in the prediction process with high computational costs. Fuzzy-logic-based models had similar accuracy, but were of higher complexity. Furthermore, Support Vector Machine (SVM)-based models were considered to have lower computational costs than ANN but with comparable accuracy. Hybrid models were found to be the most promising due to the presence of a secondary model to obtain hyperparameters for the main model. Mansouri et al. [16] explored the usage of 4 types of soft computing techniques, viz., ANN, ANFIS, MARS (Multivariate Adaptive Regression Splines) and M5Tree (M5 Model Tree), to predict FRP-confined concrete. These models were found to outperform the existing models, with ANN resulting in the best estimation of strain enhancement ratio. Sahoo et al. [17] studied fly-ash-based concrete using ANN modelling by considering two different replacement levels, i.e., one at Low-25%, the other at High-40%. Fly ash concrete resulted in a better performance than control concrete over long periods of sulphate exposure. The ANN model was developed by minimization of mean square error. Furthermore, the R^2 values ranged from 0.953 to 1.00, depicting high accuracy and reliability of the model. Recently, Khan et al. [18] discussed the performance of ANN, ANFIS and GEP models in order to estimate compressive strength of geopolymer concrete based on fly ash, with ANFIS giving the best performance of all.

As can be observed, there are only a handful of studies regarding the effects on compressive strength for partial replacement of cement by marble dust. Although a huge number of experimental studies have been carried out for investigating possible effects of waste marble powder (WMP) on concrete, there is still a lack of in-depth understanding on use of WMP in concrete. Regarding the use of soft computing, it remains in its infancy in civil engineering applications. As discussed above, soft computing has been largely applied in materials other than marble dust, such as fly ash, FRP and rubber, to name a few. This study aims to contribute by filling this gap and thus, pave the way for further research and eventual codification of marble dust in concrete. The soft computing approach makes use of various algorithms to arrive at its conclusions. There exists a relentless lack of a comprehensive understanding related to the dosage level of WMP so as to intensify the engineering properties of concrete. The WMP characteristics vary based on geological and weather conditions, and also on the methods for production of marble sheets for construction industries. In addition, thermogravimetric results along with various phases in cement paste containing WMP in cement and concrete exhibited the benign nature of the product [19,20]. Various alterations in experimentation make it difficult to generalise and hence, achieve a standard mixture design for WMP-incorporated concrete. Exploiting ML techniques can help in achieving cost- and time-effective simulation of the same, thus maximising the application of WMP in concrete industry by complementing the outcomes acquired from the already existing experimental investigations.

2. Data Collection and Modelling

Data for the present study were collected from experimental trials previously conducted by Singh et al. [21]. Table 1 shows the physical and chemical properties of cement and dried marble slurry, respectively.

Different mix design proportions for the combinations are presented in Table 2. Due the decrease in slump with the increase in dosage of marble dust a superplasticizer was used to keep a constant slump of 100 ± 10 mm. 12 different mix designs were designated for different variations in dosage of marble dust at 5 different replacement levels and superplasticizer. Thus, the data comprises of 60 instances with one associated real-value target, viz., compressive strength. We further augmented the data by replicating each instance 12 times and introducing an error to the target within the range of −10% to +10%. Thus, a total of 720 instances were considered for the experiment. 12 concrete cube samples for each variation were casted and tested thus generating a data set of 720 values. Table 3 shows the range of parameters used for developing the model.

Relationship of all input parameters with compressive strength is shown in Figure 1. Fine aggregate and waste marble powder had the strongest correlation with CS, followed by superplasticizer, water and cement. However, cement showed a weak linear relationship with CS, which is not generally the case observed in concrete. This can be owed to the poor distribution of the machine to learn from the data and that is where the expert opinion and experimental results play an important role.

Table 1. Physical and chemical properties of cement and marble dust.

Chemical Composition	OPC (%)	Marble Dust (%)	Physical Properties	OPC (%)	Marble Dust (%)
SiO_2	20.27	3.86			
Al_2O_3	5.32	4.62			
Fe_2O_3	3.56	0.78	Specific gravity	3.15	2.67
CaO	60.41	28.63			
MgO	2.46	16.9			
SO_3	3.17	-	Fineness (m^2/kg)	313	250
LOI	3.55	43.3			

Table 2. Proportions of concrete mixtures.

Water–Binder Ratio	Cement (kg/m^3)	Marble Dust (%)	Marble Dust (kg/m^3)	Coarse Aggregate (kg/m^3)	Fine Aggregate (kg/m^3)	Superplasticizer Admixture (L/m^3)	Water (kg/m^3)
0.35	422	0	0	1278	689	0.9	148
0.35	400.9	5	21.1	1278	689	1	148
0.35	379.8	10	42.2	1278	689	1.1	148
0.35	358.7	15	63.3	1278	689	1.2	148
0.35	337.6	20	84.4	1278	689	1.3	148
0.35	316.5	25	105.5	1278	689	1.4	148
0.4	394	0	0	1257.2	707.2	0.63	158
0.4	374.3	5	19.7	1257.2	707.2	0.67	158
0.4	354.6	10	39.4	1257.2	707.2	0.74	158
0.4	334.9	15	59.1	1257.2	707.2	0.84	158
0.4	315.2	20	78.8	1257.2	707.2	0.95	158
0.4	295.5	25	98.5	1257.2	707.2	1	158
0.45	351	0	0	1183	858	0.35	158
0.45	333.45	5	17.5	1183	858	0.39	158
0.45	315.9	10	35.1	1183	858	0.45	158
0.45	298.35	15	52.6	1183	858	0.52	158
0.45	280.8	20	70.2	1183	858	0.61	158
0.45	263.25	25	87.7	1183	858	0.7	158

Table 3. Range of parameters used for Modelling.

Variables	Minimum	Maximum
Cement (kg/m^3)	263.25	450
Marble dust (kg/m^3)	0	112
Water (kg/m^3)	148	200
Superlasticizer (kg/m^3)	0	1.4
Slump (mm)	84	199
Aggregate (kg/m^3)	1011.9	1278
Sand (kg/m^3)	675	858
Compressive strength (MPa)	21.23	42.67

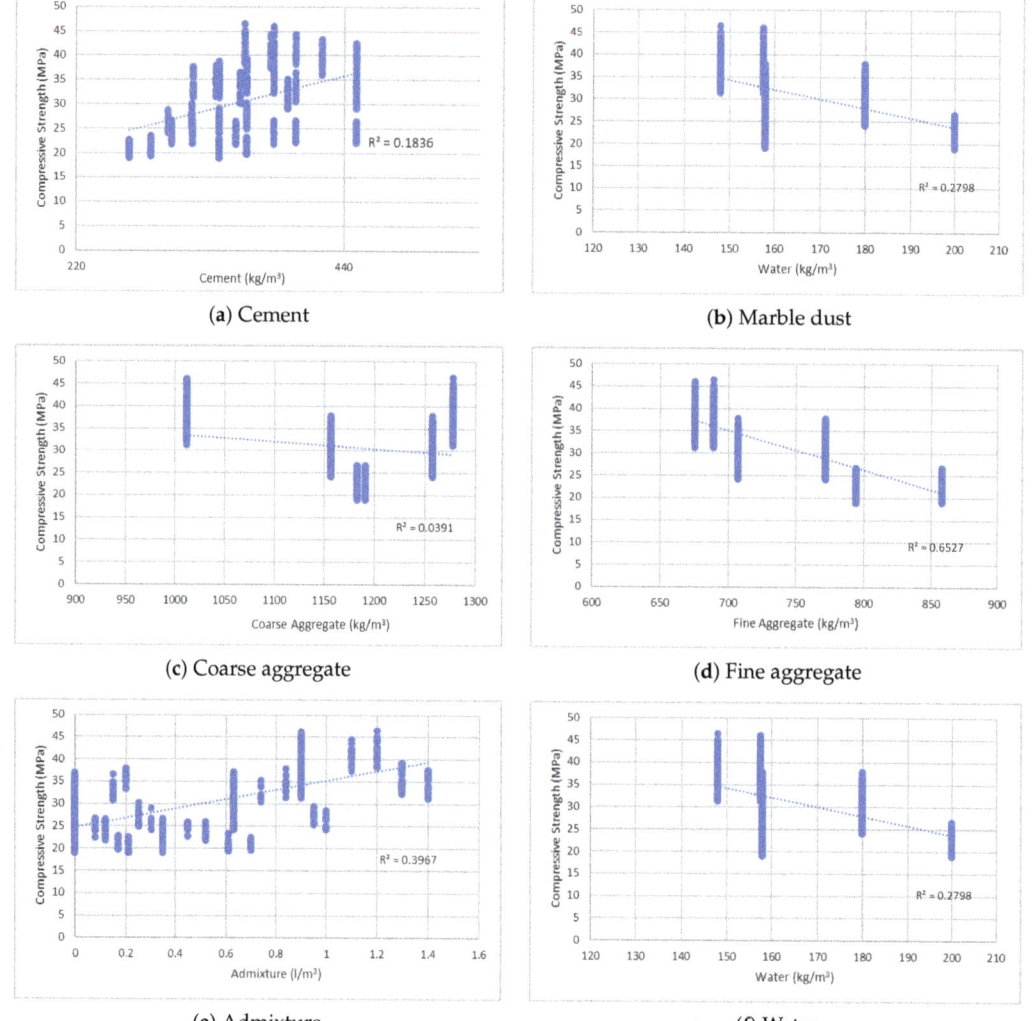

Figure 1. Cont.

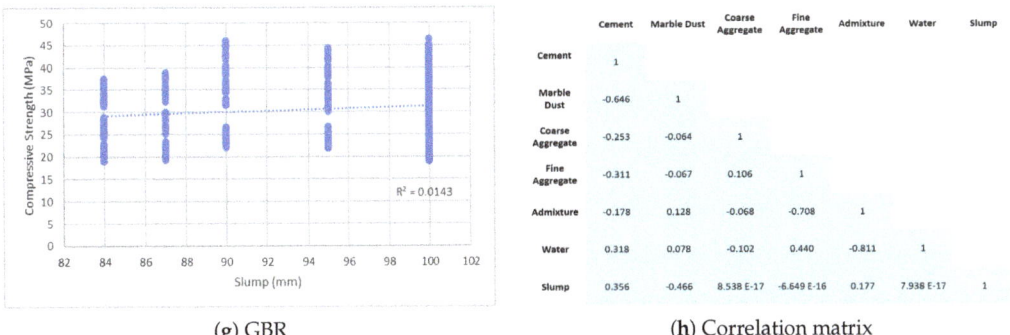

(g) GBR (h) Correlation matrix

Figure 1. (a–g) Correlation between input variables and compressive strength; (h) Correlation between various input variables.

3. Machine Learning Modelling

Different feature compositions contribute to the strength of concrete in different ways. Since the present work aims for estimating the compressive strength of concrete with partial replacement of cement by WMP at various compositions, hence, the problem is treated as that of regression. In the proposed work, we have selected different regression algorithms to analyse the effect of WMP on concrete, viz., Multiple Linear Regression (MLR), K-Nearest Neighbour (KNN), Support Vector Regression (SVR), Decision Tree (DT), Random Forest (RF), Extra Trees (ET) and Gradient Boosting (GB) [22]. For evaluating and comparing efficacy of the applied models, various statistical metrics have been used. These include R^2 Score, MAE, MSE, RMSE, MAPE, MBE which are directly computed using the first and second power of the error in prediction values. The lower value of MAE, MSE, MAPE, MBE and RMSE implies higher accuracy of a regression model. However, a higher value of R square is considered desirable. Another parameter, known as T_{stat} is also evaluated to analyse the uncertainty level during the prediction.

Considering the total number of samples as N, let y_i be the original value of ith sample with y'_i being its corresponding predicted value. Taking $\bar{y}$ as the average of all the true values, the performance parameters can be calculated as shown in Table 4. Each modelling technique was performed using in the Python programming language in i5 processor. A generalised process flow for each ML model is shown in Figure 2.

Tuning of hyperparameters for any ML algorithm is considered as a fundamental task for any ML algorithm. Manually estimating the performance of an ML algorithm can be challenging. Additionally, formation of different pairs of hyperparameters is also challenging. Hence, in our work, we have selected a Grid Search Strategy (GSS) so as to automate parameter tuning. GSS accepts manual sets of hyperparameters based on experience to form all exhaustive pairs of different hyperparametes. In order to evaluate performance of one specific pair of hyperparameters, a subset of data, known as validation data, is selected from the training data. Based on the performance on validation data, a set of hyperparameter is selected. GSS is used for all the algorithms mentioned above to generate the best combination of hyperparameters. The following section discusses the applied algorithms for the present work in detail

Table 4. Performance parameters used in the proposed method.

Metric	Formula	Description
R^2	$1 - \frac{\sum_{i=1}^{N}(y_i - y'_i)^2}{\sum_{i=1}^{N}(y_i - \bar{y}_i)^2}$	Coefficient of determination: Measure of goodness of the fit
MSE	$\frac{1}{N}\sum_{i=1}^{N}(y_i - y'_i)^2$	Mean squared error: Measures closeness of the fitted line to the data points
RMSE	$\sqrt{\frac{1}{N}\sum_{i=1}^{N}(y_i - y'_i)^2}$	Root mean squared error: Measures spread of the residuals
MAE	$\frac{1}{N}\sum_{i=1}^{N}|y_i - y'_i|$	Mean absolute error: Measures average of absolute differences between true and predicted values
MAPE	$\frac{1}{N}\sum_{i=1}^{N}\left|\frac{y_i - y'_i}{y_i}\right|$	Mean absolute percentage error: Measures average of absolute percentage differences between true and predicted values
MBE	$\frac{1}{N}\sum_{i=1}^{N}(y_i - y'_i)$	Mean bias error: Measures average of differences between true and predicted values
T_{stat}	$\sqrt{\frac{(N-1)MBE^2}{RMSE^2 - MBE^2}}$	t-statistic test: Measures significance of the differences between true and predicted values

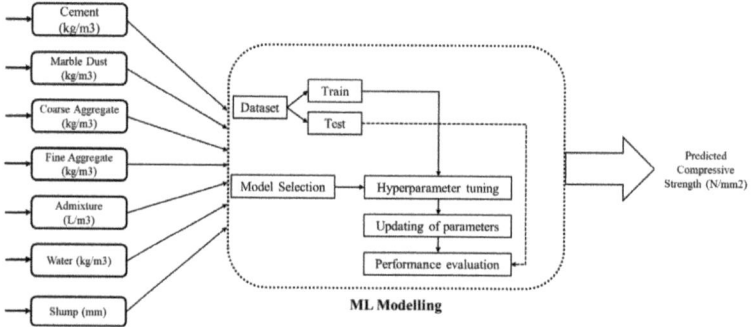

Figure 2. Process flow for various ML algorithms.

3.1. Multiple Linear Regression Model

Multiple Linear Regression MLR model assumes data points, i.e., inputs, to have a liner relationship with the outcome to be estimated. Thus, the model aims to learn a linear dependence of the output variables (compressive strength in our case) on the independent (features) variables, giving the best-fit regression line for the data. Taking into consideration Occam's razor rule [23], MLR model is applied initially to study the need of exploiting more complex data-driven regression modelling techniques. The output (i.e., concrete strength) is weighted sum of the features used. Weights used in this model are optimized using ordinary least-squares method on the estimated and actual outcomes of training data. This helps in predicting the target values such that the error difference between the predicted and true value is minimum.

3.2. K-Nearest Neighbour

K-nearest neighbour KNN [24] is a non-parametric regression method that is used to approximate the relation between input features and out variable by averaging the observations in the same neighbourhood. It exploits 'feature similarity' to predict values of any new data points. In this algorithm, all the training data is stored in memory and similarity between each test instance with the training data is calculated. The most similar K instances are chosen for prediction. The size of the neighbourhood is chosen such that the predicted value is in close proximity to the target value, resulting in minimum errors. In the present work, the KNN regression algorithm takes the number of neighbors (K) and distance metric (d) as its parameters, and considers uniform weights for all the features. In this model, similarity between a test instance with all the training instances is measured using Euclidean distance. The instances with the highest similarity, i.e., minimum distance, are chosen as the K neighbors. The value of K is chosen as 1, 3, 5, 10 and 15% of the total data, thus resulting in K = 8, 22, 36, 72, 108 and 144, respectively.

The model can be very complex for large training data, and may be infeasible to predict when there a large amounts of data.

3.3. Support Vector Regression

Support Vector Regression SVR [25] tries to learn a function that approximates the given input instance to a discrete value output. SVR aims to learn a hyperplane that can distinguish the data points with different outcomes. It is possible that data points may not be separable in lower dimensions. Therefore, a kernel function is used in order to map the data to a higher dimension. Further, there may be multiple hyperplanes that could separate the data points. However, only one particular hyperplane that demonstrate the maximum separation between the outcomes is selected. The separation margin around the hyperplane is termed as boundary. Different parameter in SVR are the type of kernel, regularization parameter (C) and regularization parameter penalty (epsilon) as its parameters. Since the strength of regularization in inversely proportional to C, hence, the value of C has been experimentally chosen to be 10. Further, the regularization parameter penalty helps the optimisation function to obtain optimal solutions by imposing a cost during the training process.

In the present work, L2-loss is used as the type of penalty. Thus, if the error value is less than 0.2 (epsilon), no penalty is associated during regularization. Further, RBF kernel is taken into consideration which assumes non-linearity in separating the training data points.

3.4. Decision Tree

Decision tree DT is a non-linear algorithm, that makes use of tree representation to solve the regression problem. DT employs a "divide and conquer" approach, where a complex task is divided into simpler, regional tasks. A tree is composed of decision nodes (features) and leaves (outcome). Commencing from the root node, each decision node applies a splitting test to the input. Based on the outcome of the test, one of the branches is chosen. The search stops upon reaching a leaf. Each path from the root to a leaf corresponds to a conjunction of different conditions in the decision nodes on the path and such a path can be written as an if–then rule. Thus, a tree can be converted to a rule base of if–then rules that are easy to interpret.

The size of the DT depends on the complexity of the problem underlying the data. Selecting an optimum size of the tree is the major hyper-parameter that can affect the efficiency of the model. Trees with less depth can lead to under-fitting, failing to reach an optimum decision owing to under-training, whereas deeper trees result in models with high complexities.

In the present work, the model considers compressive strength as its leaf node, and the input parameters as its internal nodes. The depth of the tree and splitting criteria are considered as two critical parameters that need to be tuned for best results. In this model, squared error has been selected as the splitting criteria. Tree depths of 1, 2, 3, 4, 5, and 6 have been chosen so as to search for the best parameter.

3.5. Random Forest

Random forest RF is also a non-linear model. As an improvement to the DT algorithm, RF [26] algorithm for regression was introduced, which takes into account the decision of multiple DTs. RF performs bootstrapping to construct multiple subsets of the dataset for each tree. Here, bootstrapping implies sample selection from dataset without replacement. RF is a supervised learning regression algorithm that makes use of ensemble learning by combining predictions from multiple trees so as to make more accurate predictions compared to a single model. Each tree runs individually and in parallel to each other during training time so as to make predictions. The model estimates the compressive strength based on majority voting of multiple trees.

However, similar to DT, optimum choice of tree depths is an important factor for efficient performance. Furthermore, the number of trees in a forest have to be chosen accurately so as to avoid the problem of overfitting.

In the proposed work, 100 trees are considered in the forest. Similar to DT, results for RF model have been studied for tree depths of 1, 2, 3, 4, 5, and 6. This implies that each tree in the forest bears depth.

3.6. Extra Trees

The function of Extra-trees ET regressor [27] is the same as RF, but differs in two ways, viz., selection of the splitting method and bootstrapping. A DT and RF choose the best split whereas ET chooses a random split. Moreover, unlike RF, ET does not perform bootstrapping to construct multiple subsets of the dataset.

3.7. Gradient Boosting

Gradient boosting GB algorithm [28] uses sequence of N number of DTs. A regression model is developed sequentially in order to obtain a strong regression model. First, a DT regression model is trained on using available features and real-valued regression output. Further, the residual to true and estimated real-valued regression output is used for training a new regression model but features remain unchanged. Further, the residual of second model work as the label for the third model. This process is continued until all the trees are trained. During real-time deployment, when a test instance arrives, it is fed to all the trained regression models for estimating the output values. Further multiple outputs are converted to single output values using the parameter 'shrinkage'.

4. Experimental Section and Results

The experiments using different ML algorithms were performed and analysed. Results are described in two parts. The first part helps in deciding the regression model that can be used for best prediction of compressive strength of concrete incorporated with WMP. In the latter part, the best chosen regression model is further analysed to study the relevance of each component used in manufacturing of concrete. Further, it would also help in deciding the best proportion of WMP that can replace cement in order to achieve best compressive strength.

4.1. Results Using Various Regression Models

In order to evaluate the performance of each regression model, the dataset is initially standardized in a manner such that each feature has unit variance zero mean. As described earlier, the data for the current work were collected experimentally by Singh et al. for 12 different sets of design mixes, with each mix considering 5 different marble slurry percentage replacement levels, resulting in 60 variations. A set of 720 data points was generated by casting 12 concrete cubes for each of the 60 variations [21]. From the entire data of 720 samples, 70% is considered for training, while the remaining 30% is used for testing purposes. Further, the experiment on each setup is repeated five times so as to remove any biases from the results and average results have been reported.

Figures 3–5 depict the relationships between the true and predicted values for different models. Since a higher R^2 score indicates more perfectly fit data, hence, R^2 values of each regression model have been plotted in the figure. Owing to the lowest R^2 value of 0.852 for the test data, MLR is the least-suited model for predicting compressive strength of cement incorporated with WMP. Furthermore, the model results in maximum fluctuations in prediction (Figure 4), leading to maximum error values as can be seen in Figure 5. Thus, although Occam's razor rule prioritizes simpler models, in cases where simpler models are unable to perform efficiently, complex ML models prove useful in explaining the variation of dependent variables.

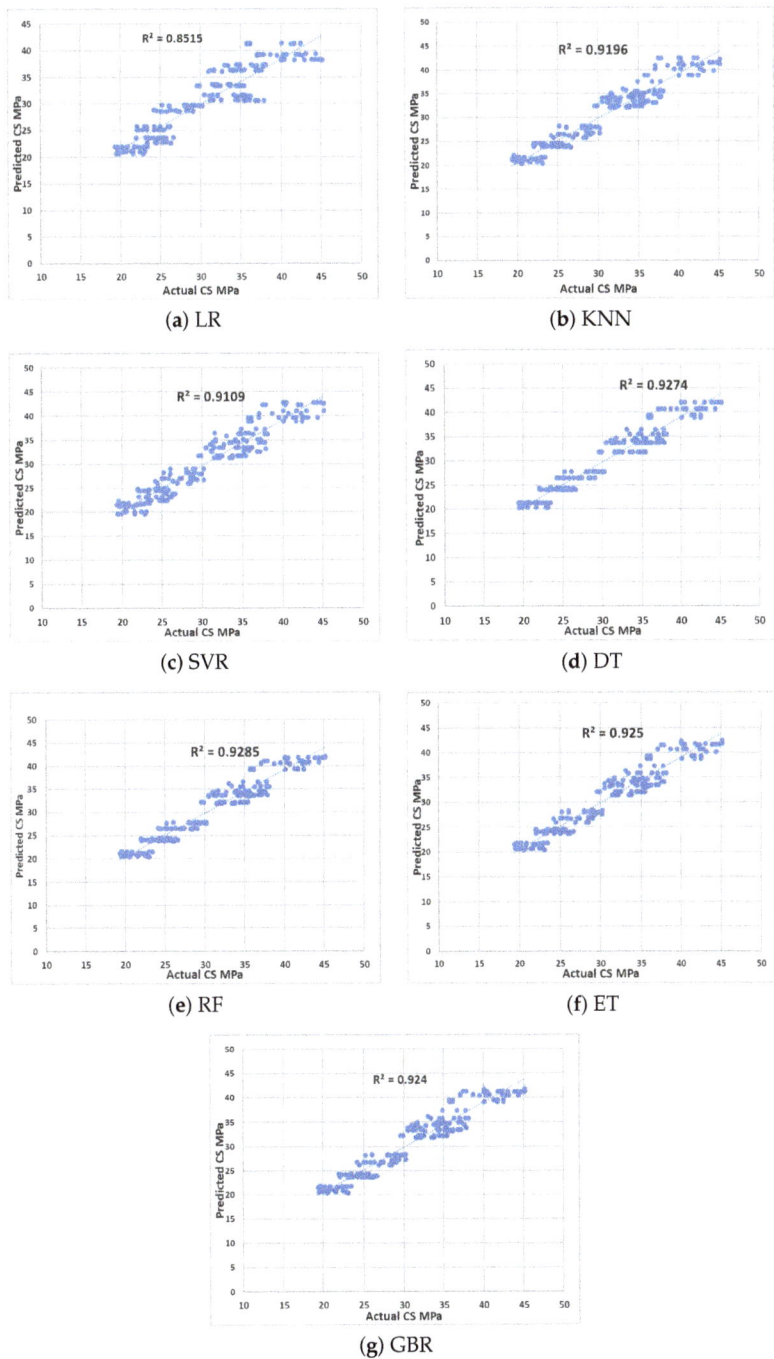

Figure 3. Performance of various ML models in CS prediction.

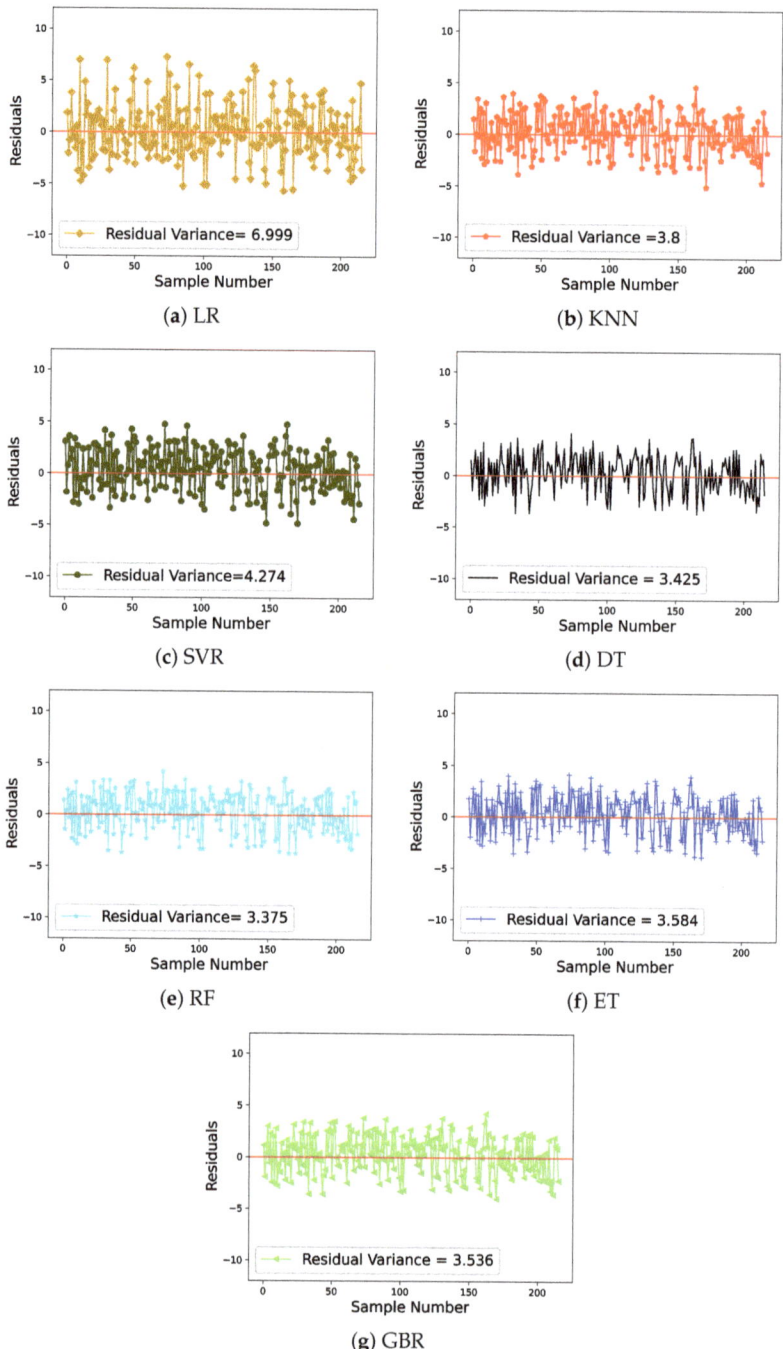

Figure 4. Fluctuations in errors, the model showing the least fluctuations represents the desirable outcome.

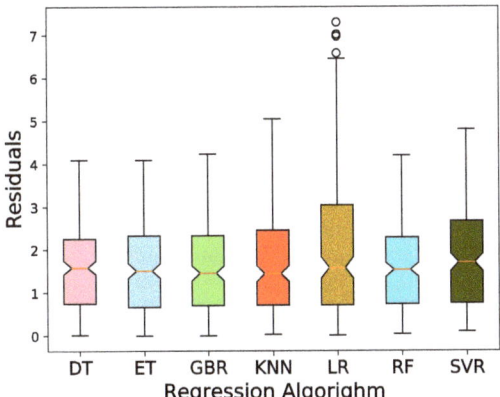

Figure 5. Error bound analysis, where the residuals are computed by taking absolute difference of true and estimated values.

From the figures, it can further be observed that RF algorithm gives the best performance out of all the models with an R^2 value of 0.926. Keeping in mind the increasing complexity of the model with increasing tree depths, a depth of 3 was found to give the best performance. Further, being flexible in nature, it is more convenient for the model to handle larger datasets more efficiently, hence, the method can be efficiently chosen for prediction applications. Moreover, the error graphs (Figures 4 and 5) show that the RF model provides the highest level of accuracy in prediction of compressive strength when compared to the other models. Similarly, for regression analysis using DT model, it was observed that a tree depth of 4 gave the best performance, taking in consideration the saturation in parametric values with increasing tree depths thereafter. For the same, the R^2 value was computed to be 0.924. However, in case of DT, a minute change in data might manifest in the structure of the tree, leading to instability. Moreover, RF algorithm solves the problem of overfitting that might occur in the case of the DT algorithm. Hence, the method is not much preferred to predict compressive strength of concrete from a given set of features.

Although ET and RF algorithms are very similar, the performance of the latter is slightly better than that of the former. Rather, in our case, performance of ET is quite similar to that of DT, with a similar R^2 value of 0.924. However, from Figure 5, it can be seen that predictions vary more from the true compressive strength values in case of ET as compared to RF algorithm. Further, for ET, best results were obtained for a tree depth of 6, which adds to the complexity of the algorithm in comparison to RF and DT, where optimal results were obtained for a tree depth of 3.

While considering GB algorithm, it was observed that the algorithm produces results in performance that is quite similar to ET algorithm with very minute difference of 0.001 in their R^2 values. Although GB is considered as one of the most powerful algorithm for regression applications, the presence of noise in the data makes it difficult for the algorithm to perform well. On the other hand, RF algorithm works efficiently even if data are missing or high noise content is present.

Further, in the case of KNN model, using GSS, the results are best obtained for 1% of neighbours (K), i.e., for K = 8 with R^2 value of 0.919. Since the model finds it difficult to handle noisy data and is sensitive to outliers, thus, with increasing number of neighbours, i.e., with increasing values of K, the error values are also increased. However, being a lazy learner owing to instance-based learning, the model requires all available data in order to make a prediction, thus making it even slower and costly for larger datasets. The results obtained for KNN are comparable to those obtained using SVR algorithm in terms of R^2. It can be observed from Figure 3 that the results for SVR model are better than the

MLR model by 6.92%, but underperforms when compared with the rest of the algorithms. Moreover, SVR model requires extensive feature scaling of variables prior to its application, thus making it computationally expensive.

Thus, the results show that the RF model is best suited to predict the compressive strength of concrete, followed by DT models. RF, being a powerful ML algorithm, can result in more accurate predictions when compared to the other algorithms, as can be seen in Figure 4. Furthermore, Figure 5 shows that the RF model gives the least variation in compressive strength values from their true values when compared with other algorithms. Further, it can handle missing data more efficiently and is usually robust to outliers.

Table 5 shows different performance measures on the applied models. From the above table, it can be observed that different performance measures consider different ML models as the best performing. The R^2 score, MAE, MSE and MAPE values are the best for RF model. On the other hand, RMSE and MBE values are best shown for MLR, whereas T_{stat} is best for GB modelling technique. Higher R^2 score for RF indicates that the model best fits the dataset compared to the rest. Further, the model shows least fluctuation in errors, as is indicated by MSE, MAE and MAPE values. This shows that there is minimal variance in residuals for RF model in comparison to the other ML models. Since the RF model is best for the majority of the performance parameters, and the other three parameters, i.e., RMSE, MBE and T_{stat} do not show any significant best performing model, hence RF was considered for further analysis. Thus, it can be seen that MLR is the least preferred model for regression since all the performance measures except RMSE and MBE are least preferred, thus leading to the need of more complex models for prediction. Furthermore, with exponentially increasing data in the current scenario, the applicability of MLR becomes minimal. Further, RF model gives the best estimation of compressive strength of concrete with partial replacement of cement by WMP. The overall performance of RF is higher when compared to the rest of the models. Other methods, compared to the rest of the methods, such as DT and ET can be considered as the next choice, but only in cases where data complexity is low.

Table 5. Performance measures of applied models.

Method	R^2 Score	MAE	MSE	RMSE	MAPE	MBE	t-Stat
MLR	0.852	2.095	7.152	9.455	6.819	0.007	0.191
KNN	0.919	1.655	3.914	9.692	5.427	0.028	0.155
SVR	0.911	1.747	4.380	9.660	5.761	0.048	0.175
DT	0.924	1.632	3.679	9.685	5.370	0.050	0.130
RF	0.926	1.608	3.561	9.679	5.291	0.051	0.126
ET	0.924	1.612	3.668	9.649	5.315	0.017	0.149
GB	0.923	1.621	3.719	9.650	5.333	0.037	0.100

4.2. Analysis of the Best Model

Each independent variable has its own contribution in deciding its effect on the compressive strength of concrete. Since RF regression technique exhibited best performance among all the models tested, the importance of features was analysed using the same. Average results are shown in the form of Variance Inflation Factor (VIF) in Figure 6. A VIF of 1 indicates that the corresponding feature has no correlation with any of the other features. Typically, a VIF value exceeding 5 or 10 is deemed too high. Any feature with such high VIF value is likely to be contributing to multicollinearity. As has been discussed in the literature and also shown by Singh et al. [21], marble dust does not explicitly affect the hydration process. Rather, it mainly works as a filler by also providing nucleation sites for enhanced hydration products. Accordingly, the contribution of marble dust on compressive strength has been found lesser as compared to the other input variables, although not entirely negligible (8%).

SHapley Additive exPlanations (SHAP) is used further for explaining the compound learned decision functions used by RF Technique as shown in Figure 7, where y-axis shows features used for the model, while the x-axis shows the impact of the corresponding feature on the model output. The position of feature between the peak and lowest values is indicated by the colour. Overlapping points show the density of Shapley values per feature. According to the figure, the RF technique is highly sensitive to WMP and fine aggregate content. Overall, mainly water, marble dust and fine aggregate contents are being used for prediction, followed by cement, superplasticizer admixture and Slump content.

Thus, from the overall analysis, it can be said that marble-slurry-incorporated concrete results in an improvement in mechanical properties at 15% replacement by weight of cement as compared to control mix for lower water–binder ratio of 0.35 and 0.40. For the water–binder ratio of 0.45, compressive strength is improved only up to a maximum replacement of 10%. Further, with the simultaneous increment in the dosage of superplasticizer and marble slurry, higher strength magnitude is observed as compared to constant dosage, owing to compactive power of superplasticizer admixture. This improvement is observed from the acceleration effect of WMP on the hydration process, which is further related to the formation of calcium carboaluminate hydrates. Furthermore, the improvement in binding capacity of carboaluminate is likely due to its compact structure as described by Bonavetti et al. [29]. Singh et al. [21] demonstrated the compaction and decreased porosity of concrete on use of marble powder using Scanning Electron Microscopy (SEM) images. Further, SHAP dependency plots help to obtain a deeper insight into the spread and variation of the predicted CS values with respect to the content of WMP as well as fine aggregate and water content being the main ingredients.

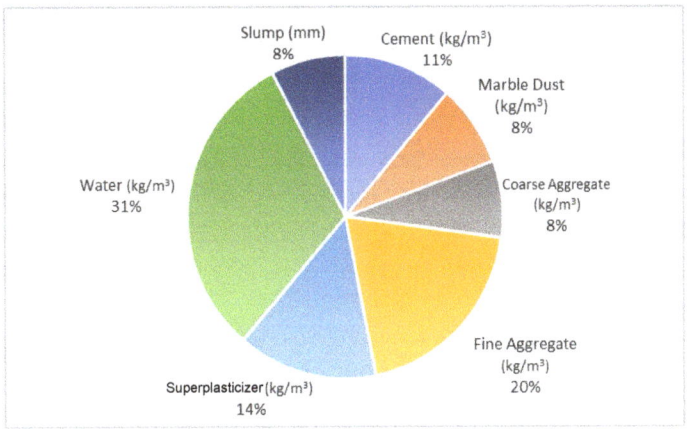

Figure 6. VIF analysis for RF.

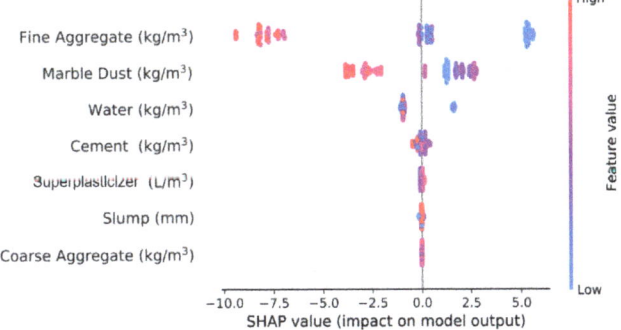

Figure 7. SHAP analysis.

5. Conclusions

On replacing cement with marble dust, there may be a dilution effect causing a reduction in the strength of concrete for varying water–binder ratios [21]. Thus, optimizing the content of marble dust is key. In this study, the compressive strength of concrete incorporated with WMP has been predicted using different ML algorithms. The estimation of compressive strength for different compositions of concrete is considered to be a regression problem. Data were collected from experimental trials using Ordinary Portland cement (OPC 43) replaced with WMP in different proportions. Performances using different regression models, viz., KNN, SVR, DT, RF, ET, GB and MLR, have been analysed and reported. Results show that the RF model is best suited to predict the compressive strength of concrete, followed by ET and DT models. Thus, RF can help in efficiently calculating the amount of WMP that can replace cement without affecting the compressive strength of concrete for practical applications. Further, on analysing the best obtained model, it can be concluded that WMP contributes approximately 8% to the total compressive strength of concrete. The data-driven models may help in predicting the strength based on input and output variables and also apply them to a large-scale dataset. However, there is no guarantee that they will explain the causality of the relationships accurately in prediction. The error may be significantly lower; however, the chemical reactions and changes taking place may not be completely predictable. These models may be used to understand the complex behaviour of marble dust which in turn would help to maximize its benefits in the construction industry. Notwithstanding, model robustness when faced with entirely new samples should be taken into consideration in this type of analysis as a calibrated model on a data point with a different structure.

Author Contributions: Conceptualization, M.S.; methodology, software and validation, P.C. and A.K.B.; formal analysis, M.S. and R.S.C.; investigation, M.S. and P.C.; resources, M.S.; data curation, M.S. and S.Y.; writing—original draft preparation, A.K.B. and S.Y.; writing—review and editing, A.K.B. and M.S.; visualization, R.S.C.; supervision, M.S. project administration, A.K.B. All authors have read and agreed to the published version of the manuscript.

Funding: This research received no external funding.

Institutional Review Board Statement: Not applicable.

Informed Consent Statement: Not applicable.

Data Availability Statement: The datasets generated during and/or analysed during the current study are available from the corresponding author on reasonable request.

Conflicts of Interest: The authors declare no conflict of interest.

Abbreviations

The following abbreviations are used in this manuscript:

WMP	Waste marble powder
ML	Machine learning
ANN	Artificial neural network
OPC	Ordinary portland cement
CS	Compressive strength
MLR	Multiple linear regression
KNN	K-nearest neighbour
SVR	Support vector regression
RF	Random forest
DT	Decision tree
ET	Extra tress
GB	Gradient boosting
GSS	Grid search strategy
MAE	Mean absolute error

MSE	Mean squared error	
RMSE	Root mean squared error	
MAPE	Mean absolute percentage error	
MBE	Mean bias error	
SHAP	SHapley Additive exPlanations	
VIF	Variance inflation factor	

References

1. Kore, S.D.; Vyas, A. Impact of marble waste as coarse aggregate on properties of lean cement concrete. *Case Stud. Constr. Mater.* **2016**, *4*, 85–92. [CrossRef]
2. Pappu, A.; Thakur, V.K.; Patidar, R.; Asolekar, S.R.; Saxena, M. Recycling marble wastes and Jarosite wastes into sustainable hybrid composite materials and validation through Response Surface Methodology. *J. Clean. Prod.* **2019**, *240*, 118249. [CrossRef]
3. Rana, A.; Kalla, P.; Csetenyi, L.J. Sustainable use of marble slurry in concrete. *J. Clean. Prod.* **2015**, *94*, 304–311. [CrossRef]
4. Shamsabadi, E.A.; Roshan, N.; Hadigheh, S.A.; Nehdi, M.L.; Khodabakhshian, A.; Ghalehnovi, M. Machine learning-based compressive strength modelling of concrete incorporating waste marble powder. *Constr. Build. Mater.* **2022**, *324*, 126592. [CrossRef]
5. Karimipour, A.; Jahangir, H.; Eidgahee, D.R. A thorough study on the effect of red mud, granite, limestone and marble slurry powder on the strengths of steel fibres-reinforced self-consolidation concrete: Experimental and numerical prediction. *J. Build. Eng.* **2021**, *44*, 103398. [CrossRef]
6. Chu, H.H.; Khan, M.A.; Javed, M.; Zafar, A.; Khan, M.I.; Alabduljabbar, H.; Qayyum, S. Sustainable use of fly-ash: Use of gene-expression programming (GEP) and multi-expression programming (MEP) for forecasting the compressive strength geopolymer concrete. *Ain Shams Eng. J.* **2021**, *12*, 3603–3617. [CrossRef]
7. al Swaidani, A.M.; Khwies, W.T. Applicability of artificial neural networks to predict mechanical and permeability properties of volcanic scoria-based concrete. *Adv. Civ. Eng.* **2018**, *2018*, 1–16. [CrossRef]
8. Chithra, S.; Kumar, S.S.; Chinnaraju, K.; Ashmita, F.A. A comparative study on the compressive strength prediction models for High Performance Concrete containing nano silica and copper slag using regression analysis and Artificial Neural Networks. *Constr. Build. Mater.* **2016**, *114*, 528–535. [CrossRef]
9. Eskandari-Naddaf, H.; Kazemi, R. ANN prediction of cement mortar compressive strength, influence of cement strength class. *Constr. Build. Mater.* **2017**, *138*, 1–11. [CrossRef]
10. Naderpour, H.; Rafiean, A.H.; Fakharian, P. Compressive strength prediction of environmentally friendly concrete using artificial neural networks. *J. Build. Eng.* **2018**, *16*, 213–219. [CrossRef]
11. Vakhshouri, B.; Nejadi, S. Prediction of compressive strength of self-compacting concrete by ANFIS models. *Neurocomputing* **2018**, *280*, 13–22. [CrossRef]
12. Duan, Z.H.; Kou, S.C.; Poon, C.S. Using artificial neural networks for predicting the elastic modulus of recycled aggregate concrete. *Constr. Build. Mater.* **2013**, *44*, 524–532. [CrossRef]
13. Deng, F.; He, Y.; Zhou, S.; Yu, Y.; Cheng, H.; Wu, X. Compressive strength prediction of recycled concrete based on deep learning. *Constr. Build. Mater.* **2018**, *175*, 562–569. [CrossRef]
14. Ly, H.B.; Nguyen, T.A.; Tran, V.Q. Development of deep neural network model to predict the compressive strength of rubber concrete. *Constr. Build. Mater.* **2021**, *301*, 124081. [CrossRef]
15. Nunez, I.; Marani, A.; Flah, M.; Nehdi, M.L. Estimating compressive strength of modern concrete mixtures using computational intelligence: A systematic review. *Constr. Build. Mater.* **2021**, *310*, 125279. [CrossRef]
16. Mansouri, I.; Ozbakkaloglu, T.; Kisi, O.; Xie, T. Predicting behavior of FRP-confined concrete using neuro fuzzy, neural network, multivariate adaptive regression splines and M5 model tree techniques. *Mater. Struct.* **2016**, *49*, 4319–4334. [CrossRef]
17. Sahoo, S.; Mahapatra, T.R. ANN Modeling to study strength loss of Fly Ash Concrete against Long term Sulphate Attack. *Mater. Today: Proc.* **2018**, *5*, 24595–24604. [CrossRef]
18. Khan, M.A.; Zafar, A.; Farooq, F.; Javed, M.F.; Alyousef, R.; Alabduljabbar, H.; Khan, M.I. Geopolymer concrete compressive strength via artificial neural network, adaptive neuro fuzzy interface system, and gene expression programming with K-fold cross validation. *Front. Mater.* **2021**, *8*, 621163. [CrossRef]
19. Aliabdo, A.A.; Abd Elmoaty, M.; Auda, E.M. Re-use of waste marble dust in the production of cement and concrete. *Constr. Build. Mater.* **2014**, *50*, 28–41. [CrossRef]
20. Santos, T.; Gonçalves, J.P.; Andrade, H. Partial replacement of cement with granular marble residue: Effects on the properties of cement pastes and reduction of CO_2 emission. *SN Appl. Sci.* **2020**, *2*, 1–12. [CrossRef]
21. Singh, M.; Srivastava, A.; Bhunia, D. An investigation on effect of partial replacement of cement by waste marble slurry. *Constr. Build. Mater.* **2017**, *134*, 471–488. [CrossRef]
22. Alpaydin, E. *Introduction to Machine Learning*; MIT Press: Cambridge, MA, United States, 2020.
23. Witten, I.H.; Frank, E.; Hall, M.A.; Pal, C.J.; DATA, M. Practical machine learning tools and techniques. In *Proceedings of the Data Mining*; Elsevier International Publishing: Amsterdam, The Netherlands, 2005; Volume 2.
24. Altman, N.S. An introduction to kernel and nearest-neighbor nonparametric regression. *Am. Stat.* **1992**, *46*, 175–185.
25. Cortes, C.; Vapnik, V. Support-vector networks. *Mach. Learn.* **1995**, *20*, 273–297. [CrossRef]

26. Breiman, L. Random forests. *Mach. Learn.* **2001**, *45*, 5–32. [CrossRef]
27. Basu, V. Prediction of Stellar Age with the Help of Extra-Trees Regressor in Machine Learning. In Proceedings of the International Conference on Innovative Computing & Communications (ICICC), West Bengal, India, 29 March 2020.
28. Zhang, Y.; Haghani, A. A gradient boosting method to improve travel time prediction. *Transp. Res. Part Emerg. Technol.* **2015**, *58*, 308–324. [CrossRef]
29. Bonavetti, V.; Rahhal, V.; Irassar, E. Studies on the carboaluminate formation in limestone filler-blended cements. *Cem. Concr. Res.* **2001**, *31*, 853–859. [CrossRef]

Disclaimer/Publisher's Note: The statements, opinions and data contained in all publications are solely those of the individual author(s) and contributor(s) and not of MDPI and/or the editor(s). MDPI and/or the editor(s) disclaim responsibility for any injury to people or property resulting from any ideas, methods, instructions or products referred to in the content.

Article

Research on the Properties of a New Type of Polyurethane Concrete for Steel Bridge Deck in Seasonally Frozen Areas

Li Li [1,2], Tianlai Yu [1,*], Yuxuan Wu [1], Yifan Wang [1], Chunming Guo [2] and Jun Li [2]

1. School of Civil Engineering, University of Northeast Forestry, Harbin 150001, China
2. Department of Highway and Bridge Engineering, Heilongjiang Institute of Construction Technology, Harbin 150001, China
* Correspondence: ytl_1965@nefu.edu.cn

Abstract: To widen the application scenarios of polyurethane concrete materials, a new type of polyurethane concrete material for steel bridge deck pavement in seasonally frozen areas was developed, and it was applied to the deck pavement engineering of steel bridges with orthotropic slabs. In this paper, we studied the properties of new polyurethane concrete through the tests of compressive strength, flexural tensile strength, and bond strength with steel plates of polyurethane concrete at different temperatures from −40 °C to 60 °C, totaling 11 temperature levels. We analyzed the elastic modulus, peak strain, and stress–strain relationship curve at the standard temperature. Then, we also conducted SEM test and IR test for the internal destruction form of polyurethane concrete, and analyzed the mechanism of its strength formation. The results show that with increasing temperature, the linear elastic range of polyurethane concrete material is shortened, the elastic modulus, compressive strength, and flexural tensile strength of the material all show a downward trend, and the peak strain and ultimate strain increase significantly. The failure state of the material is gradually transformed from brittle fracture at low temperature to plastic failure at high temperature, and the ductility of the material is significantly improved. Comparing with ones at the standard temperature, the compressive strength at 60 °C is 49.62 MPa downward by 45% and the bending tensile strength of the prism test at 60 °C is 12.34 MPa downward by 51%. Although the stress performance decreases significantly with the change of temperature, it can still meet the strength requirements of the bridge deck pavement for the pavement material. At present, the relevant research is mainly focused on the mechanical properties of new concrete under the influence of high temperature, but research on the mechanical properties of new concrete along with the temperature change is relatively limited. The proposed flexural–tensile constitutive model of polyurethane concrete for steel bridge deck pavement in seasonal freezing areas under the influence of temperature is in good agreement with the experimental results, which can provide a theoretical basis for the application of polyurethane concrete in engineering.

Keywords: polyurethane concrete; mechanical properties; constitutive relation; microscopic test; temperature effect

1. Introduction

Steel bridge deck pavement is a hot and difficult topic in the field of road engineering and road materials at home and abroad [1,2]. Compared with the ordinary road pavement, the bridge deck pavement layer must have sufficient strength, stiffness, impact resistance, wear resistance, and other mechanical properties, but also requires the pavement material to have high strength, good flexibility, and excellent durability characteristics [3–5]. Domestic bridge deck asphalt pavement technology mainly includes pouring asphalt concrete, epoxy asphalt concrete, and other pavement technologies [6,7]. Due to the defects of the existing bridge deck pavement technology, most steel bridges will produce different degrees of various diseases within ten years of operation [8,9]. Therefore, the research and

development of new concrete materials has become the focus of current steel bridge deck pavement research.

Scholars at home and abroad have had a deep understanding of the relationship between the performance and constitution of traditional concrete [10,11]. However, with the improvement in application demand, a variety of new concretes have appeared, and some scholars have conducted thorough research on the mechanical properties of these new concretes [12,13]. Polymer concrete, for example, epoxy–cement concrete and latex–cement concrete, as a new material, has been successively used in bridge pavement engineering due to its advantages of wear resistance, anti-corrosion, waterproofing and permeability prevention, and strong adhesion between layers [14–16]. At present, the relevant research is mainly focused on the mechanical properties of new concrete under the influence of high temperature, but research on the mechanical properties of new concrete along with the temperature change is relatively limited [17]. However, steel plates with good thermal conductivity will have very high temperatures, such as 60 °C in summer, and very low temperatures, such as −40 °C in winter. Therefore, the study of temperature change on the performance of new concrete is an important prerequisite to determine its application scenarios [18,19].

In recent years, polyurethane material has been gradually applied to the asphalt modification of bridge pavement engineering because of its good characteristics, and polyurethane concrete is also used in the repair and filling of steel bridge deck pavement [20–22]. To broaden the application scenario of polyurethane concrete material and solve the problem of frequent occurrence of the existing steel bridge deck pavement defects, a new type of polyurethane concrete has been developed, with good mechanical properties, durability, strong bonding with steel, early strength, low temperature work, etc. In this paper, we studied the properties of new materials through the tests of compressive strength, flexural tensile strength, and bond strength with steel plates of polyurethane concrete at different temperatures. We analyzed the elastic modulus, peak strain, and stress–strain relationship curve with temperatures. Then, we also conducted electron microscopic tests for the internal destruction form of polyurethane concrete. The constitutive model was established to provide a theoretical basis for the engineering design and application of polyurethane concrete.

2. Material Composition and Preparation

2.1. Material Composition

2.1.1. Polyurethane

The polyurethane material used in this study was obtained from a mixture of isocyanate and polyols. The isocyanate is made of BASF polyurethane black material, and the performance indexes are shown in Table 1.

Table 1. Performance indexes of isocyanate.

Detection Index	Viscosity	NCO Content, wt%	Appearance
Specifications	170–250 mPa·s	30.5%–32%	Dark brown liquid

The polyether polyols used in this study were produced by Shandong Tengzhan Polyurethane Co., Ltd., which is located in Shandong Province, China, and their performance indexes are shown in Table 2.

Table 2. Performance indexes of polyether polyols.

Detection Index	Appearance	Hydroxyl Value (mgKOH/G)	Acid Value (mgKOH/G)	Rate of Water Content (%)	PH Value	Viscosity (mPa·s)
Specifications	Colorless to pale yellow, transparent liquid	440–460	≤0.07	≤0.05	5.0–8.0	300–500

2.1.2. Cement

The cement used in this study is ordinary Portland cement (P. O 42.5) produced by Harbin Yatai Cement Co., Ltd., which is located in the city of Harbin, Heilongjiang Province, China. The cement is required to meet the specification index and should be newly produced. The moisture content of the cement is lower than 0.05%.

2.1.3. Aggregates

The aggregate used in this study adopts the first-class basalt of Harbin Acheng Quarry, and the rock index meets the factory requirements. In addition, the moisture content of the rock is lower than 0.05%. Therefore, if the rock is wet before construction, the stone needs to be dried so that its moisture content must be lower than 0.05%.

2.1.4. Retarder

According to the early retarding effect tests, the selected retarder adopts phosphate and shall meet the requirements that solid content is higher than 98%, water content is lower than 1%, viscosity should be 3000–8000 MPa·s, and free acid value is lower 0.2%.

2.1.5. Dehumidizer

Encountering water vapor or in wet environments, polyurethane concrete easily swells, producing a large number of bubbles, resulting in a pore structure and affecting the stress performance and durability. Therefore, dehumidification admixtures need to be added when used. The dehumidizer is solid silicone. Its appearance is shown in Figure 1. See Table 3 for details.

Figure 1. Appearance of the dehumidizer.

Table 3. Dehumidifier indicators.

Appearance	Diameter (mm)	Stacking Density (g/cm^3)	Balanced Water Adsorption Capacity (%)	Solid Organic Silicone Content (%)
white powder	2.3 mm	0.82	26.50%	>99

2.2. Specimen Preparation

First, the raw material composition of the polyurethane concrete material was determined, and the mix ratio was designed through the previous test. The results are shown in Table 4.

Table 4. Composition of polyurethane concrete.

Serial Number	Material	Mass Ratio to the Polyether Polyols
1	polyether polyol	1
2	Isocyanate	1
3	Retarder	0.1
4	Ordinary Portland cement (P.O.42.5)	2
5	Basalt (4.75 mm)	3.84
6	Basalt (2.36 mm)	2.16

Polyurethane concrete materials were prepared as follows:

① Cement, coarse aggregate, fine aggregate, polyether polyol, curing agent isocyanate, and other admixtures should be weighed and used according to the ratio.

② The polyether polyols and the dehumidifier were mixed for 1 min at 1500 rpm using a high speed dispersion machine, as shown in Figure 2, to ensure the full and uniform mixing of the dehumidification with the polyether polyols. After stirring, the bubbles on the surface and the material temperature should be observed. If there are more bubbles, you need to keep still until the bubbles disappear. Because the polyurethane material is very sensitive to the temperature and the temperature level directly affects the reaction speed and the curing speed, it is necessary to wait for the temperature of the material to decrease to the temperature required by the test and then to continue to add other materials to stir, and the waiting time is 5–10 min.

Figure 2. High-speed dispersion machine.

③ Then the cement was added and stirred thoroughly at 1500 rpm for 1–2 min, and the viscosity of the mixture was significantly increased. Then, isocyanate and retarder were slowly added and stirred for 1 min.

④ Finally, the fine aggregate was added to stir for 1 min; that is, the mixing was finished to obtain polyurethane concrete.

3. Test Method

3.1. Mechanical Properties Test of Polyurethane Concrete at Standard Temperature

3.1.1. Compressive Strength Test

In this study, the compressive strength test was carried out according to the "Standard for test methods of concrete physical and mechanical properties" (GB/T 50081-2019). The curing temperature was 20 ± 2 °C, the relative humidity was 50% ± 5%, and the curing age periods were 3 d, 7 d, 14 d, and 28 d. As shown in Figure 3, a 70.7 mm × 70.7 mm × 70.7 mm cube specimen was prepared, and the compressive strength of the polyurethane concrete was obtained by using a universal test machine.

Figure 3. Cube specimen of the compressive strength test and test process. (a) test specimen; (b) test process.

3.1.2. Flexural Tensile Strength Test

Flexural tensile strength tests were conducted according to the "Standard for test methods of concrete physical and mechanical properties" (GB/T 50081-2019). The curing temperature was 20 ± 2 °C, the relative humidity was 50% ± 5%, and the curing age periods were 3 d, 7 d, 14 d, and 28 d. As shown in Figure 4, a prism specimen with a size of 40 mm × 40 mm × 160 mm was prepared, and the tensile strength of the polyurethane concrete at 7 d, 14 d, and 28 d was obtained by using a universal test machine.

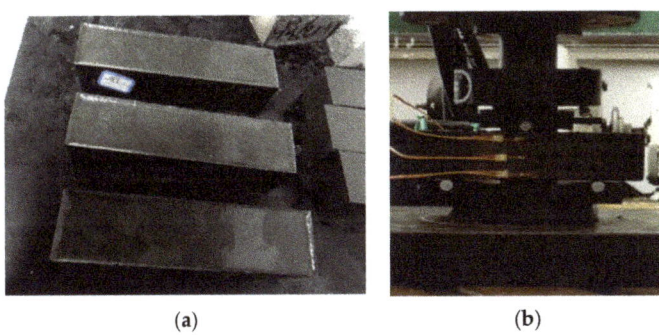

Figure 4. Prism specimen of flexural tensile strength test and the test process. (a) Test specimen; (b) Test process.

3.1.3. Test of the Bond Strength between Polyurethane Concrete and Steel

The good combination between the pavement layer and steel plate is the key to the coordinated work of bridge deck pavement and steel plate surfaces. High and durable bonding strength between the pavement layer and the steel plate and good consistency with the complex stress and strain state of the steel plate depend on the strength and toughness of the paving layer material itself and depend on the bonding effect between the paving layer and the steel plate. This paper obtains the direct shear bond strength of polyurethane concrete and steel and studies the change regulations of the bond strength between polyurethane concrete and steel with curing ages. The curing temperature was 20 ± 2 °C, the relative humidity was 50 ± 5%, and the curing age periods were 3 d, 7 d, 14 d, and 28 d. As shown in Figure 5, the treated steel plate and polyurethane concrete are poured together. After curing and forming, the interfaces of the two materials are placed at the steel pad, respectively, and a 70.7 mm × 70.7 mm × 70.7 mm square steel plate is placed between the upper bearing plate and the specimen. The steel plate surface directly in contact with the polyurethane concrete material should be derusted, roughened, and treated by a notch groove, cleanliness, and roughness to meet the design requirements of highway steel bridge deck pavement. Then, the universal experimental machine is used to put a compressive load on the top steel plate to push the specimen of polyurethane concrete out straight, and the loading speed is 0.8~1.0 MPa·s.

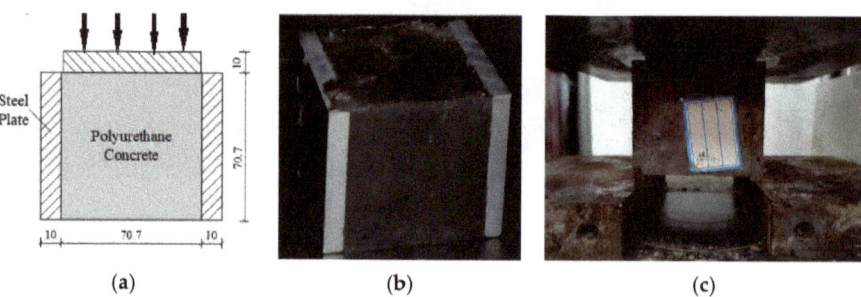

Figure 5. Test specimen of the bond strength test and test process. (**a**) Diagram; (**b**) Test Specimen (**c**) Test process.

3.1.4. Elastic Modulus Test

The polyurethane concrete elastic modulus is one of the most important mechanical indicators of polyurethane concrete material; it can not only directly reflect the stiffness characteristics of polyurethane concrete material and the deformation characteristics of polyurethane concrete structure but can also indirectly reflect the aging characteristics of polyurethane concrete material and the internal damage properties of polyurethane concrete structure.

The elastic modulus test was carried out with reference to the "Standard for test methods of concrete physical and mechanical properties" (GB/T 50081-2019). Using the standard curing method, the curing temperature was 20 ± 2 °C, and the relative humidity was 50% ± 5%. As shown in Figure 6, a standard prismatic specimen with a size of 150 mm × 150 mm × 300 mm was used, and the curing age periods were 3 d, 7 d, 14 d, and 28 d. To ensure the number of samples, 15 specimens should be prepared for each curing age, among which 3 specimens in each group determine the elastic modulus loading parameters, and the remaining specimens undergo elastic modulus tests to obtain 12 elastic modulus values and 15 axial compressive strength values.

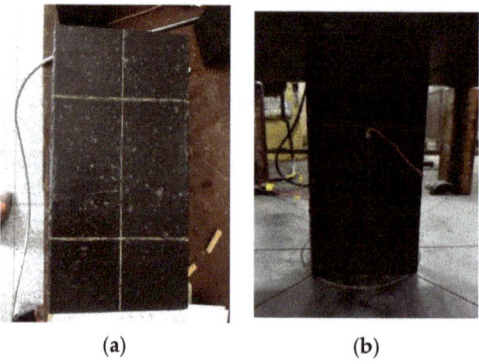

Figure 6. Test specimen and procedure of the elastic modulus test. (**a**) Test Specimen; (**b**) Test process.

3.2. Influence Test of the Vibrating Process on the Mechanical Properties of Polyurethane Concrete

The influence of the vibration process on the mechanical properties of polyurethane concrete is of practical significance for guiding the construction of steel bridge deck pavement. The influence law of different vibration forms and different vibration times on the mechanical properties of polyurethane concrete materials is studied. According to the results of the tensile flexural tensile strength test, a curing age of 14 days was used for the test. The time range of manual insertion or vibration platform vibration is considered for 10–40 s, and each 10 s is one time level, with a total of four time levels.

3.3. Test for the Effect of Temperature Change on the Mechanical Properties of Polyurethane Concrete

The steel bridge deck engineering structure is directly in the complex natural environment, so the polyurethane concrete used for the bridge deck pavement is required to adapt to the working environment of the bridge deck structure, such as the change of sunshine and temperature. As an important component of environmental factors, temperature has an important influence on the mechanical properties of polyurethane concrete. Combined with the actual temperature difference in Heilongjiang and the obvious temperature change in the steel bridge panel, to better study the mechanical properties of polyurethane concrete under the influence of temperature, the temperature control range is −40 to 60 °C, and each 10 °C is one temperature level, with a total of 11 temperature levels. As shown in Figure 7, the high- and low-temperature control chambers were used to study the compression resistance, flexural tensile resistance, bond properties, and change regulations of the polyurethane concrete material specimens at different temperature levels.

 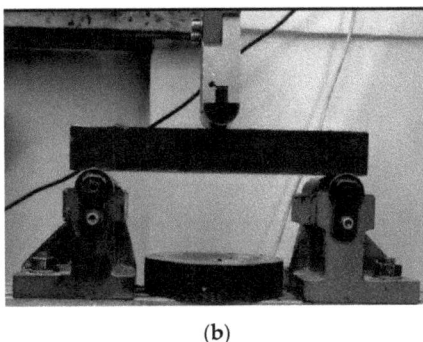

(a) (b)

Figure 7. Flexural tensile strength test piece and test process. (a) Test specimen; (b) Test process.

To make the internal and external temperatures of the polyurethane concrete material test piece reach the temperature required for the test, the test specimen is put into the test box, the target temperature of the test box is set to the required temperature of the test, and the internal temperature of the test piece reaching the set temperature is calculated by the coefficient of thermal conductivity. When the temperature inside and outside the specimen reached the set temperature, the instrument was started, and the loading test was started.

3.3.1. Effect of Temperature on the Compressive Strength of Polyurethane Concrete Material

The test was conducted according to the method in Section 2.1.1, and the test instrument adopted a high- and low-temperature control box. The compressive strength test of the polyurethane concrete cube test piece at different temperatures of −40 °C to 60 °C was conducted, and the influence of temperature on the compressive strength was analyzed from the aspects of strength change, damage form, stress and strain curve.

3.3.2. Effect of Temperature on the Flexural Tensile Resistance of Polyurethane Concrete Material

The flexural tensile test of polyurethane concrete material was conducted according to T0715-2011 in the "Standard Test Methods of Bitumen and Bituminous Mixtures for Highway Engineering" (JTG E20-2011). As shown in Figure 8, 250 mm × 30 mm × 35 mm prism small beam specimens were used, and the test instrument adopted a high- and low-temperature control box and universal material test machine. After the inside and outside temperatures of the test piece reached the test temperature, the test piece was removed and placed on the test machine for the loading test. To ensure consistent internal and outside temperatures and no temperature stress, the time of taking the test piece from the temperature box to the completion of the loading test should not exceed 60 s.

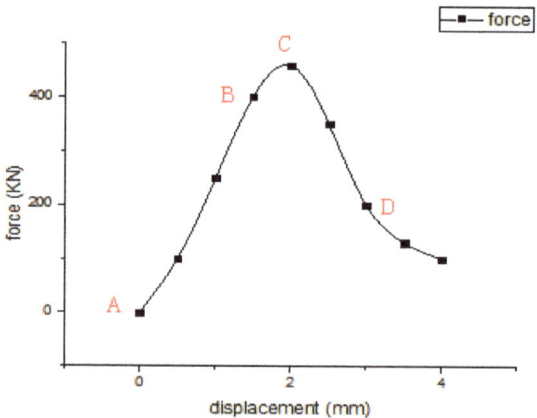

Figure 8. Twenty-eight day cube specimen failure curve.

The flexural tensile resistance test was conducted at −40–60 °C polyurethane concrete cube specimens at different temperatures. The same loading speed was used for different temperature test groups, and the vertical loading rate was 1 mm/min. The influence of temperature on the specimen flexural tensile resistance was analyzed from the aspects of strength change, failure form, failure flexural tensile strain, and stiffness modulus. The tensile strength (R_b), flexural–tensile strain (ε_b), and stiffness modulus (S_b) of the beam are calculated according to the test results.

The calculation formula is as follows:

$$R_b = \frac{3LP_b}{2bh^2}$$

$$\varepsilon_b = \frac{6hd^2}{L^2}$$

$$S_b = \frac{R_b}{\varepsilon_b}$$

In the formula:
R_b—flexural tensile strength during specimen failure (MPa);
ε_b—Maximum flexural–tensile strain during specimen failure (με);
S_b—flexural tensile stiffness modulus during specimen failure (MPa);
b—The width of the specimen (mm);
h—The height of the specimen (mm);
L—The calculating span of the beam specimen (mm);
P_b—Maximum load in case of specimen failure (N);
d—The deflection in the middle of span during specimen failure (mm).

4. Results of Mechanical Properties of Polyurethane Concrete Material

4.1. Mechanical Properties Test Results of Polyurethane Concrete Material at Standard Temperature

4.1.1. Cube Compressive Strength Test

1. Cube compressive strength test results analysis

The results of the cube compressive strength test at standard temperature are shown in Table 5.

Table 5. Record table of compressive strength test results of cube specimens at standard temperature.

The Specimen Number	Age (d)	Failing Load F (KN)	Compression Strength (MPa)	Average Compression Strength (MPa)
PUC-1		400	80.02	
PUC-2	3 d	381.1	76.24	78.19
PUC-3		391.5	78.32	
PUC-4		398.6	79.74	
PUC-5	7 d	402.3	80.48	80.35
PUC-6		404	80.82	
PUC-7		443.98	88.82	
PUC-8	14 d	465.02	93.03	89.86
PUC-9		438.5	87.73	
PUC-10		457.6	91.54	
PUC-11	28 d	458	91.63	91.75
PUC-12		460.2	92.07	
PUC-13		456.2	91.26	
PUC-14	60d	459.6	91.94	91.77
PUC-15		460.3	92.08	

According to the test data, polyurethane concrete has the performance of early strength and fast strength, which can reach 85% of the final strength in 3 d, 87.6% of the final strength in 7 d, and 97.9% of the final strength in 14 d. After 14 d, the strength stabilized and reached the final strength.

Based on the cube compressive strength test data of the 28-day polyurethane concrete test specimen, the test force–displacement curve is shown in Figure 8.

According to the test results, the deformation and destruction process of polyurethane concrete are divided into five stages:

Stage I (changes process from O to A)

With the increase in the displacement of the test machine, the stress gradually increases. Before the displacement does not reach 0.4 mm, the stress and strain curve increases rapidly, and the slope of the curve increases with increasing load. In this stage, the macro performance is as follows: as the load increases, the specimen transverse expansion is small, and the specimen volume decreases with increasing load.

Stage II (changes process from A to B)

With the increasing displacement of the test machine, the stress gradually increases, and then the stress-strain curve basically changes linearly during the 0.4 mm-1.3 mm period. In this stage, the macro performance is that the elastic deformation gradually increases as the load increases, and the unloading deformation can be recovered.

Stage III (changes process from B to C)

As the displacement of the test machine increases, the stress increases gradually. When the displacement is in the range of 1.3–1.9 mm, the stress-strain curve slows down rising, and the slope of the curve decreases with increasing load. At this stage, the macroscopic performance is as follows: the test part starts from the end. With the increase in the load, the macroscopic crack develops from the end to the middle. When the load reaches the limit load of the test part, the crack runs through, forming multiple open joints parallel to the direction of the compressive stress.

Stage IV (changes process from C to D)

With the increasing displacement of the test machine, the stress gradually decreased. When the displacement is greater than 1.9 mm, the stress–strain curve slowly decreases and gradually decreases gently after point D. The macro performance of this stage is as follows: with the increase in displacement, many split cracks and inclined cracks are produced parallel to the direction of compressive stress, and the specimen is divided into multiple small columns, resulting in concrete instability but still maintaining the basic shape.

Stage V (changes process after D)

After point D, polyurethane concrete is finally completely damaged and forms a network fracture failure form.

2. Polyurethane concrete SEM (scanning electron microscope) test analysis

In the process of the compressive strength test of the cube specimen, the four change points of the compressive stress–strain relationship curve were sampled, namely before loading, elastic–plastic stage, plastic stage, and destruction point. Combining the microscopic cracks and macro-cracks of the specimen, the formation mechanism of concrete strength was analyzed by the scanning electron microscope.

Stage I (changes process from O to A)

The microscopic performance is the tension structural surface and the microcrack closure in the specimen with increasing load. The destruction pattern is shown in Figure 9a.

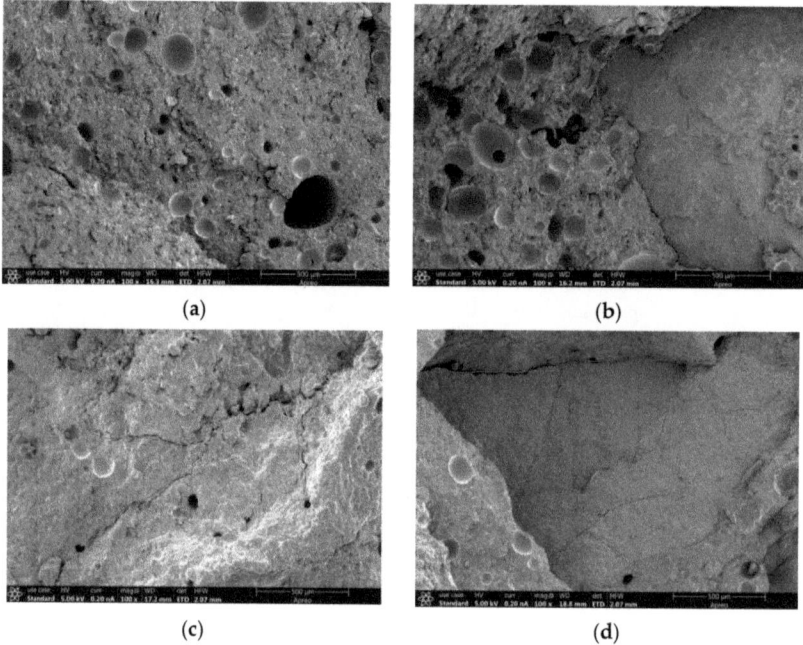

Figure 9. Microscopic form of 28-day cube specimen destruction. (**a**) Stage I (changes process from O to A); (**b**) Stage II (changes process from A to B); (**c**) Stage III (changes process from B to C); (**d**) Stage IV (changes process from C to D).

Stage II (changes process from A to B)

The microscopic performance is as follows: with the increase in displacement, the microcracks along the junction of stone and substrate longitudinal development, not deep into the substrate interior. The crack width is almost unchanged. The destruction morphology is shown in Figure 9b.

Stage III (changes process from B to C)

With the increase in load, microcracks develop rapidly, and the development of microcracks undergoes qualitative change. Microcracks develop along the junction of the stone and the substrate, forming multiple mesh cracks inside the substrate. The destruction morphology is shown in Figure 9c.

Stage IV (changes process from C to D)

The microscopic performance of the specimen is that multiple mesh cracks inside the substrate continue to develop; with the increase in displacement, the mesh density increases and cannot be recovered after damage. The destruction morphology is shown in Figure 9d.

Stage V (changes process after D)

After point D, the fracture cross forms a macroscopic fracture surface. As the displacement increases, the overall slip of the deformation extension fracture surface finally becomes completely unstable and forms a network fracture failure form. The disruption morphology is shown in Figure 10.

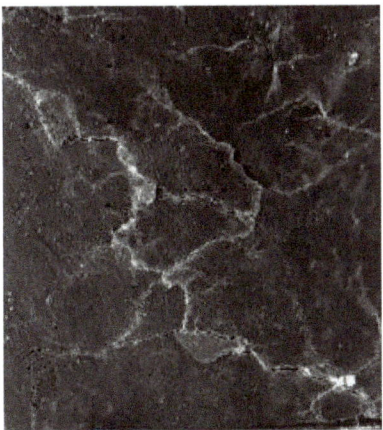

Figure 10. Macro-cracks of 28-day cube specimen destruction.

3. Polyurethane concrete IR (infrared spectroscopy) test analysis

After the cube specimen compressive strength test, the damaging cube specimen test of the polyurethane concrete was taken as the IR test sample to confirm the composition structure and destruction form of the polyurethane concrete.

As can be seen in Figure 11, the absorption peak of C-S-H (922.85 cm^{-1}) and Ca(OH)$_2$ (3567.34 cm^{-1}) exists in the IR test graph in addition to the main group of polyurethane.

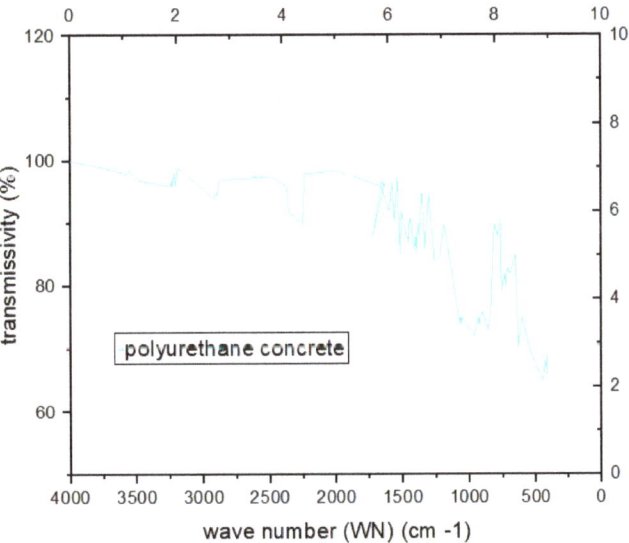

Figure 11. IR test graph of polyurethane concrete.

This is due to internal cement particles contacting the water in the air to form the crystallization reaction when the damage of the polyurethane concrete occurred at the same time, as the hydration reaction product will produce a certain strength which will close the subtle cracks. It proves that the polyurethane concrete has a certain self-healing ability. Specific test results are shown in Table 6.

Table 6. Categories of functional groups represented by the peaks in the IR test graph of polyurethane concrete.

Sections	Peak (cm^{-1})	Categories of Functional Groups	Force Form
Soft segment area	2971.76	C—H	stretching vibration
	2917.23	C—H	stretching vibration
	2863.73	C—H	stretching vibration
	1449.86	—CH$_2$—	scissoring vibration
	1373.58	—CH$_3$	symmetrical bending vibration
	1070.12	C—O	stretching vibration
Hard section area	3245.93	N—H	symmetrical and asymmetrical stretching vibration
	2254.69	—N=C=O	
	1723.62	C=O	stretching vibration
	1597.53	C=C	
	1517.24	N—H	in-plane bending vibration
	1412.33	—CO—O—	
	1220.81	C—O	asymmetrical stretching vibration
	931.4	—NH—CO—O—	symmetrical stretching vibration
	817.23	C—H	out-of-planebending vibration
	756.41	C—H	out-of-planebending vibration
Groups after cement particle hydration	451.38	O—Si—O	
	1450.24	O—Ca—O	
	922.85	Si—O	

4.1.2. Test of Flexural Tensile Strength of Prisms

The results of the prism flexural tensile strength test at standard temperature are shown in Table 7.

Table 7. Results of the flexural tensile strength test of prisms at standard temperature.

The Specimen Number	Age (d)	Failure Load P (KN)	Rupture Strength (MPa)	Average Rupture Strength (MPa)
PUC-3-1	3	6.99	16.38	17.06
PUC-3-2		7.62	17.86	
PUC-3-3		7.23	16.94	
PUC-7-1	7	8.64	20.25	20.28
PUC-7-2		8.72	20.44	
PUC-7-3		8.60	20.16	
PUC-14-1	14	12.04	28.22	28.58
PUC-14-2		12.64	29.6	
PUC-14-3		11.9	27.89	
PUC-28-1	28	12.25	28.71	28.82
PUC-28-2		12.26	28.73	
PUC-28-3		12.38	29.01	
PUC-60-1	60	12.55	29.41	29.06
PUC-60-2		12.46	29.20	
PUC-60-3		12.19	28.57	

According to the test results, the specimen can reach 59% of the final strength by 3 days, 70.4% of the final strength in 7 d, and 99.2% of the final strength at 14 d, and the strength will stabilize to reach the final strength after 14 d.

The bottom tensile strain of the flexural tensile strength test specimen was further measured to obtain the tensile stress-strain curve of each test specimen. The average stress-strain curve was synthesized according to the test results of the three test specimens, as shown in Figure 12.

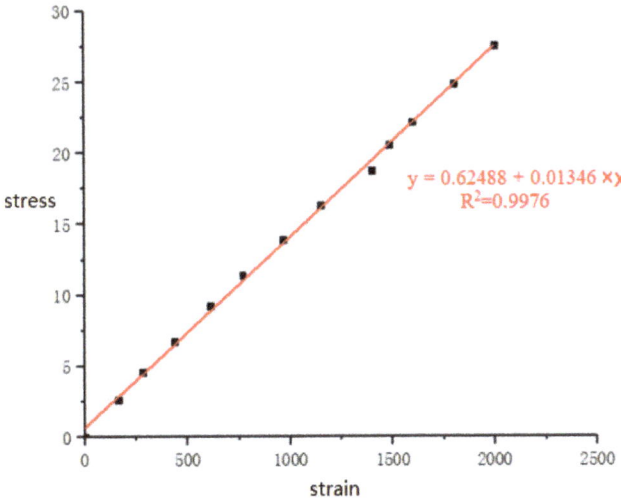

Figure 12. Tensile stress–strain curve diagram.

According to the figure, the tensile stress-strain curve of polyurethane concrete is basically linear. Therefore, the flexural tensile elastic modulus of polyurethane concrete can be obtained from the curve slope, namely, 1.35×10^4 MPa.

4.1.3. Test Results of the Bond Strength between Polyurethane Concrete and Steel

The results of the bond strength test at standard temperatures are shown in Table 8.

Table 8. Results of the bond strength of polyurethane concrete specimens and steel plates at different ages.

The Specimen Number	Age (d)	Failure Loads P (KN)	Bond Strength (MPa)	Average Bond Strength (MPa)
PUC-3-1	3	22.8	4.56	
PUC-3-2	3	23.4	4.68	4.66
PUC-3-3	3	23.7	4.74	
PUC-7-1	7	26.99	5.39	
PUC-7-2	7	26.78	5.36	5.41
PUC-7-3	7	27.32	5.47	
PUC-14-1	14	40.22	8.05	
PUC-14-2	14	39.56	7.91	7.78
PUC-14-3	14	36.88	7.38	
PUC-28-1	28	43.64	8.73	
PUC-28-2	28	44.32	8.87	8.89
PUC-28-3	28	45.31	9.06	
PUC-60-1	60	44.7	8.94	
PUC-60-2	60	45.1	9.02	8.91
PUC-60-3	60	43.8	8.76	

According to the test results, from Figure 13, 52% of the final bond strength arrives at 3 d, 60.79% at 7 d, 8 7.42% at 14 d, and the final bond strength at 28 d tends to stabilize.

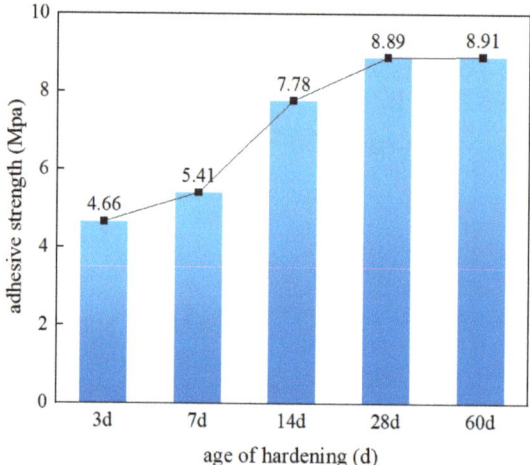

Figure 13. Change curve of the bond strength of polyurethane concrete material and steel plate at different ages.

According to the current design requirements of steel bridge deck pavement, the test results of the interface bonding strength between the pavement material and the steel plate should meet the requirements of the following formula:

$$\overline{\sigma_{rm}} \geq \sigma_d + Z_a S$$

In the formula:

$\overline{\sigma_{rm}}$—The average interface bond strength of the actual measurement of the combined structural specimen (MPa);

σ_d—Interface bond strength design value (MPa)

$$\sigma_d = K_C K_J \sigma_{st}$$

In the formula:

K_C—Highway grade coefficient, the coefficient of highway is 1.4;

K_J—Correction coefficient of traffic load grade, super extremely heavy and extremely heavy traffic value are 1.4~1.5, heavy traffic value is 1.3~1.4, and medium traffic value is 1.1~1.2;

σ_{st}—Under the action of the standard axle load, standard value of bonding strength of protective interface, MPa; standard value of bonding strength of modified asphalt mixture protective layer is 0.3 MPa; standard value of bonding strength of epoxy asphalt mixture protective layer is 0.6 MPa;

c—Coefficients that change with the guaranteed rate in the standard normal distribution table, Highway and primary highway guarantee rate are 95%, namely, $Z_a = 1.645$; Other highway guarantee rates is 90%, namely, $Z_a = 1.282$;

S—Measured standard deviation of combined structural specimen strength.

As polyurethane concrete material as a steel bridge deck pavement layer is still in the exploratory stage and there are no issued national industry specifications and standards, the evaluation parameters of the designed bond strength are calculated according to the highest value, and the design value of the interface bond strength (σ_d) is:

$$\sigma_d = K_C K_J \sigma_{st} = 1.4 \times 1.5 \times 0.6 = 1.26 \text{ MPa}$$

$$\sigma_d + Z_a S = 1.26 + 1.645 \times 0.17 = 1.54$$

As seen from the results, the average value of the bond strength between the 3-day curing age polyurethane concrete material and the steel plate is also much higher than 1.54 MPa, which meets the relevant requirements.

4.1.4. Elastic Modulus Test

After judging and removing the suspicious data from the test value of each group of tests by means of the 3S rule, the elastic modulus and axial compressive strength test results obtained are shown in Table 9.

Table 9. Test values of the elastic modulus and axial compressive strength.

AGE (d)	Modulus of Elasticity (10^4 N/mm^2)		Axial Compressive Strength (N/mm^2)	
	Average	Ratio to the 28 d Test Values/%	Average	Ratio to the 28 d Test Values/%
3	1.18	81.37	80.23	86.91
7	1.31	90.97	82.54	89.41
14	1.42	98.61	90.12	97.62
28	1.44	100	92.31	100
60	1.45	100.69	92.44	100.14

According to the test results, both the elastic modulus and the axial compressive strength at 3 d of age reached more than 80%, with a large increase within 7 d and a gradual increase thereafter. The elastic modulus of polyurethane concrete is smaller and more flexible than that of concrete materials of the same design strength. Combined with the test results, it can be seen that the polyurethane concrete material combines the flexibility of asphalt concrete materials and the stiffness of concrete materials, which belongs to the material with good stiffness, good deformation resistance, and excellent ductility, and is suitable for steel bridge deck pavement materials.

4.2. Mechanical Properties Test Results of Materials under Different Vibration Processes

The test results of different forms of vibration are shown in Figure 14.

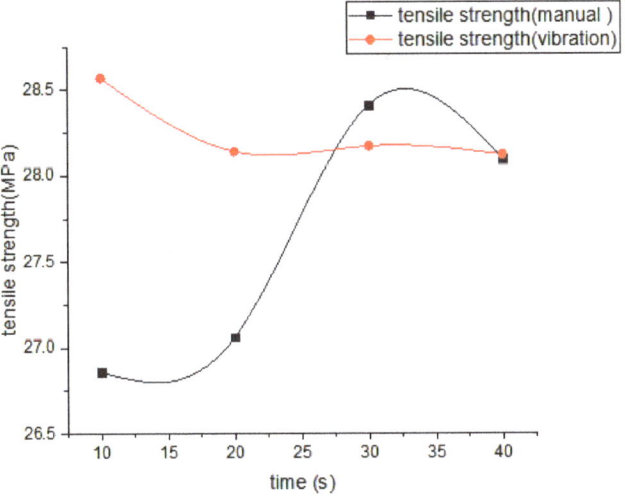

Figure 14. Results of different vibration forms.

It can be concluded from the test results that the tensile strength of polyurethane concrete is significantly lower than that of the polyurethane concrete prismatic test formed by the vibration platform. The flexural tensile strength of the polyurethane concrete prism changes with the vibration time and gradually decreases after 30 s. The highest strength is 28.49 MPa at 32 s. The flexural tensile strength of the polyurethane concrete prism specimen decreases gradually, and the highest strength is 28.57 MPa at 10 s. This is because an excessively long vibration time will lead to the segregation of polyurethane cement slurry and gravel, affecting the final fracture resistance strength. According to the test results, to facilitate construction, a vibration of 10 s was applied during the construction process.

4.3. Mechanical Properties Test Results of Materials at Different Temperatures

4.3.1. Effect of Temperature on the Compressive Strength of Polyurethane Concrete Material

The compressive strength results of the polyurethane concrete cubes at different test temperatures are shown in Figure 15.

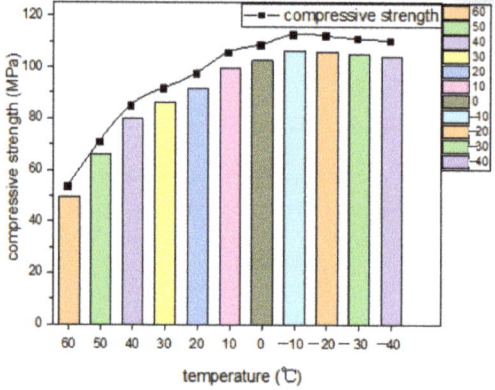

Figure 15. Compressive strength of polyurethane concrete cubes at different temperatures.

It can be seen from the figure that the compressive strength of the cube concrete cube increases significantly with decreasing temperature, which is basically consistent with the change trend of the elastic modulus with temperature. The compressive strength of the cube test specimen at 60°C is 49.62 MPa, which can still meet the strength requirements of pavement materials on bridge deck pavement. The test temperatures were 40 °C, 20 °C, 10.8%, 84.9%, 106.7%, and 109.5%, respectively. The temperature change also greatly affects the compressive strength of polyurethane concrete.

From the analysis of the material chemical composition, polyurethane concrete can have the characteristics of high hardness and good flexibility at ordinary temperatures, mainly because polyurethane contains urethane in its molecular structure (molecular chain). Temperature can affect the activity of the molecular chain of polyurethane material, which then makes the material different in macroscopic strength. At the structural level, polyurethane at low temperature presents high hardness, changing strength change stable characteristics, and polyurethane heated expansion reduces the stiffness shows plastic characteristics, after compression compared to room temperature will appear greater deformation, lead to aggregate and binder is not easy to coordinate deformation, affect the two-interface bond relative slip, eventually affecting the material strength.

4.3.2. Influence of Temperature on the Flexural Tensile Strength of Polyurethane Concrete Material

The flexural tensile strength results of polyurethane concrete at different test temperatures are shown in Figure 16.

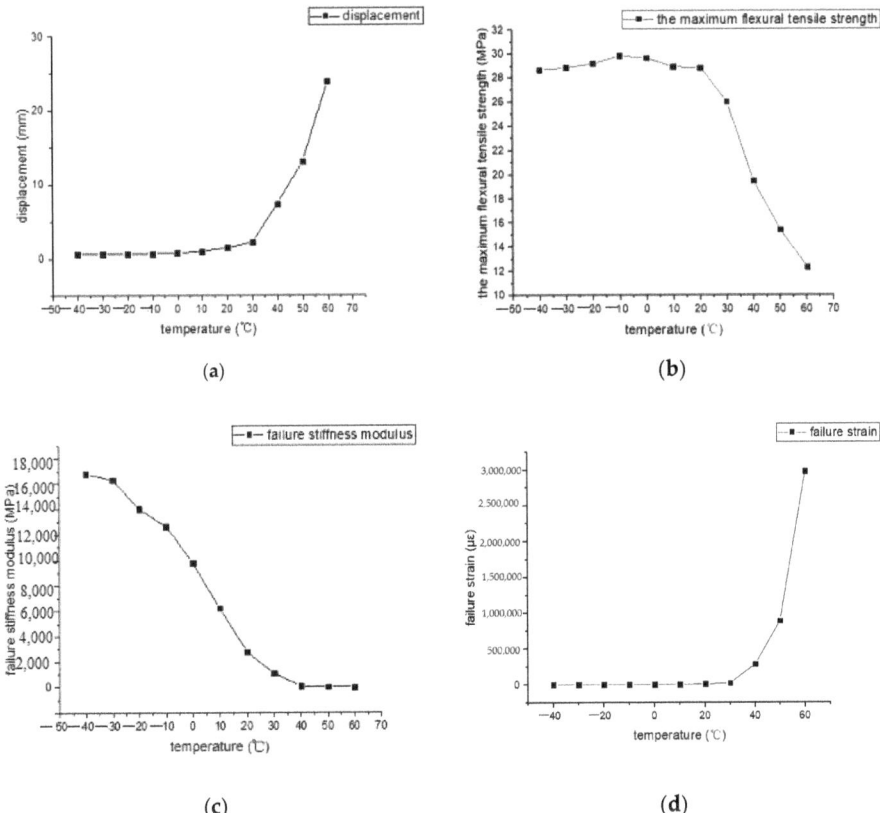

Figure 16. The flexural tensile strength of polyurethane concrete at different temperatures. (**a**) Deflection; (**b**) Flexural tensile strength (**c**) Stiffness modulus (**d**) Flexural tensile strain.

It can be seen from the test results that as the test temperature decreases, the maximum deflection of the polyurethane concrete material in the middle of the span during bending failure continuously decreases. During the reduction from 60 °C to 30 °C, the maximum deflection decreased gradually and rapidly, which decreased from 23.83 mm to 2.14 mm and the decreasing amplitude was 91%; the maximum deflection decreased slowly during the temperature reduction process from 30 °C to −10 °C, which decreased from 2.14 mm to 0.67 mm and the decreasing amplitude was 69%. The change in the maximum deflection in the middle of the span was even gentler from −10 °C to −40 °C, which decreased from 0.67mm to 0.57 mm and the decreasing amplitude was 15%.

The flexural tensile strength of polyurethane concrete gradually increases with decreasing temperature. In the process from 60 °C to 30 °C, the flexural tensile strength and stiffness modulus increase rapidly. From 30 °C to −40 °C, the flexural tensile strength changes smoothly, but the stiffness modulus gradually increases rapidly, the rigidity gradually increases, and the deformation resistance is enhanced. Therefore, the higher the temperature is, the less the polyurethane concrete material can withstand the load; the smaller the strength modulus is, the smaller the deformation resistance, but it still exceeds the standard value, and there is a large safety reserve.

4.3.3. Effect of Different Temperatures on the Bond Strength of Polyurethane Concrete and Steel Plate

The test results of bond strength between polyurethane concrete and steel plate at different temperatures are shown in Figure 17.

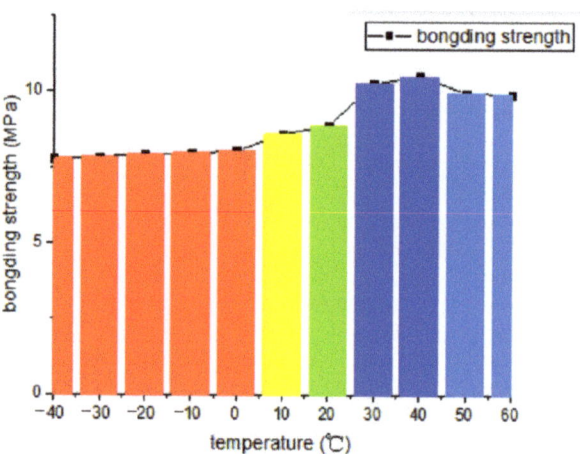

Figure 17. The test results of bond strength between polyurethane concrete and steel plate at different temperatures.

From the results of the test, the bond strength of polyurethane concrete and steel plate changes greatly with the temperature, and with stable variation in a temperature range of 10–20 °C, it is 8.89 MPa. The higher the temperature, the higher the bond strength of the polyurethane material with the steel plate. The bond strength at 40 °C reached 10.52 MPa. The bond strength was decreased while rising to 50 °C, but was still higher than the bond strength at 20 °C. There was a sharp reduction in bond strength in the temperature range of 10–0 °C, and the bond strength at low temperatures was in a stably varying state. The bond strength is not less than 7.78 MPa. It is still higher than the bond strength between the other steel bridge deck pavement materials and the steel bridge deck.

5. Conclusions

The present study mainly yielded the following conclusions:

The mechanical properties of the polyurethane concrete test at ordinary temperatures gradually grow and finally tend to be constant as the curing age increases. Polyurethane concrete has the performance of early strength and fast strength, which can reach 85% of the final strength in 3 d, 87.6% of the final strength in 7 d, and 97.9% of the final strength in 14 d. After 14 d, the strength stabilized and reached the final strength.

The strength and elastic modulus decrease with increasing temperature, the peak strain and limit strain increase with increasing temperature, and the mechanical property index of polyurethane concrete at different temperatures meets the requirements of paving materials. The compressive strength of the cube test specimen at 60°C is 49.62 MPa, which can still meet the strength requirements of pavement materials on bridge deck pavement. The test temperature was 40 °C, 20 °C, 10.8%, 84.9%, 106.7%, and 109.5%, respectively. The temperature change also greatly affects the compressive strength of polyurethane concrete.

The flexural tensile strength of the polyurethane concrete prisms formed by the vibrating table is better than the manual vibration, but the long vibration time will lead to the segregation of the polyurethane cement slurry and the gravel, which will affect the final tensile strength. According to the test results, vibration of the 10 s construction process is recommended.

The destructive state of polyurethane concrete is obviously affected by the temperature. With increasing temperature, the specimen changes from brittle destruction at low temperature to plastic destruction at high temperature. Under the action of high temperature, the crack width of the specimen during the destruction is smaller, the crack expansion speed is slower, and the specimen has good integrity after the destruction.

In recent years, the cost of raw materials increased quickly, especially the polyether polyols and isocyanate, which limits the study of other properties of new materials to a certain extent. In the future, we should study the durability and pavement properties of polyurethane concrete as a pavement material in the seasonal frozen areas.

Author Contributions: Conceptualization, T.Y. and L.L.; methodology, L.L.; software, L.L.; validation, T.Y., L.L. and Y.W. (Yuxuan Wu); formal analysis, Y.W. (Yuxuan Wu); investigation, C.G.; resources, J.L.; data curation, L.L.; writing—original draft preparation, L.L.; writing—review and editing, L.L.; visualization, Y.W. (Yuxuan Wu); supervision, T.Y.; project administration, Y.W. (Yifan Wang); funding acquisition, T.Y. All authors have read and agreed to the published version of the manuscript.

Funding: 1. Project of Research on the New Type of Steel Bridge Deck Materials for Beijing-Harbin Expressway Reconstruction and Expansion from Heilongjiang Province Transport Department; 2. Eagle Project from Harbin Science and Technology Bureau.

Institutional Review Board Statement: Not applicable.

Informed Consent Statement: Not applicable.

Data Availability Statement: Data sharing is not applicable to this article.

Conflicts of Interest: There are no conflict of interest.

References

1. Liu, J.; Yan, Y.; Li, X. Study on properties of epoxy asphalt mixture in steel deck. *J. China Foreign Highway* **2013**, *33*, 296–299.
2. Jiang, Z.; Tang, C.; Yang, J.; You, Y.; Lv, Z. A lab study to develop polyurethane concrete for bridge deckpavement. *Int. J. Pavement Eng.* **2020**, *23*, 1404–1412. [CrossRef]
3. Chen, Z.; Xu, W.; Zhao, J.; An, J.; Wang, F.; Du, Z.; Chen, Q. Experimental Study of the Factors Influencing the Performance of the Bonding Interface between Epoxy Asphalt Concrete Pavement and a Steel Bridge Deck. *Buildings* **2022**, *12*, 477. [CrossRef]
4. He, Q.; Zhang, H.; Li, J.; Duan, H. Performance evaluation of polyurethane/epoxy resin modified asphalt as adhesive layer material for steel-UHPC composite bridge deck pavements. *Constr. Build. Mater.* **2021**, *291*, 123364. [CrossRef]
5. Moslemi, A.; Navayi Neya, B.; Davoodi, M.-R.; Dehestani, M. Proposing and finite element analysis of a new composite profiled sheet deck – Applying PU and PVC and stability considerations. *Structures* **2021**, *34*, 3040–3054. [CrossRef]
6. Zhang, H.; Pan, Y.; Zhang, J. Experimental study on the properties of large span epoxy asphalt concrete. *Highway* **2013**, *58*, 36–39.
7. Zhao, G.; Jiang, Z. Study on Paving Technology and Performance of Pting Steel Decks. *J. Chongqing Jiaotong Univ. (Nat. Sci. Ed.)* **2020**, *39*, 46–51.
8. Ma, R.; Chen, F. Investigation and cause analysis of ERS steel bridge deck pavement disease. *J. China Foreign Highway.* **2019**, *39*, 175–178.
9. Zhao, J.; Chen, L.; Liu, G.; Meng, X. Compressive Properties and Microstructure of Polymer-Concrete Under Dry Heat Environment at 80 °C. *Arab. J. Sci. Eng.* **2022**, 1–16. [CrossRef]
10. El-Helou, R.G.; Koutromanos, I.; Moen, C.D.; Moharrami, M. Triaxial Constitutive Law for Ultra-High-Performance Concrete and Other Fiber-Reinforced Cementitious Materials. *J. Eng. Mech.* **2020**, *146*, 04020062. [CrossRef]
11. Ye, Z.; Li, Y.; Zhao, K. A constitutive relation of steel fiber concrete under low strain rate. *Exp. Mech.* **2019**, *34*, 284–288.
12. Xinzhi, Z.; Shuang, Z.; Yuanhong, Z. Constitutive relation of stiffened square CFST Columns. IOP conference series. *Earth Environ. Sci.* **2021**, *643*, 12014.
13. Zhong, C.; Ye, Z.; Wang, Y. A new form of nonlinear steel fiber concrete constituterelation. *Bull. Chin. Ceram. Soc.* **2018**, *37*, 1583–1588.
14. Hadigheh, S.A.; Ke, F.; Fatemi, H. Durability design criteria for the hybrid carbon fibre reinforced polymer (CFRP)-reinforced geopolymer concrete bridges. *Structures* **2022**, *35*, 325–339. [CrossRef]
15. Ding, H.; Sun, Q.; Wang, Y.; Jia, D.; Li, C.; Ji, C.; Feng, Y. Flexural Behavior of Polyurethane Concrete Reinforced by Carbon Fiber Grid. *Materials* **2021**, *14*, 5321. [CrossRef]
16. Fan, G.; Sha, F.; Yang, J.; Ji, X.; Lin, F.; Feng, C. Research on working performance of waterborne aliphatic polyurethane modified concrete. *J. Build. Eng.* **2022**, *51*, 104262. [CrossRef]
17. Ding, Z.; Ni, F.; Li, S. Analysis of key high temperature performance indexes of commonly used modified asphalt in steel bridge deck pavement. *J. Build. Materials* **2021**, *24*, 833–841.

18. Lee, K.S.; Choi, J.I.; Park, S.E.; Hwang, J.S.; Lee, B.Y. Damping property of prepacked concrete incorporating coarse aggregates coated with polyurethane. *Cem. Concr. Compos.* **2018**, *93*, 301–308. [CrossRef]
19. Lei, J.; Feng, F.; Xu, S.; Wen, W.; He, X. Study on Mechanical Properties of Modified Polyurethane Concrete at Different Temperatures. *Appl. Sci.* **2022**, *12*, 3184. [CrossRef]
20. Zhang, H.; He, Q. Progress on Flexible Improvement of Epoxy Asphalt for Steel Bridge Paving. *J. Highw. Transp. Res. Dev.* **2021**, *38*, 52–64.
21. Fu, D.; Hu, D.; Qian, Z. Research on Material Performance Index and Material Development of Steel Bridge deck Paving. *Highway* **2013**, 188–193.
22. Li, L.; Yu, T. Curing comparison and performance investigation of polyurethane concrete with retarders. *Constr. Build. Mater.* **2022**, *326*, 126883. [CrossRef]

Article

Research on Dynamic Stress–Strain Change Rules of Rubber-Particle-Mixed Sand

Yunkai Zhang [1], Fei Liu [1,*], Yuhan Bao [2,†] and Haiyan Yuan [3,†]

1. Beijing Advanced Innovation Center for Future Urban Design, Beijing 100044, China
2. College of Civil Engineering, Nanjing Forestry University, Nanjing 210018, China
3. Chang'an Dublin International College of Transportation, Chang'an University, Xi'an 710064, China
* Correspondence: liufei@bucea.edu.cn
† These authors contributed equally to this work.

Abstract: We conducted GDS dynamic triaxial tests to study the change rules of the hysteresis curve morphology, axial strain, dynamic elastic modulus, and damping ratio of waste tire rubber-mixed sand-based subgrade model samples with different rubber particle sizes, rubber mixing amounts, and loading times. The research revealed the developmental rule of the hysteresis curves of waste tire rubber-mixed-sand samples under cyclic loading. From the analyses and comparison of the dynamic stress–strain change rules of rubber-particle-mixed-sand samples under different test conditions, it was concluded that the dynamic elastic modulus and shear stiffness of rubber-mixed-sand samples were smaller than those of pure sand samples under cyclic loading while their damping ratios were greater than that of pure sand samples, promoting vibration resistance and reduction to a larger extent. Therefore, this conclusion is of guiding significance for engineering practice.

Keywords: waste tire particle; dynamic characteristic; hysteresis curve

Citation: Zhang, Y.; Liu, F.; Bao, Y.; Yuan, H. Research on Dynamic Stress–Strain Change Rules of Rubber-Particle-Mixed Sand. *Coatings* **2022**, *12*, 1470. https://doi.org/10.3390/coatings12101470

Academic Editor: Ofelia-Cornelia Corbu

Received: 25 July 2022
Accepted: 28 September 2022
Published: 4 October 2022

Publisher's Note: MDPI stays neutral with regard to jurisdictional claims in published maps and institutional affiliations.

Copyright: © 2022 by the authors. Licensee MDPI, Basel, Switzerland. This article is an open access article distributed under the terms and conditions of the Creative Commons Attribution (CC BY) license (https://creativecommons.org/licenses/by/4.0/).

1. Introduction

Waste tire particles have damping characteristics because rubber has high elasticity and good stability. Rubber particles have a smaller elastic modulus than other modified reinforced materials, but they can recover from larger deformation and have greater elastic deformation ability when external forces are removed. In addition, there is no obvious yielding point on the stress–strain curves of rubber particles. All these indicate that rubber particles enjoy an excellent vibration reducing performance. At present, research on the static characteristics of waste tire particle-mixed soil dominates compared with that of dynamic characteristics in China and abroad. Some scholars have drawn valuable conclusions through tests.

Feng et al. [1] studied the dynamic stress–strain change rules of a mixture of sand and rubber particles via dynamic triaxial tests. The test results were then compared with the Hardinen conventional soil dynamic characteristics formula. As a result, an improved Hardinen formula of the dynamic characteristics of rubber-mixed sand was obtained.

Pamukcue et al. [2] studied the dynamic characteristics of a sand–rubber particle mixture through the resonant column test, from which it was concluded that a smaller stiffness of this mixture compared with pure sand results from the different dynamic modulus of the two materials and the increased contact area between particles. If there is a big difference in the expansion coefficients of the two materials, the damping ratio of mixed samples will be larger than that of pure sand.

Nakeh [3] studied the influences of dynamic stress and estimated the confining pressure on the dynamic elastic modulus of rubber-mixed sand in large-scale dynamic triaxial tests using the exponential function model.

Anasta et al. [4–6] studied the change rules of the dynamic elastic modulus and damping ratio of rubber-mixed sand under different test conditions in dynamic triaxial tests. The theoretical formulas for the dynamic elastic modulus and damping ratio of rubber-mixed sand were obtained through function fitting based on a large number of tests.

Sui X X [7] mixed quartz particles into sand and studied the dynamic elastic modulus and damping ratio of mixed sand samples under different test conditions via dynamic triaxial tests. It was concluded that the dynamic elastic modulus and damping ratio of mixed sand samples are proportional to the dynamic stress applied in the tests.

In dynamic triaxial tests, Li L H [8] found that the dynamic elastic modulus of waste tire particle-mixed sand is smaller than that of pure sand, and the equivalent damping significantly rises with the increase in rubber content.

Shang S P [9] studied the change rules of the dynamic elastic modulus of glass sand under different test conditions in dynamic triaxial tests. It was concluded that the dynamic shear modulus of rubber-glass-mixed sand decreases after rubber particles are added.

The application of discarded tires in geotechnical engineering, particularly soil reinforcement technology, has been extensively studied. The application of soil reinforcement technology dates back to antiquity. In ancient times, basic materials were combined with dirt to create rudimentary soil reinforcements. Recently, reinforcing technology has been implemented in geotechnical engineering, and reinforced materials have evolved from natural textiles and steel bars to geosynthetics such as geomembranes and geocells. Diverse soil-reinforcement materials are often lightweight, simple to assemble, inexpensive, and simple to produce and transport. In addition, reinforcing technology may raise soil strength, boost soil stability, and significantly enhance soil mechanical qualities [10–13].

To sum up, research on the dynamic characteristics of saturated sand dominates while that on the dynamic characteristics of waste tire rubber particles mixed with sand is inadequate. In this paper, GDS dynamic triaxial tests were conducted to study the change rules of the hysteresis curve, axial strain, dynamic elastic modulus, and damping ratio of waste tire rubber-mixed sand-based subgrade model samples with different rubber particle sizes, rubber mixing amounts, and loading times. Finally, conclusions of guiding significance for engineering practice were drawn from tests.

2. Test Content and Scheme

2.1. Test Instruments and Functions

2.1.1. Composition of Dynamic Triaxial Apparatus System DYNTTS

The dynamic triaxial apparatus system *DYNTTS* made in England was adopted in the tests, whose composition is shown in Figure 1. Specifically, the data acquisition system consisted of a distributed control system (*DCS*) signal receiver, an axial force sensor, and a pore water pressure sensor, and the external control system was composed of an axial force controller, a pressure chamber, a confining pressure controller, and a back pressure controller.

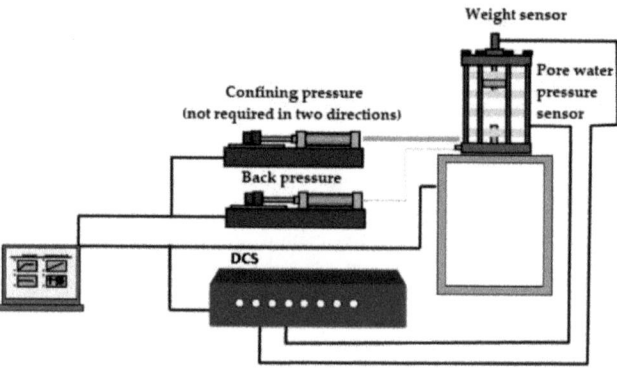

Figure 1. The basic framework of the *DYNTTS*.

2.1.2. Experimental Principle of GDS Dynamic Triaxial Apparatus

The test principle of the GDS dynamic triaxial apparatus is that the confining pressure value σ_3 and the back pressure value P^0 are preset via the confining pressure controller and the back pressure controller. Dynamic loads or static loads are applied to control force or displacement by operating GDSLAP when each set of confining pressure and back pressure reaches the test target values. GDSLAP coordinates and controls the relationship among the confining pressure controller, the back pressure controller, and the axial displacement force control system and processes the data collected by each sensor according to the set calculation mode. Then, the difference between the actual and target values is identified during the test before each set of data generated during the test is output. The details are as follows:

The initial diameter and height of the triaxial sample were set to be D_0 and H_0, respectively. The initial area A_0 and initial volume V_0 of the sample can be obtained according to Formulas (1) and (2).

$$V_0 = \frac{\pi D_0^2 H_0}{4} \quad (1)$$

$$A_0 = \frac{\pi D_0^2}{4} \quad (2)$$

D, H, A, and V represent the sample's diameter, height, area, and volume, respectively. The volume change of the sample is obtained as $\Delta V = V_0 - V$, the volume change data of the back pressure controller reflect the pore water pressure changes, and the derivation formulas are obtained as follows: the axial force of actual sample $\sigma = F/A + \sigma_3$, the effective stress $\sigma_3' = \sigma_3 - u_w$ derived from pore water pressure, the effective axial stress $\sigma' = \sigma - u_w$, the axial strain $\varepsilon_a = (\Delta H/H_0) \times 100\%$, the radial strain $\varepsilon_r = [(-D + D_0)/D_0] \times 100\%$, and the deviator stress $\sigma_1 = \sigma_3' \times K_c - \sigma_3'$ derived from effective consolidation confining pressure σ_3' and consolidation ratio K_c.

2.2. Test Scheme and Damage Standard

2.2.1. Test Scheme

A sinusoidal load with a stress amplitude of 30 kPa, a frequency of 1 Hz, and a vibration frequency of 300 times was applied via stress control in the dynamic pore pressure test of saturated rubber-mixed sand samples, and the parameters of dynamic stress, dynamic strain, and dynamic pore pressure were recorded in GDSLAP. The test was stopped when the dynamic pore pressure reached the damage standard, or the vibration frequency reached 300 s, the scheme of which is shown in Table 1.

Table 1. Experimental scheme for research on dynamic pore pressure of rubber-mixed sand.

Mixing Amount/%	Particle Size/mm	Consolidation Stress Ratio	Confining Pressure/kPa	Dynamic Stress/kPa
10%	0.75	1	50, 100, 150	30
	1.5	1, 1.5, 2	50, 100, 150	30
	2.5	1	50, 100, 150	30
20%	0.75	1	50, 100, 150	30
	1.5	1, 1.5, 2	50, 100, 150	30
	2.5	1	50, 100, 150	30
30%	0.75	1	50, 100, 150	30
	1.5	1, 1.5, 2	50, 100, 150	30
	2.5	1	50, 100, 150	30

2.2.2. Damage Standard

Pore pressure standard was adopted in the research on dynamic pore pressure and hysteresis curve of samples; that is, the samples were damaged when the over-consolidation stress ratio $u_d/\sigma_3 = 1$. The pores in soil mass were compressed, and the dynamic pore water pressure between particles continuously increased under the cyclic action of dynamic loads. When the dynamic pore water pressure was equal to the effective confining pressure ($u_d = \sigma_3$), the effective stress between particles of soil mass in the samples approached zero, leading to liquefaction damage of the soil mass.

2.3. Test Procedures

2.3.1. Test Materials and Sample Preparation

(1) Rubber particles. The rubber particles used in this test were produced via the normal temperature crushing method. Screened by the vibrating screen in the laboratory, the following three kinds of rubber particles with average particle sizes of 0.75 mm, 1.5 mm, and 2.5 mm, respectively, were mainly utilized, as shown in Figures 2–5.

Figure 2. Rubber particles of 0.75 mm.

Figure 3. Rubber particles of 1.5 mm.

Figure 4. Rubber particles of 2.5 mm.

Figure 5. Sand used in the test.

(2) Sand particles. ISO Fujian standard sand (as shown in Figure 5) was used as sand material in the tests to ensure the consistency of each group of tests. In addition, the density bottle method was adopted to measure the specific density (2.60) of sand in a dry state.

(3) Preparation of raw materials for dynamic triaxial test samples: waste tire rubber particles (three kinds of particle sizes) and Fujian standard sand. Tools for sample preparation: rubber membranes, three-segment dies, permeable stones, and filter paper.

To ensure the consistent arrangement of sand particles in each group of samples, the dry density of sand in rubber-mixed sand samples was required to be equal to that of pure sand samples when different contents of rubber particles were mixed into sand particles. The calculation formula for the corresponding proportion is shown as follows.

$$\rho = \frac{M_s}{V_s + V_k} = \frac{M_o - M_x}{V_o - V_x} \tag{3}$$

where V_o, V_x, V_s, and V_k represent the total volume of rubber-mixed sand, the rubber volume, the sand volume, and the pore volume between particles, respectively, while M_o, M_s, and M_x stand for the total mass of samples, the mass of sand particles, and the mass of rubber particles, respectively.

2.3.2. Installation, Saturation, and Consolidation of Samples

(1) Sample installation: The rubber-mixed sand was sampled via the moist sampling method, and the rubber-mixed sand particles were weighed according to the proportion which was boiled in water and cooled for 12 h for later use. The apparatus was normally plugged in, and all pipes were not blocked before the sample was installed. First, a permeable stone covered with moist single-layer filter paper was put on the base of the apparatus. Then, a three-segment die was installed, and the boiled and cooled rubber-mixed sand was put into the rubber membrane evenly and slowly. After the sand loading, the back pressure was set to be −30 kPa. When the back pressure loading was stable, the three-segment die was removed. At this moment, the sample was upright due to negative pressure, as shown in Figure 6. After that, the external pressure hood was closed, the apparatus was filled with water, and the axial force sensor value was set to 0.005 kN when the apparatus base rose and contacted the top sensor. When the axial force reached 0.005 kN, the apparatus no longer rose. The sample was contacted with the axial force sensor as shown in Figure 7. Then, the vent plug at the top of the pressure chamber was opened, and the inlet valve was opened to fill the pressure chamber with water. Finally, the inlet valve and the vent plug were closed after all steps were conducted.

Figure 6. Installed sample.

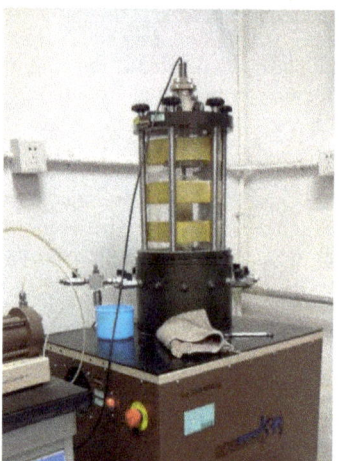

Figure 7. Pressure hood installed.

(2) Sample saturation: Water head saturation and back pressure saturation were utilized to saturate the samples, thus improving the saturation of the rubber-mixed sand samples. First, water head saturation was performed. Specifically, the drain valve of the apparatus base was connected with that of the water container. Then, the drain valve of the container was opened so that the boiled and cooled distilled water flowed into the bottom from the apparatus base. In this way, the pores between sample particles were filled with water, and the air in the samples was exhausted, thus gradually elevating the saturation of the samples. In addition, excess distilled water flowed out from the top of the samples to the container prepared in advance. After the water head saturation, a small amount of air remained in the samples. However, the air in the samples was melted into water by applying five-level back pressure saturation. Figure 8 shows the five-level back pressure loading change curve. In the process of back pressure loading, the effective confining pressure remained at 20 kPa (the confining pressure remained greater than the back pressure to prevent the sample preparation from being damaged due to the overload of the dynamic pore pressure in the samples, and the five-level loading

process could minimize the disturbance of the test apparatus loading on the samples). After the five-level back pressure loading, the saturation of the samples was detected. If the B value $\Delta u / \Delta \sigma_3 > 0.95$, the saturation condition was met. If not, saturation detection was conducted after the samples were re-installed and the saturation process was reloaded.

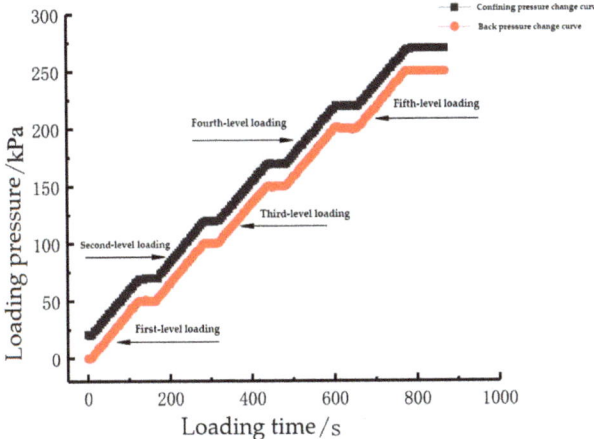

Figure 8. Five-level loading curve back pressure.

3. Research on Dynamic Stress–Strain Change Rules of Rubber-Particle-Mixed Sand

3.1. Change Rules of the Dynamic Stress–Strain Hysteresis Curve

The dynamic stress–strain relationship of the soil mass presents two characteristics under cyclic loading, namely nonlinearity and hysteresis. As shown in Figure 9, the stress–strain data points recorded in the dynamic triaxial tests were connected into a curve called a stress–strain hysteresis curve. The stress–strain hysteresis curve is a developmental curve where loading and unloading repeat cyclically, reflecting the relationship between the dynamic stress σ_d and the dynamic strain ε_d at every moment during the cycle of the dynamic stress cyclic loading. The hysteresis of the dynamic stress–strain hysteresis curve was mainly displayed as the dynamic stress–strain hysteresis curve would gradually move towards the direction where strain magnified with the increase in cycle time since the soil mass was an elastoplastic material. The encircled central point shifted from the original point to the positive X-axis direction and the soil mass was subjected to unrecoverable permanent strain.

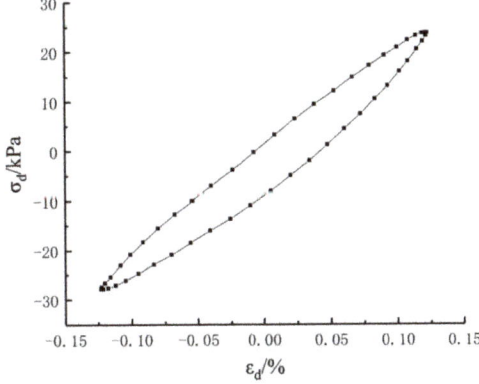

Figure 9. Dynamic stress–strain hysteresis curve.

Research on Hysteresis Curve Morphology of Rubber-Particle-Mixed Sand

As the dynamic loading frequency in the test was 1 Hz and the interval of the data recording was set as 0.02 s, a total of 50 data points was acquired in the interval of 1 s. In order to analyze the change rules of the dynamic stress–strain hysteresis curve of the soil mass more meticulously, the hysteresis curves of the first 100 times of the cyclic loading recorded by the GDS operating system in the dynamic triaxial tests are exhibited in Figure 9.

Figure 10a–j is hysteresis curves of the soil samples in various groups with different mixing amounts of rubber particles. It was observed from the hysteresis curve of the pure sand in Figure 10a that cumulative strain took place in the pure sand samples under cyclic loading and the strain reached 0.54% after 100 times of vibration. The morphology of the hysteresis curves coincided at each time.

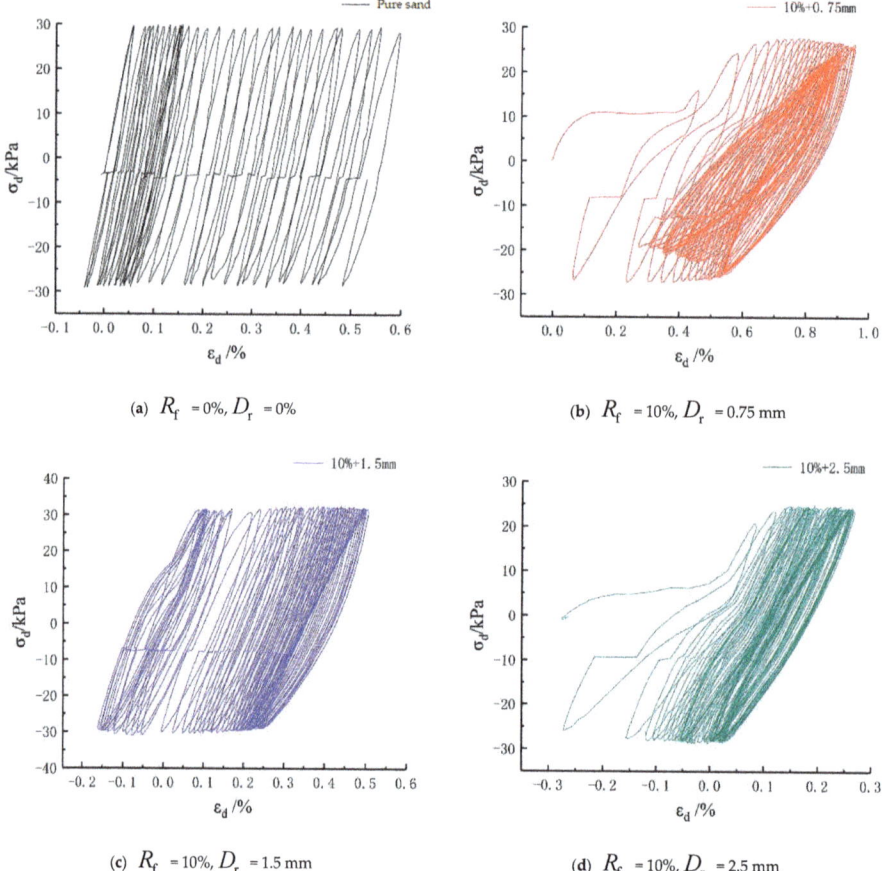

(a) R_f = 0%, D_r = 0%

(b) R_f = 10%, D_r = 0.75 mm

(c) R_f = 10%, D_r = 1.5 mm

(d) R_f = 10%, D_r = 2.5 mm

Figure 10. Cont.

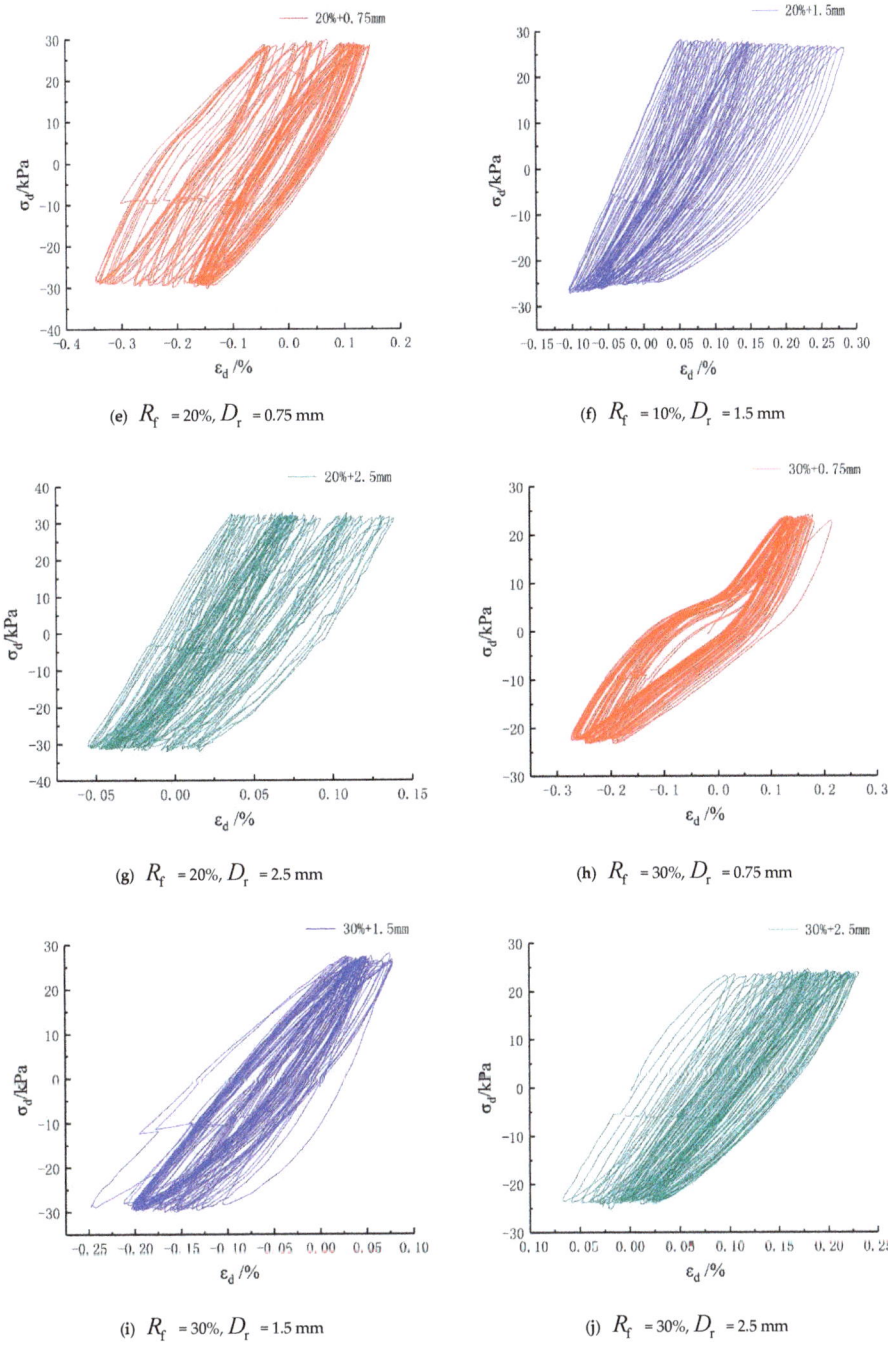

Figure 10. Relationship curve of mixed sand σ_d-ε_d with different rubber content. (**a**) $R_f = 0\%$, $D_r = 0\%$; (**b**) $R_f = 10\%$, $D_r = 0.75$ mm; (**c**) $R_f = 10\%$, $D_r = 1.5$ mm; (**d**) $R_f = 10\%$, $D_r = 2.5$ mm; (**e**) $R_f = 20\%$, $D_r = 0.75$ mm; (**f**) $R_f = 20\%$, $D_r = 1.5$ mm; (**g**) $R_f = 20\%$, $D_r = 2.5$ mm; (**h**) $R_f = 30\%$, $D_r = 0.75$ mm; (**i**) $R_f = 30\%$, $D_r = 1.5$ mm; (**j**) $R_f = 30\%$, $D_r = 2.5$ mm.

Figure 10b–d refers to the images of the hysteresis curves when the mixing amount of rubber R_f was 10%, and it was seen that the developmental rule of the hysteresis curves in three groups of samples all presented a trend from looseness to tightness. The samples showed a maximum strain of 0.95% within the range of vibration time while mixed with 0.75 mm rubber particles and the cumulative strain displayed an increasing trend. The strain range of the samples mixed with 1.5 mm rubber particles was −0.15%–0.47% and that of the samples mixed with 2.5 mm rubber particles was −0.25%–0.26%.

Figure 10e–g shows the images of the hysteresis curves with 20% mixing amount of rubber R_f. The strain ranges of the samples mixed with rubber particles in the three sizes with the three mixing amounts were −0.41%–0.14%, −0.12%–0.27%, and −0.06%–0.13%, respectively. It was found that when the mixing amount of rubber increased to 20%, the range of the dynamic strain reduced correspondingly, and the central symmetry points of the hysteresis curves swung around the original points.

Figure 10h–j represents the images of the hysteresis curves with 30% mixing amount of rubber R_f. The strain ranges of the samples mixed with rubber particles in the three sizes with the three mixing amounts were −0.28%–0.21%, −0.25%–0.06%, and −0.09%–0.23%, respectively. It was never discovered that a relatively large cumulative strain was produced in three groups of samples under cyclic loading. The morphology of the hysteresis curves was similar at each time and the central symmetry points of the hysteresis curves also swung around the original points.

The stress–strain hysteresis curve of the pure sand samples was taken as a reference object. As shown in Figure 1, with 100 times of vibration N, the strain ε_d of the pure sand samples was 0.54% and the compressive strain of the samples increasingly enlarged until its damage as the test went on. Due to the high-resilience compressibility of rubber, the three groups of samples mixed with rubber particles in the size of 2.5 mm showed a different changing form in the stress–strain curve from the pure sand samples. In this case, the curve was presented as a reciprocating form of tension and compression and the strain fluctuated between −0.33% and 0.27%. It was figured out that the pure sand samples mixed with rubber particles acquired a greatly enhanced non-deformability. It was also observed that the strain change rates of the pure sand mixed with the rubber particles in the sizes of 1.5 mm and 0.75 mm were higher than that of the pure sand mixed with the rubber particles in the size of 2.5 mm. The strain of the samples mixed with the rubber particles in the size of 1.5 mm was between −0.26% and 0.42% and that of the samples mixed with the rubber particles in the size of 0.75 mm was between −0.36% and 0.95%. The rubber particles of a relatively large size could improve the non-deformability of the samples and the hysteresis curve of the samples mixed with such rubber particles presented a reciprocating form of tension and compression.

The influence of the mixing amount of rubber particles on the stress–strain relationship is expressed below. Under the cyclic loading, the compressive strain of the pure sand samples continuously increased until it was damaged. The stress–strain hysteresis curve of the rubber-mixed sand samples was similar to that of the pure sand samples because of a small mixing amount of rubber ($R_f = 10\%$). However, the samples with $R_f = 20\%$ and $R_f = 30\%$ had a stressed skeleton, in which the rubber particles were primary, on account of the large mixing amount of rubber particles, so the stress–strain hysteresis curve showed a totally different changing form from the pure sand. From the specific data in Figure 4, the strain range of the rubber particles in the three sizes was between −0.36% and 0.24% and the tensile and compressive strain was distributed. The strain of the samples with $R_f = 10\%$ changed from −0.19% to 0.9542% and the compressive strain dominated in the hysteresis curve, which was similar to that of the pure sand.

3.2. Change Rules of Hysteresis Curves under Graded Vibration Times

The dynamic stress–strain hysteresis curves of four times at 10 s, 100 s, 200 s, and 300 s under the cyclic loading were selected as shown in Figures 11–13 below to more specifically study the dynamic stress–strain curve change rules of the rubber-particle-mixed-sand samples under cyclic loading. The developmental rule of the hysteresis curves of specified time was researched according to the hysteresis curve strain and the encircled area of each group of samples chosen at different times.

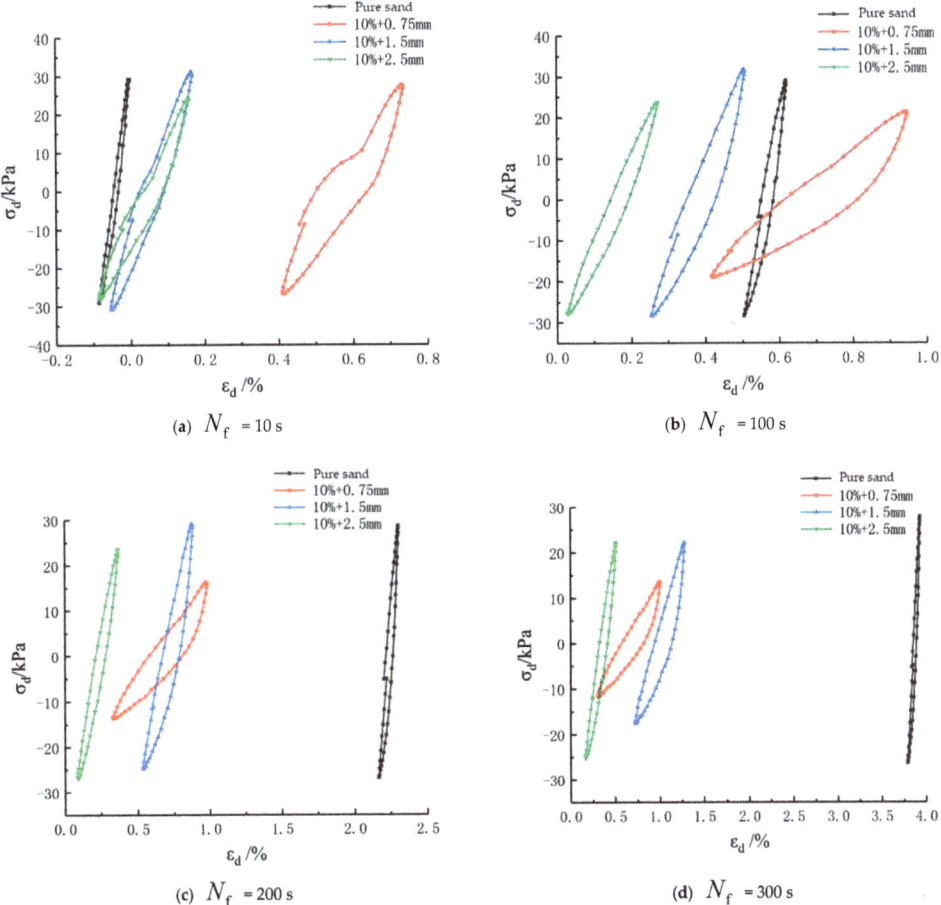

Figure 11. Dynamic stress–strain curve of a specified time (R_f = 10%. (**a**) N_f = 10 s; (**b**) N_f = 100 s; (**c**) N_f = 200 s; (**d**) N_f = 300 s.

Figure 11a–d is the dynamic stress–strain hysteresis curves of the rubber-particle-mixed-sand samples in the three particle sizes under graded vibration times when the mixing amount of rubber R_f was 10%. The maximum strains of the samples mixed with 0.75 mm rubber particles achieved 0.73%, 0.95%, 0.97%, and 1.12%, respectively, under the cyclic loading at four times (10 s, 100 s, 200 s, and 300 s). The rubber-mixed sand samples in the size of 0.75 mm presented obvious softening under cyclic loading and the encircled area by the dynamic stress hysteresis curve decreased with the increase in vibration time, indicating a reduced energy consumption of the samples. The samples mixed with 1.5 mm rubber particles showed the maximum strains, namely 0.16%, 0.47%, 0.88%, and 1.27%, respectively, under cyclic loading at four times (10 s, 100 s, 200 s, and 300 s). The samples mixed with 2.5 mm rubber particles exhibited the maximum strains, namely 0.15%, 0.26%, 0.36%, and 0.51%, respectively, under cyclic loading at four times (10 s, 100 s, 200 s, and 300 s).

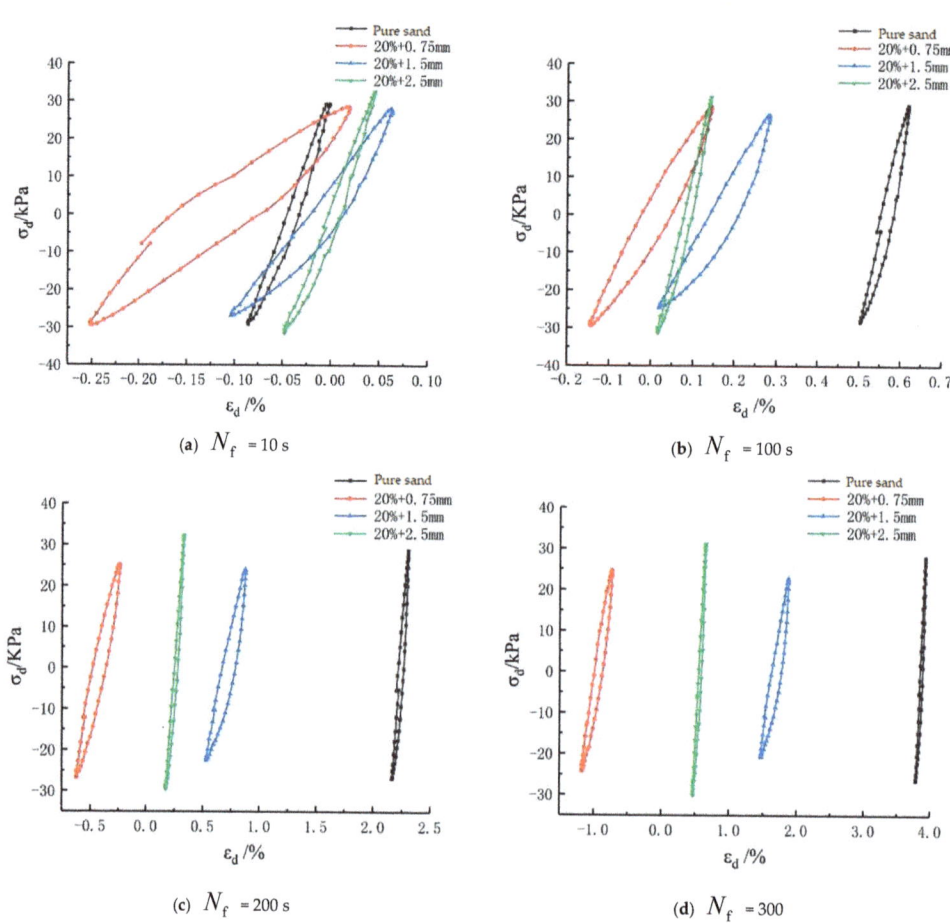

Figure 12. Dynamic stress–strain curve of a specified time (R_f = 20%). (**a**) N_f = 10 s; (**b**) N_f = 100 s; (**c**) N_f = 200 s; (**d**) N_f = 300 s.

To further study the impacts of a 20% mixing amount of rubber particles on the stress–strain relationship curve, the hysteresis curves of various groups under graded vibration times are listed as shown in Figure 12. Specifically, the samples mixed with 0.75 mm rubber particles had the maximum strains, namely 0.02%, 0.14%, −0.24%, and −0.73%, respectively, under cyclic loading at four times (10 s, 100 s, 200 s, and 300 s), which presented a reciprocating strain form of tension and compression and did not produce a large cumulative strain. The samples mixed with 1.5 mm rubber particles reached the maximum strains, namely 0.06%, 0.27%, 0.87%, and 1.87%, respectively, under cyclic loading at four times (10 s, 100 s, 200 s, and 300 s), whose dynamic strain enlarged as the number of cycles increased and which generated a large cumulative strain. The samples mixed with 2.5 mm rubber particles reached the maximum strains, namely 0.04%, 0.13%, 0.33%, and 0.64%, respectively, under cyclic loading at four times (10 s, 100 s, 200 s, and 300 s), whose dynamic strain developmental rule was similar to that of the samples mixed with 1.5 mm rubber particles and whose dynamic strain was augmented with the increase in the number of cycles. Under 300 times of vibration, the latter only achieved a maximum strain of half of that of the former. Since rubber particles of large size possess a great energy absorption and vibration reducing performance, they could effectively decrease the strain of the samples under cyclic loading when mixed in the sand.

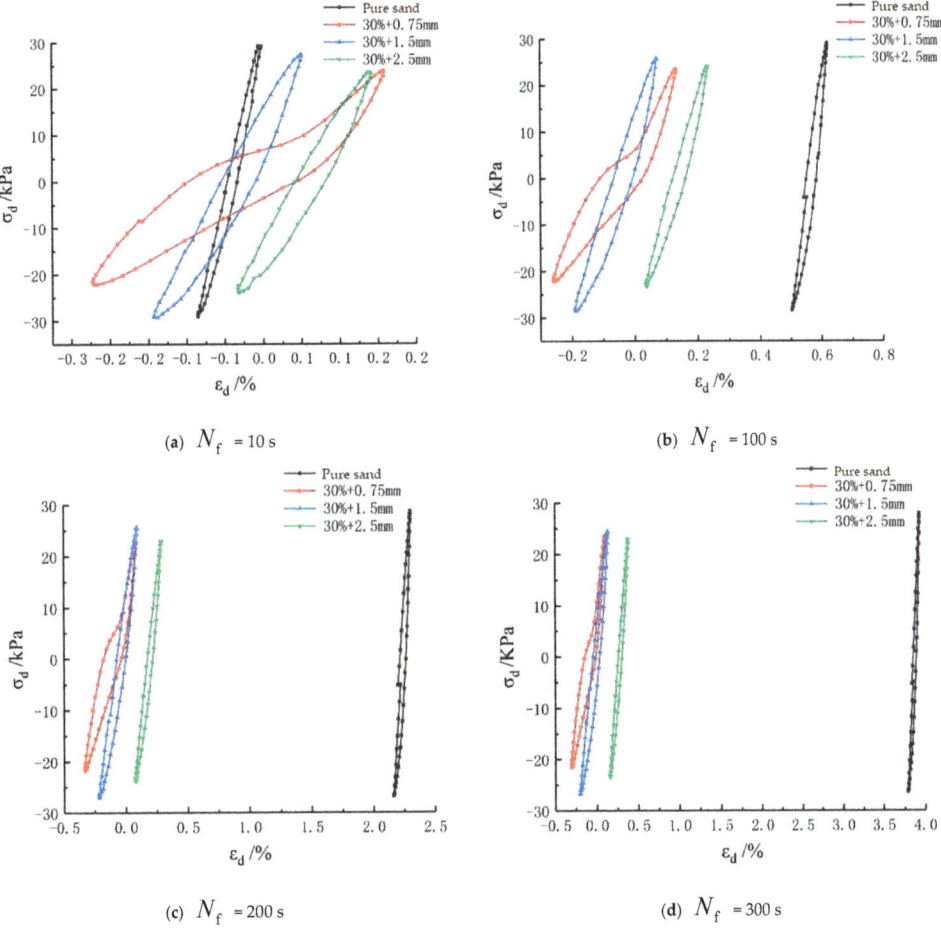

Figure 13. Dynamic stress–strain curve of a specified time ($R_f = 30\%$). (**a**) $N_f = 10$ s; (**b**) $N_f = 100$ s; (**c**) $N_f = 200$ s; (**d**) $N_f = 300$ s.

Furthermore, when the mixing amount of rubber particles was lifted to 30%, it could be seen from Figures 3–5 that the samples primarily showed a reciprocating form of tension and compression. When the mixing rubber particle size was 0.75 mm, the maximum strains under cyclic loading at four times (10 s, 100 s, 200 s, and 300 s) were 0.16%, 0.21%, 0.08%, and 0.09%, respectively. When the mixing rubber particle size was 1.5 mm, the maximum strains under cyclic loading at four times (10 s, 100 s, 200 s, and 300 s) were 0.05%, 0.06%, 0.09%, and 0.13%, respectively. When the mixing rubber particle size was 2.5 mm, the maximum strains under cyclic loading at four times (10 s, 100 s, 200 s, and 300 s) were 0.14%, 0.23%, 0.29%, and 0.38%, respectively. As can be observed from the abovementioned three groups of data, as the mixing amount of rubber was 30%, the change in the samples under cyclic loading was mainly a reciprocating form of tension and compression, and prominent strain cumulation was not found in the samples. With a high mixing amount of rubber, the stressed skeleton inside the samples experienced a change in its form, which was dominated by the rubber particles, so the samples presented a reciprocating form of tension and compression under cyclic loading.

3.3. Change Rules of Dynamic Elastic Modulus and Damping Ratio of Hysteresis Curves

3.3.1. Change Rules of Dynamic Elastic Modulus

The hysteresis curve of the soil mass is the basis of studying the dynamic triaxial characteristics. The characteristic dynamic parameters of the soil mass, including the dynamic elastic modulus and the damping ratio, could be obtained by the quantitative characteristic analysis [14] of the hysteresis curves.

Figure 14 shows a standard dynamic stress–strain curve. A quantitative analysis of the hysteresis curve was conducted to gain the dynamic elastic modulus E_d and the damping ratio λ. The dynamic elastic modulus of the soil mass was calculated using the slope of the two endpoints A and C on the hysteresis curve in Figures 4–6 according to Formula (4).

$$E_d = \frac{\sigma_{Amax} - \sigma_{Cmin}}{\varepsilon_{Amax} - \varepsilon_{Cmin}} \quad (4)$$

where: σ_{Amax} and σ_{Cmin} denote the maximum value and the minimum value of the dynamic stress of the samples in each hysteresis curve, respectively.

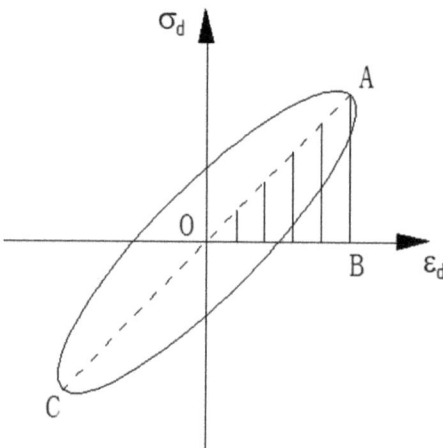

Figure 14. Standard dynamic stress–strain curve.

ε_{Amax} and ε_{Cmin} denote the dynamic strain values, which correspond to the maximum value and the minimum value of the dynamic stress of the samples in each hysteresis curve, respectively.

Figure 15 displays a curve of the relationship between vibration time N and dynamic elastic modulus E_d of the samples mixed with the rubber particles with different mixing amounts in different sizes. It was observed from the figure that the dynamic elastic modulus E_d of the pure sand samples and the samples with 20% and 30% mixing amounts of rubber R_f presented a decreasing trend with the increase in vibration time. When the mixing amount of rubber R_f was 30%, the dynamic elastic modulus of the mixed samples remained at a basically stable state as the vibration time increased. The dynamic elastic modulus E_d of all the rubber-mixed sand was smaller than that of the pure sand samples. Specifically, with a 10% mixing amount of rubber, the dynamic elastic modulus was reduced by 0.53 MPa (76%) compared with the pure sand samples to the greatest extent. With a 20% mixing amount of rubber, the dynamic elastic modulus was reduced by 0.49 MPa (70%) compared with that of the pure sand samples to the greatest extent. With a 30% mixing amount of rubber, the dynamic elastic modulus was reduced by 0.58 MPa (82.8%) compared with the pure sand samples to the greatest extent. After adding rubber particles to the pure sand, the samples had a decreased dynamic elastic modulus, illustrating that the sand mixed with rubber particles could reduce the shear stiffness of the soil mass, giving full play to the vibration resistance and reduction.

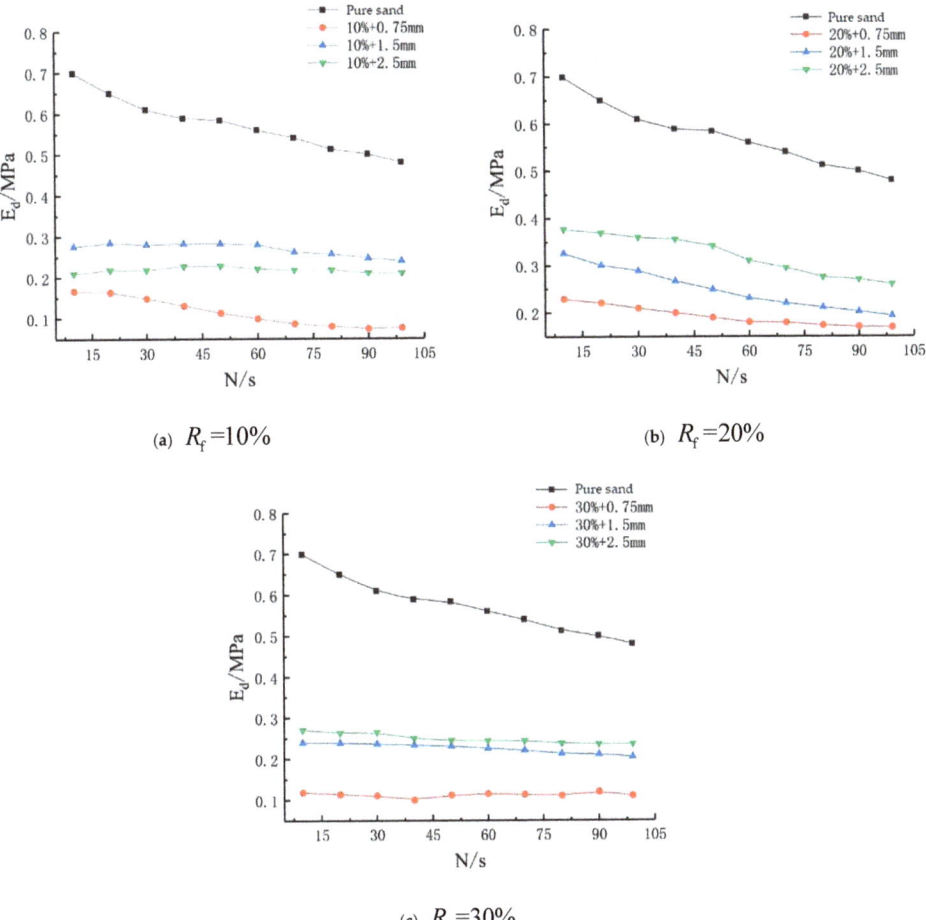

Figure 15. Rubber-mixed sand elastic modulus with different rubber content and size. (**a**) R_f = 10%; (**b**)R_f = 10%; (**c**) R_f = 30%.

3.3.2. Change Rules of Damping Ratio

For geomaterials, the damping ratio of soil is one of the important parameters to reflect the dynamic characteristics of soil mass, which reveals the hysteresis of the dynamic stress–strain relationship of soil mass under cyclic loading.

At present, soil is regarded as a viscoelastic body to calculate the damping ratio, which is the ratio of the energy consumed to the total energy stored during a cycle of stress cyclic loading, and the calculation formula is shown in Figure 10.

$$\lambda = \frac{\Delta W}{4\pi W}$$

where: λ refers to the damping ratio,

ΔW is the area of the hysteresis curve, representing the energy consumed in loading and unloading, and W is the area of ΔOAB, representing the total energy of loading and unloading.

The area encircled by the hysteresis curve needs to be accurately acquired for the calculation of the damping ratio. Currently, the most commonly used method is to fit the hysteresis curve using the elliptic curve. The calculated area of the ellipse is considered to be the area encircled by the hysteresis curve. However, the actual hysteresis curve is not a standard elliptic curve. Taking this test as an example, the GDS system was set to collect 50 data points in each vibration cycle, so the hysteresis curve was a curve formed by connecting 50 data points. Therefore, there were some errors in the method of fitting the stress–strain hysteresis curve with an elliptic curve, especially when the coincidence rate between the curve shape formed by connecting the test data points and the elliptic curve shape was low. In this case, the calculated damping ratio was quite different from the actual results. As a result, the polygon accumulation method proposed by Chen W [15] was applied to calculate the damping ratio in this paper. First, the stress–strain data points corresponding to the selected specific time were connected into a curve in sequence. At this time, the graph encircled by the curve was a polygon, and then the area of the hysteresis curve was obtained by calculating the polygon area.

In Figure 16, two points A and B are two adjacent data points. The OA side of the triangle was supposed as vector a and the OB side as the vector b, and the area of the triangle encircled by the two vectors was half of the vector value of their cross product.

$$a \times b = \begin{vmatrix} a_y & a_z \\ a_y & b_z \end{vmatrix} i - \begin{vmatrix} a_x & a_z \\ b_x & b_z \end{vmatrix} j + \begin{vmatrix} a_x & a_y \\ b_x & b_y \end{vmatrix} k \tag{5}$$

$$S_{OAB} = \frac{1}{2}|OA \times OB| = -\frac{1}{2}\begin{vmatrix} \varepsilon_{ad} & \sigma_{ad} \\ \varepsilon_{bd} & \sigma_{bd} \end{vmatrix} \tag{6}$$

where: ε_{ad}, σ_{ad}, ε_{bd}, and σ_{bd} mean the dynamic strain and dynamic stress, which correspond to the adjacent data points A and B in Figure 16.

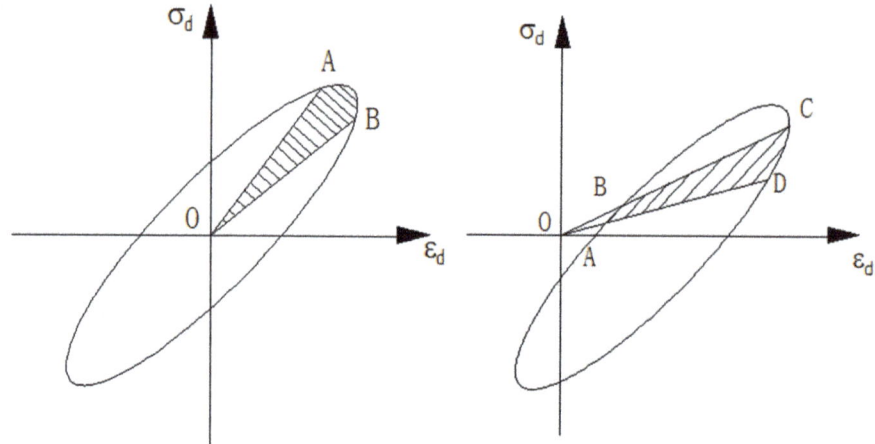

Figure 16. Approximate calculation method of the dynamic stress–strain curve.

Following the polygon accumulation method, the data in the GDS test recording system were utilized to calculate the area encircled by the hysteresis curve with Formula (7).

$$\Delta W = -\frac{1}{2}(\begin{vmatrix} \varepsilon_{1d} & \varepsilon_{1d} \\ \varepsilon_{1d} & \varepsilon_{1d} \end{vmatrix} + \begin{vmatrix} \varepsilon_{2d} & \varepsilon_{2d} \\ \varepsilon_{3d} & \varepsilon_{3d} \end{vmatrix} + \cdots + \begin{vmatrix} \varepsilon_{(n-1)d} & \varepsilon_{(n-1)d} \\ \varepsilon_{nd} & \varepsilon_{nd} \end{vmatrix} + \begin{vmatrix} \varepsilon_{nd} & \varepsilon_{nd} \\ \varepsilon_{1d} & \varepsilon_{1d} \end{vmatrix}) \tag{7}$$

where: ε_{nd} and σ_{nd} represent the dynamic strain and dynamic stress corresponding to the adjacent data points in the hysteresis curve.

Figure 17 shows that the damping ratio of both the mixed sand samples and the pure sand samples exhibited an increasing trend as the vibration time grew and the damping ratio of the rubber-mixed sand samples was larger than that of the pure sand samples. With a 10% mixing amount of rubber, the damping ratio was augmented by 6.54 times to the maximum extent. With a 20% mixing amount of rubber, the damping ratio was augmented by 5.05 times to the maximum extent. With a 30% mixing amount of rubber, the damping ratio was augmented by 5.34 times to the maximum extent. Since the dry density ρ_d of all the samples was kept same in this test, the pore characteristics of the particles were the same. When the samples are mixed with rubber particles, energy absorption will occur on account of the elasticity of the rubber particles. Thus, the hysteresis of the soil mass is strengthened under cyclic loading. This was why the damping ratio of the rubber-mixed sand samples was greater than that of the pure sand samples.

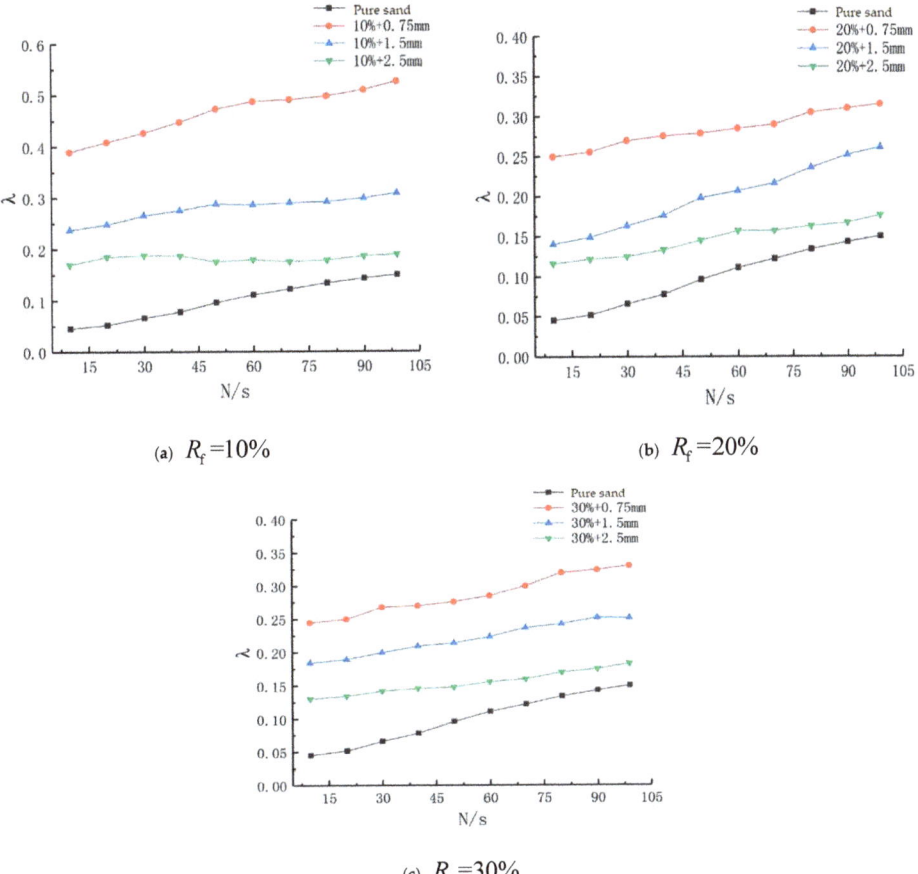

Figure 17. Rubber-mixed sand damping ratio with different rubber content and size. (a) R_f = 10%; (b) R_f = 10%; (c) R_f = 30%.

4. Conclusions

The dynamic stress–strain hysteresis curve, dynamic elastic modulus, and damping ratio of saturated rubber-mixed sand samples were studied in this paper, and the conclusions were as follows:

(1) Under the same confining pressure and dynamic stress, the dynamic elastic modulus of the rubber-mixed sand samples was smaller than that of the pure sand samples. With a 30% mixing amount of rubber in the particle size of 0.75 mm, the dynamic elastic

modulus of the samples showed a maximum decrease of 0.58 MPa. The sand mixed with rubber particles could decrease the shear stiffness of the soil mass, making full use of the vibration resistance and reduction. The behavior of sand became more ductile with an increasing granulated rubber content. Adding granulated rubber led to greater yielding strain and less tangent stiffness of the sand. The maximum dilation angle decreased with the decrease in granulated rubber content [16].

(2) Under the same confining pressure and dynamic stress, the damping ratio of the rubber-mixed sand samples was larger than that of the pure sand samples. When the samples were mixed with rubber particles, the rubber particles presented energy absorption because of their elasticity and thus the hysteresis of the soil mass was strengthened under cyclic loading. Therefore, the damping ratio of the rubber-mixed sand samples was greater than that of the pure sand samples.

(3) The sand samples mixed with rubber particles could prominently reduce the cumulative strain generated under the cyclic loading. When the mixing amount of rubber particles was large, the cumulative strain produced in the samples was far smaller than that of the pure sand samples. As rubber particles are featured by high resilience, they were regarded as stressed skeletons in the rubber-mixed sand with a high mixing amount to bear the external loading. At this moment, the strain produced due to the cyclic loading was a partially recoverable elastic strain, equivalent to adding a spring with recoverable strain in the pure sand samples. Hence, the cumulative strain generated in the samples under the cyclic loading was obviously decreased after adding rubber particles into the sand.

There were some limitations to this study. A study with larger sample size is required in future study. Moreover, a geotechnical engineer is required to check the particle size before prescribing a suitable quantity for the application of crumb rubber in the field.

Author Contributions: Conceptualization, Y.Z., F.L., Y.B. and H.Y.; Data curation, Y.Z., F.L., Y.B. and H.Y.; Formal analysis, Y.Z., F.L., Y.B. and H.Y.; Funding acquisition, Y.Z., F.L. and Y.B.; Investigation, Y.Z., F.L. and Y.B.; Methodology, Y.Z., F.L., Y.B. and H.Y.; Project administration, Y.Z., F.L., Y.B. and H.Y.; Resources, Y.Z., F.L., Y.B. and H.Y; Software, Y.Z., F.L., Y.B. and H.Y.; Supervision, Y.Z., F.L., Y.B. and H.Y.; Validation, Y.Z., F.L., Y.B. and H.Y.; Visualization, Y.Z., F.L., Y.B. and H.Y.; Writing—original draft, Y.Z., F.L., Y.B. and H.Y.; Writing—review & editing, Y.Z., F.L., Y.B. and H.Y. All authors have read and agreed to the published version of the manuscript.

Funding: Basic Scientific Research Business Expenses Project of Municipal Universities (X18199); Scientific Research Project of Beijing Municipal Education Commission-General Project of Science and Technology Plan (General Project) (Z18028); University Scientific Research Fund Natural Science Project-Special Fund (ZF17067); National College Students' Innovation and Entrepreneurship Training Program (202110016019; 202210016015).

Institutional Review Board Statement: Not applicable.

Informed Consent Statement: Not applicable.

Data Availability Statement: Not applicable.

Conflicts of Interest: The authors declare no conflict of interest.

References

1. Feng, Z.Y.; Sutter, K.G. Dynamic properties of granulated rubber-sand mixtures. *Geotech. Test. J.* **2000**, *23*, 338–344.
2. Pamukcu, S.; Akbulut, S. Thermoelastic enhancement of damping of sand using synthetic ground rubber. *J. Geotech. Geoenviron. Eng.* **2006**, *132*, 501–510. [CrossRef]
3. Senetakis, K.; Anastasiadis, A.; Pitilakis, K. Dynamic properties of dry sand/rubber (SRM) and gravel/rubber (GRM) mixtures in a wide range of shearing strain amplitudes. *Soil Dyn. Earthq. Eng.* **2012**, *33*, 38–53. [CrossRef]
4. Anastasiadis, A.; Senetakis, K.; Pitilakis, K. Small-strain shear modulus and damping ratio of sand-rubber and gravel-rubber mixtures. *Geotech. Geol. Eng.* **2012**, *30*, 363–382. [CrossRef]
5. Anastasiadis, A.; Senetakis, K.; Pitilakis, K.; Gargala, C.; Karakasi, I.; Edil, T. Dynamic behavior of sand/rubber mixtures. Part I: Effect of rubber content and duration of confinement on small-strain shear modulus and damping ratio. *J. ASTM Int.* **2011**, *9*, 1–17.

6. Senetakis, K.; Anastasiadis, A.; Pitilakis, K.; Souli, A.; Edil, T.; Dean, S.W. Dynamic behavior of sand/rubber mixtures, part II: Effect of rubber content on G/G0-γ-DT curves and volumetric threshold strain. *J. ASTM Int.* **2011**, *9*, 103711–103712. [CrossRef]
7. Sui, X.X. *The Study on Seismic Isolation Performance of Granulated Rubber-Sand Mixture*; Huanan University: Changsha, China, 2009.
8. Li, L.H.; Xiao, H.L.; Tang, H.M.; Hu, Q.Z.; Sun, M.J.; Sun, L. Dynamic properties variation of tire shred-soil mixtures. *Rock Soil Mech.* **2014**, *35*, 359–364.
9. Shang, S.-P.; Sui, X.-X.; Zhou, Z.-J.; Liu, F.-C.; Xiong, W. Study of dynamic shear modulus of granulated rubber-sand mixture. *Rock Soil Mech.* **2010**, *31*, 377–381.
10. Akbarimehr, D.; Eslami, A.; Aflaki, E. Geotechnical behaviour of clay soil mixed with rubber waste. *J. Clean. Prod.* **2020**, *271*, 122632. [CrossRef]
11. Mase, L.Z.; Likitlersuang, S.; Tobita, T. Cyclic behaviour and liquefaction resistance of Izumio sands in Osaka, Japan. *Mar. Georesour. Geotechnol.* **2018**, *37*, 765–774. [CrossRef]
12. Mase, L.Z. Shaking Table Test of Soil Liquefaction in Southern Yogyakarta. *Int. J. Technol.* **2017**, *8*, 747–760. [CrossRef]
13. Mojtahedzadeh, N.; Ghalandarzadeh, A.; Motamed, R. Experimental evaluation of dynamic characteristics of Firouzkooh sand using cyclic triaxial and bender element tests. *Int. J. Civ. Eng.* **2021**, *20*, 125–138. [CrossRef]
14. Cetin, H.; Söylemez, M. Soil-particle and pore orientations during drained and undrained shear of a cohesive sandy silt clay soil. *Can. Geotech. J.* **2004**, *41*, 1127–1138. [CrossRef]
15. Chen, W.; Kong, L.W.; Zhu, J.Q. A simple method to approximately determine the damping ratio of soils. *Rock Soil Mech.* **2007**, *28* (Suppl. 1), 789–791.
16. Anvari, S.M.; Shooshpasha, I.; Kutanaei, S.S. Effect of granulated rubber on shear strength of fine-grained sand. *J. Rock Mech. Geotech. Eng.* **2017**, *9*, 936–944. [CrossRef]

Article

The Cracking Resistance Behavior of Geosynthetics-Reinforced Asphalt Concrete under Lower Temperatures Using Bending Test

Qiaoyi Li [1,2], Yonghai He [3], Guangqing Yang [1,4,*], Penghui Su [1,2] and Biao Li [5]

[1] State Key Laboratory of Mechanical Behavior and System Safety of Traffic Engineering Structures, Shijiazhuang Tiedao University, Shijiazhuang 050043, China; liqiaoyi@stdu.edu.cn (Q.L.); suphsph@163.com (P.S.)
[2] School of Traffic and Transportation, Shijiazhuang Tiedao University, Shijiazhuang 050043, China
[3] Hebei Provincial Communications Planning, Design and Research Insititute Co., Ltd., Shijiazhuang 050043, China; yonghaihe@126.com
[4] School of Civil Engineering, Shijiazhuang Tiedao University, Shijiazhuang 050043, China
[5] School of Management, Shijiazhuang Tiedao University, Shijiazhuang 050043, China; libiao_hb1990@163.com
* Correspondence: yanggq@stdu.edu.cn

Citation: Li, Q.; He, Y.; Yang, G.; Su, P.; Li, B. The Cracking Resistance Behavior of Geosynthetics-Reinforced Asphalt Concrete under Lower Temperatures Using Bending Test. *Coatings* 2022, 12, 812. https:// doi.org/10.3390/coatings12060812

Academic Editors: Ofelia-Cornelia Corbu and Ionut Ovidiu Toma

Received: 7 May 2022
Accepted: 8 June 2022
Published: 10 June 2022

Publisher's Note: MDPI stays neutral with regard to jurisdictional claims in published maps and institutional affiliations.

Copyright: © 2022 by the authors. Licensee MDPI, Basel, Switzerland. This article is an open access article distributed under the terms and conditions of the Creative Commons Attribution (CC BY) license (https:// creativecommons.org/licenses/by/ 4.0/).

Abstract: Asphalt is a kind of temperature-sensitive material. With the decrease of temperature, the deformation capacity of an asphalt mixture will be significantly reduced. When the temperature is greatly reduced, the asphalt layer will produce large shrinkage tensile stress and strain, resulting in cracking. Therefore, the cracking resistance behavior is essential for the asphalt. In order to study the cracking resistance behavior of geosynthetics-reinforced asphalt under lower temperatures, the bending tests were carried out indoors at a temperature of -10 °C. The results showed that compared with the unreinforced asphalt sample, the flexural tensile strength at failure of the geogrid-reinforced sample was increased by 14.1% and 12.3%, corresponding to AC-13C and AC-20C. Additionally, the geotextile-reinforced sample was reduced by 2.5% and 3.6%, corresponding to AC-13C and AC-20C. The values of the bending stiffness modulus of the geogrid- and geotextile-reinforced samples were reduced by 6% and 1%. The cracking energy of the geogrid-reinforced asphalt provides by 45.2% and 30.8% more than unreinforced asphalt, corresponding to AC-13C and AC-20C. The cracking energy of the geotextile-reinforced asphalt is increased by 4.5% and 0.6% compared with unreinforced asphalt, corresponding to AC-13C and AC-20C. The cracking resistance behavior of geogrid-reinforced asphalt is better than unreinforced and geotextile-reinforced asphalt. The asphalt shows obvious brittleness at a temperature of -10 °C, and the existence of the geosynthetics does not change the shape of the load–deflection curves.

Keywords: geosynthetics; geogrid; geotextile; reinforcement; asphalt concrete; the bending test

1. Introduction

Geosynthetics-reinforced asphalt has attracted much focus in recent years. Geosynthetics in asphalt can effectively improve fatigue life and decrease rutting [1–6]. Meanwhile, it also can improve the cracking resistance of the asphalt. There are many studies on the cracking resistance of geosynthetics-reinforced asphalt. Canestrari et al. [7] carried out a three-point bending test at a temperature of 20 °C on a geogrid-reinforced asphalt beam. Results show that the reinforced interfaces lead to higher peak load and deflection values. Zofka et al. [8] also conducted a three-point bending test on three kinds of different asphalt beams at a temperature of 13 °C. He pointed out that four times more energy is necessary for the crack to propagate through the carbon-reinforced beam than the unreinforced beam. The effect of reinforcement, in particular the CF geogrid, is significant. Ingrassia et al. [9] studied the cracking resistance of two geomembranes-reinforced asphalts and one geogrid-reinforced asphalt compared with unreinforced asphalt at 20 °C. Results show that the unreinforced system exhibits a higher value of flexural strength (P_{max}) than the reinforced systems. However, after reaching P_{max}, unreinforced asphalt rapidly loses its resistance until complete failure, without any residual flexural resistance. On the contrary, the reinforced systems show a significant post-peak dissipative phase. The geocomposites increased the energy necessary for the crack propagation by three to eight times compared to the unreinforced pavement. Ragni et al. [10] assessed the effectiveness of asphalt pavement rehabilitation with geocomposites to limit fatigue cracking, reflective cracking, and rutting. Kumar et al. [11] evaluated the cracking resistance potential of geosynthetic-reinforced asphalt overlays by using a direct tensile strength test at 20, 30, and 40 °C. He pointed out that the performance of specimens conditioned at temperatures of 20 °C is superior to those at 30 and 40 °C. Spadoni et al. [12] assessed the influence of four different geocomposites, obtained by combining a reinforcing geosynthetic with a bituminous membrane, regarding the crack propagation and interlayer bonding of asphalt pavements. The three-point bending test was carried out on double-layered asphalt specimens at temperatures of 20 °C. Results showed that the main contribution of the geocomposites consisted in increasing the crack propagation energy in the layer above the reinforcement (from five to ten times with respect to the unreinforced system). Ram Kumar et al. [13] reviewed the flexural fatigue properties, interfacial shear characteristics, and mechanical properties of geogrids embedded with asphalt layers. From their summary and the literature reviewed, it can be summarized that many scholars mainly research the cracking resistance of the geosynthetics-reinforced asphalt at a temperature of 20 °C. There are few studies on the cracking resistance of the geosynthetics-reinforced asphalt under lower temperatures.

In this paper, the objective of this study was to investigate the cracking resistance of reinforced asphalt and unreinforced asphalt at a temperature of −10 °C. The obtained results were then compared with previously published results to obtain a more comprehensive understanding of the geosynthetics-reinforced asphalt.

2. Materials and Methods

2.1. Geosynthetics

In this study, geogrids and geotextiles were used in asphalt reinforcement. The transverse ribs coated with bitumen of the geogrid were made of 12K carbon fiber, the longitudinal ribs made of 1100 Tex glass fiber, and the geotextile were made of polyester fiber. Table 1 shows the properties of the geosynthetics used in this study.

Table 1. Characteristics of geosynthetics.

Geosynthetic	Direction	Material	Thickness (mm)	Elongation at Rupture (%)	Tensile Force (kN/m)
Geotextile	Longitudinal	Polyester glass fiber	1.2	4.6	9.40
	Transversal	Polyester glass fiber	1.2	4.7	9.28
Geogrid	Longitudinal	Glass fiber	0.7	3–4.5	45.00
	Transversal	Carbon fiber	0.7	2–2.5	76.92

2.2. Asphalt Mix Design

There were two kinds of double-layered hot-mix asphalt slab specimens made in the study. One double-layered asphalt was made of the AC-20C asphalt mixture. The AC-20C asphalt mixture was composed of crushed limestone aggregates and 4.3% bitumen content by weight. The aggregates were divided into 5 specifications: 16–22 mm limestone aggregate, 11–16 mm limestone aggregate, 6–11 mm limestone aggregate, 3–6 mm limestone aggregate, and 0–3 (machine-made sand) limestone aggregate. The density of the aggregates is shown in Table 2. The nominal maximum size of the AC-20C asphalt mixture was 22 mm. Another double-layered asphalt was made of two asphalt mixtures, AC-13C and AC-20C. The lower layer consisted of AC-20C, the same as the first type of asphalt, and the upper layer was an AC-13C asphalt mixture. The AC-13C asphalt mixture was composed of crushed limestone aggregates and 4.8% bitumen content by weight. The nominal maximum size of the AC-13C asphalt mixture was 15 mm. The asphalt mix design was based on JTGF40-2004. The mineral aggregate gradation of asphalt mixture is presented in Table 3. Before fabrication of reinforced beam sample, asphalt mixture density test and Marshall stability test was conducted indoors according to the JTG E20-2011. The property of the asphalt was shown in Table 4. The bitumen was SBS-modified bitumen, penetration in 25 °C (0.1 mm) of 55, softening point of 81 °C, flashpoint of 270 °C, and Brookfield viscosity 135 °C of 1.9 Pa/s.

Table 2. The density of the aggregates.

Size	Bulk Density(g/cm^3)	Apparent Specific Gravity
16–22 mm	2.680	2.745
11–16 mm	2.687	2.743
6–11 mm	2.681	2.748
3–6 mm	2.683	2.747
0–3 mm (machine-made sand)	2.617	2.737

Table 3. Mineral aggregate gradation of asphalt mixture.

Sieve Size (mm)	AC-13C Passing (%)	AC-20C Passing (%)
26.5	-	100.0
19	-	99.5
16	100.0	90.5
13.2	98.1	75.9
9.5	76.8	61.0
4.75	50.8	40.5
2.36	36.4	30.0
1.18	26.8	22.0
0.6	17.2	14.1
0.3	11.8	9.6
0.15	8.3	6.6
0.075	6.3	5.0

Table 4. The property of the asphalt.

-	Void Ratio/%	Aggregate Clearance Rate/%	Asphalt Saturation/%	Stability/kN	Flow Value/mm	Marshall Modulus (kN/mm)
AC-13C	3.1	11.0	77.2	14.2	3.6	3.944
AC-20C	3.5	14.2	75.0	14.4	3.4	4.364

2.3. Sample Production

The preparation process of geosynthetics-reinforced asphalt trabecular specimen mainly includes five processes: asphalt mixture production, lower-layer rolling forming, laying geosynthetics, upper-layer rolling forming, and plate cutting. Asphalt mixture production: The dried mineral aggregate and bitumen mixed to homogenous in the asphalt mixture mixer. Lower-layer rolling forming: The homogenous asphalt mix was put into the mold of 300 mm × 300 mm × 50 mm size. Compaction carried out by a roller to form a base plate. Laying geosynthetics: When lower-asphalt slab cooled to approx. 25 °C, placed the lower slab to mold (300 mm × 300 mm × 100 mm), and brushed the tack coat on the surface of the plate before the installation of geogrid. The tack coat was painted at 0.5 L/m^2. Plate cutting: Five beams were cut from the double-layered asphalt slab. The dimension of beam sample was L/H/W = 250/50/47 (unit: mm). The preparation of the test specimen is presented in Figure 1.

Figure 1. The preparation of test specimens. (**a**) Asphalt-mixture mixing; (**b**) layer rolling forming; (**c**) laying geosynthetics and brushing the tack coat; (**d**) double-layered asphalt slab; (**e**) plate cutting; (**f**) beam sample of the bending test.

2.4. Bending Tests

Before the bending test, the beam samples must be placed in a low temperature environment of $-10\ °C$ for no less than 4 h. The bending test was carried out at an ambient temperature of $-10\ °C$. The loading rate was 50.8 mm/min. Many research projects used this displacement rate [7–10]. When the load value reached 60% of the maximum load, the beam sample was considered to be damaged and the test ended. The bending test apparatus is shown in Figure 2. Five repetitions were performed for each test condition.

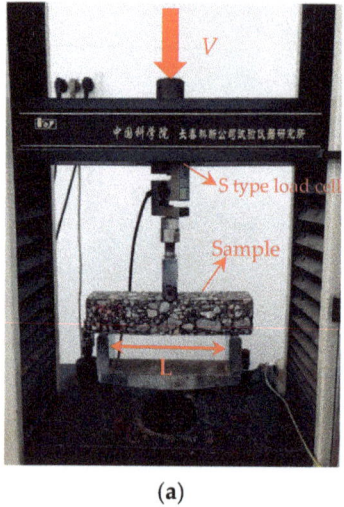 

(a) (b)

Figure 2. Test apparatuses. (a) The bending test apparatus; (b) temperature control box. Chinese in the figure: Changchun Kexin Company Test Instrument Research Institute, Chinese Academy of Sciences, Changchun, China.

The flexural strength point (P_{max}, $\delta_{P,max}$) was obtained by the bending test. It is noticed that when the difference between a certain data in a group of measured values and the average value is greater than 1.67 times of the standard deviation, the measured value shall be discarded (JTG E20-2011) and the average value of other measured values shall be taken as the test result. The values of R_B, ε_B, and S_B can be calculated from the following Equations:

$$R_B = \frac{3 \times L \times P_{max}}{2 \times b \times h} \quad (1)$$

$$\varepsilon_B = \frac{6 \times h \times \delta}{L^2} \quad (2)$$

$$S_B = \frac{R_B}{\varepsilon_B} \quad (3)$$

R_B—Flexural tensile strength of specimen at failure, MPa.
ε_B—Maximum bending tensile strain of specimen at failure, $\mu\varepsilon$.
S_B—Bending stiffness modulus of specimen at failure, MPa.
b—Mid-span width of specimen, mm.
h—Mid-span height of specimen, mm.
L—The span length of specimen, mm.
P_{max}—Maximum load of specimen, N.
δ—Mid-span deflection of specimen in failure, mm.

3. Results and Discussion

3.1. The Bending Test Results

The load–deflection curves obtained in the bending test are reported in Figure 3 and the average values of the corresponding characteristic parameters are summarized in Table 5.

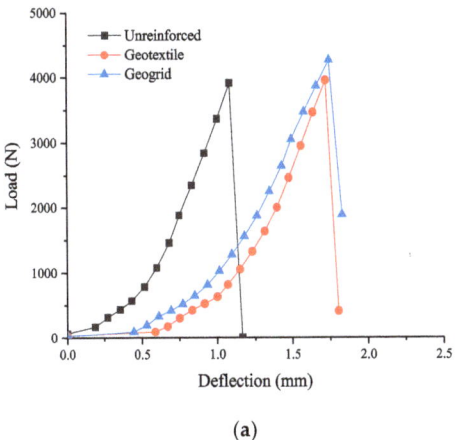

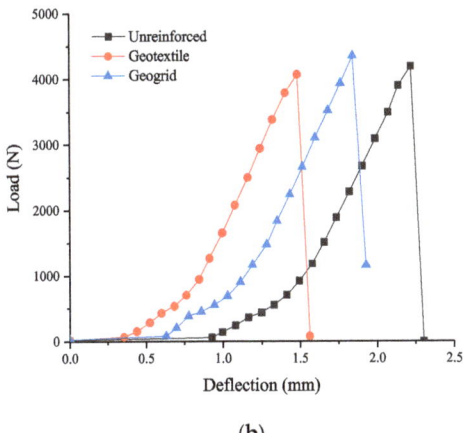

Figure 3. The load vs. deflection curves. (**a**) AC-13C and (**b**) AC-20C.

Table 5. The average values of the corresponding characteristic parameters.

Asphalt Mix	Interface Type	P_{max} (N)	δ (mm)	R_B (MPa)	ε_B (με)	S_B (MPa)
	Unreinforced	3780.45	0.72	9.65	5381.250	1793.67
AC-13C	Geotextile	3683.78	0.73	9.41	5484.375	1714.94
	Geogrid	4311.11	0.87	11.01	6528.075	1686.12
	Unreinforced	4012.11	0.80	10.24	6034.200	1697.60
AC-20C	Geotextile	3865.55	0.75	9.87	5608.475	1759.75
	Geogrid	4504.67	0.91	11.50	6853.325	1678.02

Table 5 shows that the order of the flexural tensile strength R_B value from large to small was $R_B^{CF} > R_B^{UN} > R_B^{GT}$. Compared with the flexural tensile strength R_B^{UN} of the unreinforced asphalt sample, the value of the flexural tensile strength R_B^{CF} of the geogrid-reinforced sample was increased by 14.1%, and the value of the flexural tensile strength R_B^{GT} of the geotextile-reinforced asphalt sample was reduced by 2.5%, corresponding to AC-13C. For the AC-20C, the value of the flexural tensile strength R_B^{CF} of the geogrid-reinforced sample was increased by 12.3%; the value of the flexural tensile strength R_B^{GT} of the geotextile-reinforced asphalt sample was reduced by 3.6%. The order of the maximum bending tensile strain value at failure was $\varepsilon_B^{CF} > \varepsilon_B^{UN} > \varepsilon_B^{GT}$, corresponding to AC-20C. However, for the AC-13C, the order is $\varepsilon_B^{CF} > \varepsilon_B^{GT} > \varepsilon_B^{UN}$. This indicated that the geogrid can effectively improve the maximum bending tensile strain and alleviate the brittleness of asphalt at lower temperatures. When the cracks develop from the bottom of the asphalt beam to the interface, due to the network structure of the geogrid, the expansion of the cracks will be limited, and it is reflected in the test results that the asphalt beam shows a high tensile strain.

The order of the bending stiffness modulus value at failure was $S_B^{UN} > S_B^{GT} > S_B^{CF}$, corresponding to AC-13C. Compared with the bending stiffness modulus value S_B^{UN} of the unreinforced asphalt sample, the values of the bending stiffness modulus S_B^{CF}, S_B^{GT} were reduced by 6% and 1%, respectively, for the geogrid-reinforced asphalt and geotextile-reinforced asphalt. However, for the AC-20C, the order of the bending stiffness modulus value at failure was $S_B^{GT} > S_B^{UN} > S_B^{CF}$. It indicated that the geogrid is beneficial to improve the stress relaxation performance of asphalt, so as to inhibit the generation of cracks and prolong the failure time, and improve the low-temperature crack resistance of asphalt.

The stress is transferred through the mineral aggregate particles, and the geogrid laid on the interface of the asphalt mixture layer is equivalent to forming a "stress absorption layer", which changes the transfer mode of the interlayer force. In the process of stress transfer from top to bottom, part of the stress is dissipated when it is transferred to the "stress absorption layer", so that the remaining stress can be uniformly transferred to the bottom, thus delaying the generation of cracks. When the tensile stress at the bottom of the specimen exceeds the ultimate tensile strength of the asphalt mixture, cracks will occur at the bottom of the specimen and expand rapidly. When the crack extends to the geosynthetics, the presence of geosynthetics changes the stress at the crack tip, effectively reduces the stress concentration phenomenon, and is conducive to preventing the extension of the crack. At the same time, the tensile force, interlayer adhesion, and the friction of geosynthetics will restrict the opening deformation of cracks [7–9,13].

From the flexural tensile strength and strain to the flexural tensile strength and the bending stiffness modulus, the geogrid-reinforced asphalts have a better behavior on the cracking resistance under lower temperature. Figure 3 shows that the load–deflection curves are smooth in the whole test process and that the existence of the geosynthetics does not change the shape of the curve. For the AC-13C, the initial deflections of the geogrid and geotextile asphalt layer are higher than the unreinforced sample. However, the initial deflections of the geogrid and geotextile asphalt layer are lower than the unreinforced sample for the AC-20C.

3.2. Comparative Analysis with Previous Studies

Figure 4 shows the load–deflection curves obtained by other scholars. All results show that reinforcement has no effect on crack initiation at a temperature of 20 or 13 °C. There is, however, a significant impact on the softening region; that is, geosynthetics decrease the crack propagation after the crack has been initiated. Compared to the previously published results at a temperature of 20 or 13 °C (Figure 4), with the results at a temperature of −10 °C in this paper (Figure 3), we found that when the load reached the maximum value, the transformation trend of load deflection curve will be different. At the temperature of −10 °C, the crack developed to the top, and the double-layer asphalt beam was destroyed rapidly. However, at a temperature of 20 or 13 °C, the geogrid decreases the crack propagation. It shows that in a low-temperature environment, the effect of a geogrid on crack development is weaker than that in normal temperature environment. The flexural tensile strain of asphalt at a temperature of −10 °C is lower, the thickness of the upper asphalt is thin, and the beam is damaged too fast.

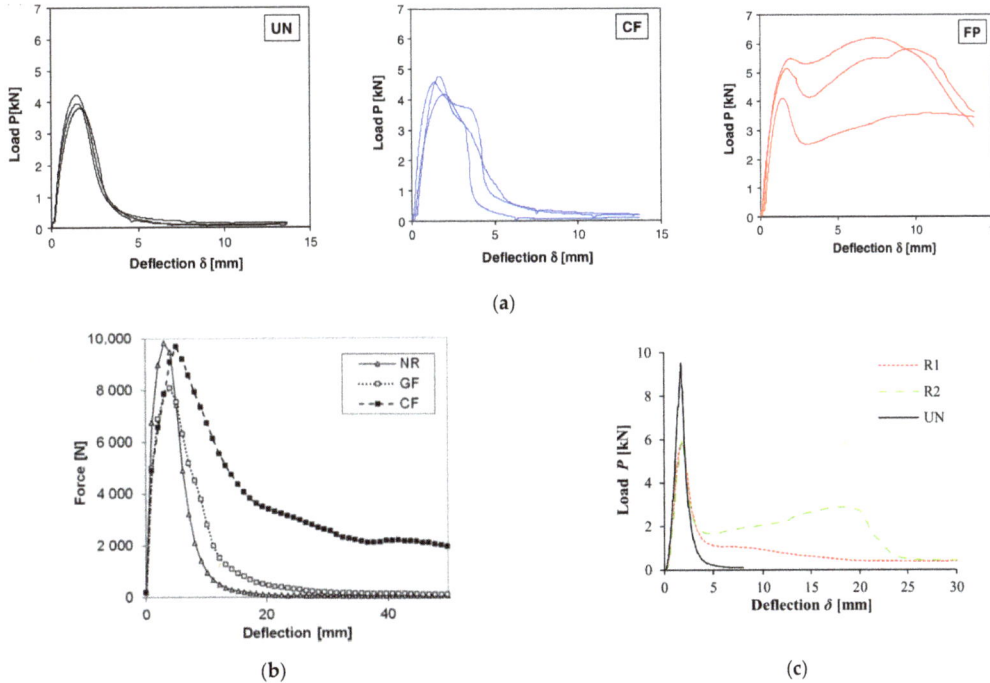

Figure 4. The load–deflection curves obtained by other scholars. (**a**) Reprinted with permission from [7]; Copyright 2013 Springer Nature; (**b**) Reprinted with permission from [8]; Copyright 2018 Taylor & Francis; (**c**) Reprinted with permission from [9].

Figure 5 shows the cracking energy of the six types of asphalt samples. The cracking energy of the geogrid-reinforced asphalt beam is higher than the unreinforced and geotextile-reinforced asphalt beam. The cracking energy of the geogrid-reinforced asphalt is 45.2% and 30.8% higher than unreinforced asphalt, corresponding to AC-13C and AC-20C. The cracking energy of the geotextile-reinforced asphalt is increased by 4.5% and 0.6% compared with unreinforced asphalt, corresponding to AC-13C and AC-20C. Comparing AC-20C and AC-13C asphalt beams, it can be found that the cracking energy of AC-20C asphalt samples are almost the same as the AC-13C asphalt beam. There is little difference between AC-20C and AC-13C samples.

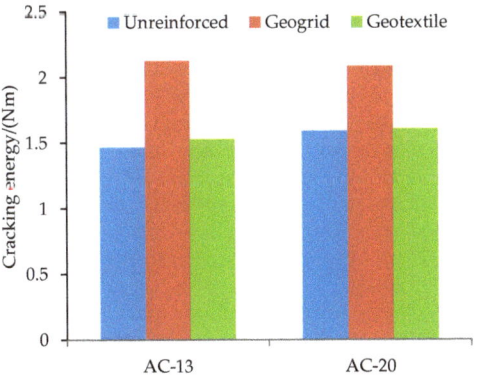

Figure 5. The cracking energy of the six types of asphalt samples.

3.3. Crack Propagation Analyses

Figures 6 and 7 show that only one obvious crack occurs in the beam at a temperature of −10 °C. The crack initiation position is close to the load application point, and the strike is basically perpendicular to the upper and lower surfaces of the beam. The crack opening size of unreinforced asphalt and geotextile-reinforced asphalt was relatively large, while the crack opening size of geogrid-reinforced asphalt was relatively small. When the crack at the bottom of the beam extends to the interlayer, it mainly depends on geosynthetics to inhibit the further expansion of the crack. However, due to the lower tensile strength of the geotextile, the geotextile will be damaged over time, and will finally form a crack with a large opening. The tensile strength of the geogrid is relatively high and will not be damaged, but the geogrid will be deformed to a certain extent. The deformation is really small; however, the asphalt has reached the cracking limit. The crack will continue to expand upward and eventually form a through crack with a small opening.

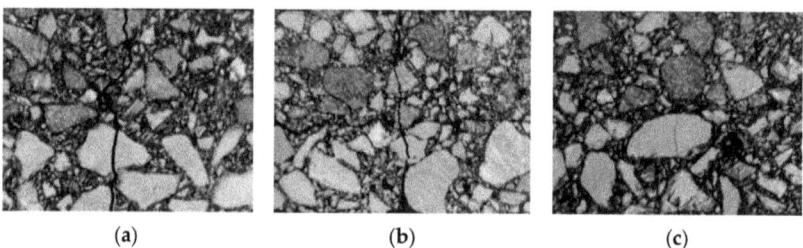

Figure 6. Crack propagation diagram of the AC-13C. (**a**) Unreinforced, (**b**) geotextile, (**c**) geogrid.

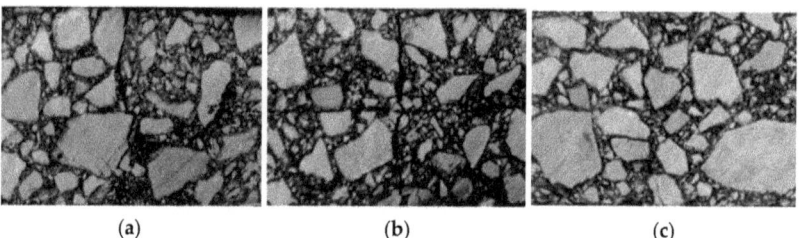

Figure 7. Crack propagation diagram of the AC-20C. (**a**) Unreinforced, (**b**) geotextile, (**c**) geogrid.

4. Conclusions

This article has discussed the cracking resistance behavior of geosynthetics-reinforced asphalt at −10 °C using the bending test. The following conclusions can be drawn regarding the results presented:

1. Compared with the unreinforced asphalt sample, the flexural tensile strength at failure of the geogrid-reinforced sample was increased by 14.1% and 12.3%, and the geotextile-reinforced sample was reduced by 2.5% and 3.6%, corresponding to AC-13C and AC-20C. The values of the bending stiffness modulus of the geogrid and geotextile-reinforced samples were reduced by 6% and 1%;

2. Through the analysis of the maximum load, flexural tensile strength, and maximum bending tensile strain, it is shown that the cracking resistance behavior of geogrid-reinforced asphalt is better than unreinforced and geotextile-reinforced asphalt at a temperature of −10 °C;

3. The flexural tensile strength at failure of the AC-20C asphalt samples is higher than the AC-13C asphalt samples. The initial deflections of the geogrid and geotextile asphalt layer are higher than the unreinforced sample of the AC-13C. However, the initial deflections of the geogrid and geotextile asphalt layer are lower than the unreinforced

sample for the AC-20C. The reason may be that the maximum aggregate size is different between the AC-20C and the AC-13C;

4. The asphalt shows obvious brittleness at a temperature of $-10\ °C$, and the existence of the geosynthetics does not change the shape of the load–deflection curves. Compared to the previously published results at a temperature of 20 or 13 $°C$, geosynthetics have no obvious inhibition effect on crack propagation at a temperature of $-10\ °C$;

5. The cracking energy of the geogrid-reinforced asphalt is 45.2% and 30.8% higher than unreinforced asphalt, corresponding to AC-13C and AC-20C. The cracking energy of the geotextile-reinforced asphalt is increased by 4.5% and 0.6% compared with unreinforced asphalt, corresponding to AC-13C and AC-20C. The cracking energy of AC-20C asphalt samples are almost the same as the AC-13C asphalt samples.

Author Contributions: Conceptualization, Q.L.; formal analysis, Q.L.; investigation, Q.L., P.S. and B.L.; data curation, Q.L., P.S. and B.L.; writing—original draft preparation, Q.L.; writing—review and editing, Y.H. and G.Y.; supervision, Y.H. and G.Y.; project administration, G.Y.; methodology P.S. and B.L.; validation P.S. and B.L. All authors have read and agreed to the published version of the manuscript.

Funding: This research was funded by the Key Research and Development Program of Hebei Province (Grant no. 20375504D).

Institutional Review Board Statement: Not applicable.

Informed Consent Statement: Not applicable.

Data Availability Statement: Not applicable.

Conflicts of Interest: The authors declare no conflict of interest.

Nomenclature

Notations
UN	Unreinforced asphalt
CF	Carbon geogrid-reinforced asphalt
GT	Geotextile-reinforced asphalt
FP	Glass fiber-reinforced polymer geogrid
NR	No-reinforcement asphalt
GF	Reinforced with glass geogrid
R1	Reinforced with continuous fiberglass fabric
R2	Reinforced with a non-woven polyester fabric and multidirectional fiberglass

References

1. Ferrotti, G.; Canestrari, F.; Virgili, A.; Grilli, A. A Strategic Laboratory Approach for the Performance Investigation of Geogrids in Flexible Pavements. *Constr. Build. Mater.* **2011**, *25*, 2343–2348. [CrossRef]
2. Ferrotti, G.; Canestrari, F.; Pasquini, E.; Virgili, A. Experimental Evaluation of the Influence of Surface Coating on Fiberglass Geogrid Performance in Asphalt Pavements. *Geotext. Geomembranes* **2012**, *34*, 11–18. [CrossRef]
3. Saride, S.; Kumar, V.V. Influence of Geosynthetic-Interlayers on the Performance of Asphalt Overlays on Pre-Cracked Pavements. *Geotext. Geomembranes* **2017**, *45*, 184–196. [CrossRef]
4. Lee, J.H.; Baek, S.B.; Lee, K.H.; Kim, J.S.; Jeong, J.H. Long-Term Performance of Fiber-Grid-Reinforced Asphalt Overlay Pavements: A Case Study of Korean National Highways. *J. Traffic Transp. Eng.* **2019**, *6*, 366–382. [CrossRef]
5. Lee, J.; Kim, Y.R.; Lee, J. Rutting Performance Evaluation of Asphalt Mix with Different Types of Geosynthetics Using MMLS3. *Int. J. Pavement Eng.* **2015**, *16*, 894–905. [CrossRef]
6. Partl, M.N.; Sokolov, K.; Kim, H. Evaluating and Modelling the Effect of Carbon Fiber Grid Reinforcement in a Model Asphalt Pavement. In Proceedings of the Fourth International Conference on FRP Composites in Civil Engineering (CICE2008), Zurich, Switzerland, 22 July 2008; pp. 22–24.
7. Canestrari, F.; Belogi, L.; Ferrotti, G.; Graziani, A. Shear and Flexural Characterization of Grid-Reinforced Asphalt Pavements and Relation with Field Distress Evolution. *Mater. Struct. Constr.* **2015**, *48*, 959–975. [CrossRef]
8. Zofka, A.; Maliszewski, M.; Maliszewska, D. Glass and Carbon Geogrid Reinforcement of Asphalt Mixtures. *Road Mater. Pavement Des.* **2017**, *18*, 471–490. [CrossRef]

9. Ingrassia, L.P.; Virgili, A.; Canestrari, F. Effect of Geocomposite Reinforcement on the Performance of Thin Asphalt Pavements: Accelerated Pavement Testing and Laboratory Analysis. *Case Stud. Constr. Mater.* **2020**, *12*, e00342. [CrossRef]
10. Ragni, D.; Montillo, T.; Marradi, A.; Canestrari, F. *Fast Falling Weight Accelerated Pavement Testing and Laboratory Analysis of Asphalt Pavements Reinforced with Geocomposites*; Springer International Publishing: New York, NY, USA, 2020; Volume 48, ISBN 9783030297794. [CrossRef]
11. Kumar, V.V.; Saride, S. Evaluation of Cracking Resistance Potential of Geosynthetic Reinforced Asphalt Overlays Using Direct Tensile Strength Test. *Constr. Build. Mater.* **2018**, *162*, 37–47. [CrossRef]
12. Spadoni, S.; Ingrassia, L.P.; Paoloni, G.; Virgili, A.; Canestrari, F. Influence of Geocomposite Properties on the Crack Propagation and Interlayer Bonding of Asphalt Pavements. *Materials* **2021**, *14*, 5310. [CrossRef] [PubMed]
13. Ram Kumar, B.A.V.; Jallu, H. *Performance of Geogrid Reinforced Asphalt Layers—A Review*; Springer: Singapore, 2022; Volume 172, ISBN 9789811643958.

Review

Use of Clay and Titanium Dioxide Nanoparticles in Mortar and Concrete—A State-of-the-Art Analysis

Georgiana Bunea [1,*], Sergiu-Mihai Alexa-Stratulat [1], Petru Mihai [2] and Ionuț-Ovidiu Toma [1,*]

[1] Department of Structural Mechanics, Faculty of Civil Engineering and Building Services, "Gheorghe Asachi" Technical University of Iasi, 700050 Iasi, Romania

[2] Department of Concrete Structures, Building Materials, Technology and Management, Faculty of Civil Engineering and Building Services, "Gheorghe Asachi" Technical University of Iasi, 700050 Iasi, Romania

* Correspondence: georgiana.bunea@academic.tuiasi.ro (G.B.); ionut.ovidiu.toma@tuiasi.ro (I.-O.T.)

Abstract: In the past decades, nanomaterials have become one of the focal points in civil engineering research. When added to cement-based construction materials (e.g., concrete), it results in significant improvements in their strength and other important properties. However, the final mix characteristics depend on many variables that must be taken into account. As such, there is no general consensus regarding the influence upon the original material of certain nano-sized additives, the optimum dosage or the synergistic effect of two or more nano-materials. This is also the case for titanium dioxide (TiO_2) and nanoclay (NC). The paper focuses on reporting the existing research data on the use of the above-mentioned materials when added to mortar and concrete. The collected data is summarized and presented in terms of strength and durability properties of cement mortar and concrete containing either TiO_2 or NC. Both nano-materials have been proven, by various studies, to increase the strength of the composite, at both room and elevated temperature, when added by themselves in 0.5%~12% for TiO_2 and 0.25%~6% for NC. It can be inferred that a combination of the two with the cementitious matrix can be beneficial and may lead to obtaining a new material with improved strength, elastic and durability properties that can be applied in the construction industry, with implications at the economic, social and environmental levels.

Keywords: nanomaterials; nanoclay; titanium dioxide; cementitious material; sustainability

Citation: Bunea, G.; Alexa-Stratulat, S.-M.; Mihai, P.; Toma, I.-O. Use of Clay and Titanium Dioxide Nanoparticles in Mortar and Concrete—A State-of-the-Art Analysis. *Coatings* 2023, *13*, 506. https://doi.org/10.3390/coatings13030506

Academic Editor: Andrea Nobili

Received: 22 January 2023
Revised: 22 February 2023
Accepted: 23 February 2023
Published: 24 February 2023

Copyright: © 2023 by the authors. Licensee MDPI, Basel, Switzerland. This article is an open access article distributed under the terms and conditions of the Creative Commons Attribution (CC BY) license (https:// creativecommons.org/licenses/by/ 4.0/).

1. Introduction

Various global-sized revolutions have changed previously held ideas, encouraging both research and industry. Nowadays, the growth and general evolution within societies is significantly accelerated by technological breakthroughs. This has resulted in new problems which are currently unsolved, namely pollution and the threat of natural resource depletion. In the field of civil engineering, researchers are compelled to mitigate the influence the construction industry has, considering that in 2020 it was responsible for approximately 37% of the global process-related carbon dioxide emissions. Of this percentage, 10% was generated by the manufacturing processes of building materials [1]. Nanotechnology has influenced almost all of nowadays advancements in terms of mechanical properties and durability of materials, construction materials included. The term "nanomaterial" has been defined by the European Union Commission [2] in the Official Journal of the European Union as being an artificially-created or a natural material, of which at least 50% of the particles have one, or more, external dimensions between 1 nm and 100 nm. In cementitious composites, many components fit this description, and their study is now possible due to the progress of material characterization techniques at this level.

To this day, several nanomaterials have been studied with respect to their applications in the construction industry. In addition to strength, other properties were also considered within the framework of nanotechnology. A recent study [3] highlighted the positive effect of several nanomaterials on the thermal resistance of cementitious composites, of which nano-silica stands out as being one the most studied materials. Additionally, several other nanomaterials were mentioned as increasing the high-temperature strength of the cement-based composites: carbon-based nanomaterials (carbon nanotubes, carbon nanofibers, graphene oxide, graphene nanoplatelets), nanoclay, nano-alumina, nano-iron oxides and nano-titania. Other studies [4,5] focused on the environmental impact of using nanoparticles in concrete, concluding that TiO_2 has a positive influence from this perspective [4].

The present work aims at creating a structured report encompassing the existing information about the impact of titanium dioxide (TiO_2) and nanoclay (NC) on cementitious composites from the point of view of strength enhancements and improvement of durability characteristics. A definite remark upon the influence of nanoclay on cement mortar or concrete is difficult to advance, as the variables change from one study to another. An important change is the type of nanoclay used in the study, the number of curing days or the additives introduced in the mix, e.g., fly ash, polypropylene fibers, superplasticizer, silica fume. These influence the final measured values for the strengths of the composite. Titanium dioxide nanoparticles succeed in improving both mechanical properties and photocatalytic reactions of cement mortar and concrete due to their chemical and physical properties. They are usually used in mortar/concrete mixes combined with superplasticizers in order to obtain a higher workability. Considering that TiO_2 is a non-reactive powder, there are studies in which pozzolans were added with the purpose of promoting the cement hydration reaction, e.g., fly ash, silica fume.

The mechanisms of improving the strength and durability properties of cementitious materials by using each of the two nanomaterials are different from one another but the end results are similar: NC is a highly pozzolanic material with significant influences on strength and durability at later ages while TiO_2 is inert and plays a filler effect. However, due to the very small particle sizes of nano-TiO_2, it serves as nucleation sites in the cement matrix with benefits in terms of strength and durability at earlier ages compared to NC. The information summarized in this work could serve as the starting point for future research works investigating the influence of combining TiO_2 and NC on the strength, elastic and durability properties of cement mortar and concrete.

2. The Influence of Nanoclay on Cementitious Materials

The use of nanoclay as a component of other materials started in the late 20th century when researchers observed that, by using this nanomaterial, the properties of the new composite material improved as compared to the material without the addition of nanoclay [6–9]. Among its first applications were the polymer matrices. The next observable research stage involved combining nanoclays with other materials of the same size (e.g., carbon nano-tubes [10,11]) as well as using them in cementitious mixes [12–14].

Nanoclay is defined as a layered mineral silicate, which, due to its filler and pozzolanic characteristics, succeeds in enhancing the mechanical and durability properties of several types of materials such as polymers or cement-based materials [15–18]. The use of such layered crystals, i.e., clays, in polymer composites resulted in improved physical properties and in the heat deflection temperature mainly due to their high surface area [19]. Moreover, the benefits of using nanoclay did not go unnoticed and their field of application soon extended to construction materials.

2.1. Types of Nanoclay Used in Combination with Cementitious Materials

There are several types of natural clays that were used as raw materials for the manufacturing of nanoclays: bentonite, montmorillonite, kaolin, illite, halloysite. They all have the same base crystalline structure, with the difference coming from the types of bonds between the stacked layers. This leads to different properties with direct effect in terms of the obtained results when added in the composition of another material [20,21]. Most of the research in terms of nanoclays used in cement-based materials is conducted with montmorillonite or kaolin clays, usually in a modified form, in order to obtain a significant improvement of the physical properties of the resulting material.

Montmorillonite (MMT) nanoclay has a 2:1 layer structure and very weak Van der Waals forces keeping the outer layers together. This means that when combined with water the particles absorb water molecules and the distance between layers increases, leading to swelling [21]. A more detailed explanation of this phenomenon was given in [20]. Based on the available information at that time, one way to decrease the swelling was to add exchangeable cations with a lower hydration energy, thus obtaining a more hydrophobic nano-montmorillonite. Some researchers based their studies on this method and used ammonium cations to decrease the hydrophilicity of MMT obtaining an organo-modified montmorillonite [12,17,22–27]. Other studies prepared the nanoclay particles by subjecting them to very high temperatures, in the range of 750–900 °C for 2 h, obtaining a nano-calcined montmorillonite clay [28–31].

Unlike montmorillonite clays, kaolin ones have hydrophobic properties, as they are composed only of two layers connected not only by Van der Waals forces but also by hydrogen bonds which do not allow water molecules to enter between them [21]. Therefore, no additional treatment is needed before using nano-kaolin particles in the cement-based composites. However, when subjected to high temperatures, due to the calcination process, the hardness of the kaolin clay nano-particles is increased together with their degree of pozzolanicity. The particles change their shape and size, become more hydrophobic, as in the case of MMT nanoclay, and increase their whiteness. Due to its improved afore mentioned properties, the calcined kaolin nanoclay, i.e., metakaolin, became the focus for researchers instead of the raw kaolin nanoclay [32–36].

2.2. Chemical Structure and Properties

Nanoclay is a general term for the nanoparticles, which, as mentioned above, are comprised of layered mineral silicates [37] and, depending on the arrangement of the layers in the crystal lattice and the existing types of bonds, they result in different types. As mentioned before, MMT and kaolin are the most used types of nanoclays in civil engineering. Kaolin is a 1:1- type of clay mineral consisting of one silica tetrahedral sheet connected to the alumina octahedral sheet by means of the hydrogen atoms, having the chemical formula $Al_2Si_2O_5(OH)_4$ [38]. Montmorillonite, on the other hand, is a 2:1-type of clay mineral having a sandwich-like structure and it consists of two silica tetrahedral sheets with a single alumina or magnesium octahedral sheet in between, having the formula $(Na,Ca)_{0.33}(Al,Mg)_2(Si_4O_{10})(OH)_2 \cdot nH_2O$ [17,39]. The dimension of a phyllosilicate sheet is approximately 1 nm, whereas the nanoparticle size is in the interval 30–100 nm [40].

The benefits of using them in combination with cement are two-fold. On one hand, their very small particle size leads to a higher specific surface area than the micro-sized particles, on which the hydration products can form. Moreover, because of this, nanoclay particles succeed in entering the existing voids between the cement paste and the aggregates, at a nanoscale level, thus obtaining a decrease in porosity. Combining the aforementioned phenomena occurring inside the cement-based material, the resulted strength and durability increases [18,37]. On the other hand, the increase in strength is given also by the nanoclay chemical reactivity with calcium hydroxide (CH) during the hydration process. Having in its composition two layers of silica, nanoclay promotes the formation of calcium-silicate-hydrates (CSH) in a pozzolanic reaction. The increase in the concentration of CSH leads to an increased strength of the material [14,41].

2.3. Nanoclay Particles in Cementitious Materials—Technological Flow

Because of their very small size, nanoclay particles behave differently when mixed with water than the micro- or macro-sized particles usually used in civil engineering. The effect of electrostatic attraction is greater in case of nanoparticles leading to the occurrence of flocs. When water is added, the flocculation effect can become more prominent, especially in case of MMT clays, where the water molecules can intercalate with the outer layers of the crystal and cause swelling [21,42]. This agglomeration of particles within the mortar or concrete has a negative impact upon the strength. Therefore, there is a stringent need for particle dispersion before the start of cement hydration.

The scientific literature gives two main ways in which this dispersion can be performed. The first one, which is the most recommended one, is to introduce the nanoclay particles in the water and then to subject the solution to ultrasonic waves. In this manner, the vibration caused within the molecules due to the impact of ultrasonic waves will prevent nanoclay particles to flocculate and will disperse them. Then the nanoclay-water solution can be used in two different ways to create the mortar or concrete samples: mechanical mixing or ultrasonic mixing. The mechanical mixing follows the steps given in ASTM C305 norm [18,43]. Using the ultrasonic mixing, the dry mixed materials (cement, sand, additives) are added to the nanoclay suspension already obtained and the ultrasonication is started again for the whole mix [44]. A previous study [45] concluded that in case of nanoclay, the sonication process has a positive effect on the final mechanical properties of the concrete samples. They also stated that the maximum allowable percentage of nanoclay used in cement-based materials is 5% [22,45].

The second method of dispersion is a simpler one and does not require any additional equipment than the one needed for preparing the mortar/concrete, namely, the dry mixing method. In this case, the mixing is performed in steps. The cement and the additives are dry mixed together with the nanoclay particles. In this manner, the nanoclay particles can be easily dispersed, due to the lack of a solving agent. Because the quantity of nanoclay is very small compared to that of cement, the distance between the nanoclay particles tends to increase during the dry mixing. This leads to a decrease in the flocculation probability when adding water. Afterwards, the aggregates are added and the mixing is restarted. Finally, the water is introduced in the composite [29,46].

A comparison between the previously mentioned methods of mixing nanoclay with cement-based materials, i.e., ultrasonication and dry mixing, showed that for a substitution of cement of 1–3% nanoclay, no differences were registered in terms of mechanical strengths. It was therefore concluded that both methods were equally reliable in properly dispersing nanoclay particles within the cement matrix [42].

2.4. Results of Laboratory Analyses (Microstructural Analyses)

When analyzing the nanoclay-cement-based material composite, the laboratory results vary depending on several factors, e.g., the nanoclay particle size, the nanoclay/cement ratio, the type of cement, the water/cement ratio, type and quantity of additives, type and size of aggregates. However, there are some common properties which are further discussed in this paper. Taking into account that when using sand or gravel, an important quantity of silica is introduced in the composite and the properties change significantly. The analyses will be presented considering this factor.

2.4.1. Scanning Electron Microscopy (SEM) Analysis

The layered structure of the organo-modified montmorillonite nanoclay (OMMT), obtained by cation exchange with quaternary ammonium cations, was extensively studied by means of SEM analysis [22]. Because MMT has a very high hydrophilicity, when it is introduced in the cement matrix, it begins to absorb the water from the cement matrix. In this manner, at first, the hydration of cement particles is slowed down and the workability is significantly reduced. The advantage of montmorillonite particles is that they begin to release the accumulated water at a later period during the hydration process of cement, leading to an improvement in strength. The phenomenon was later observed in other supplementary cementitious materials, such as zeolites, and it is now known as internal curing. However, the major problem with MMT particles, from this point of view, is that because of the initial occurring phenomenon, they introduce a water to cement ratio gradient near them as the quantity of water increases in that area. By increasing the water, the porosity is automatically increased, reducing the cohesion between the cement matrix and MMT particles [17]. When adding OMMT instead of MMT the water/cement ratio gradient around the nanoclay particles is decreased and the development of cracks is hindered [17].

The effects of high temperatures on mortar containing 5 wt% nano-calcinated montmorillonite clay (NCMC) was also studied [28] by firstly analyzing the SEM images, presented in Figure 1, for samples with and without NCMC at 25 °C, 250 °C and 900 °C, at 28 days of curing. The investigations highlighted the positive effect the NCMC has on the mortar matrix, especially at the level of the interfacial transition zone (ITZ) between the aggregates and the cement paste (#2 in Figure 1d). It was therefore concluded that by adding NCMC to the mortar, a denser matrix was obtained [28] mainly due to the filler effect of the nanoparticles. Additionally, an increase in the pozzolanic activity was detected. When the temperature was increased to 250 °C, the cement hydration was accelerated and more CSH were obtained in case of NCMC mortar than in case of control mortar (#3 in Figure 1b,e) [28]. After 300 °C, the development of microcracks began and at 450 °C the cracks widened and were clearly visible as Portlandite started to degrade and CSH particles lost their structural integrity [47]. At 900 °C, wide cracks were present in the control sample (#4 in Figure 1c) leading to a significant strength loss. However, when adding NCMC in the mix, the cement matrix experienced fewer cracks (#4 in Figure 1f), thus leading to higher strength values [28].

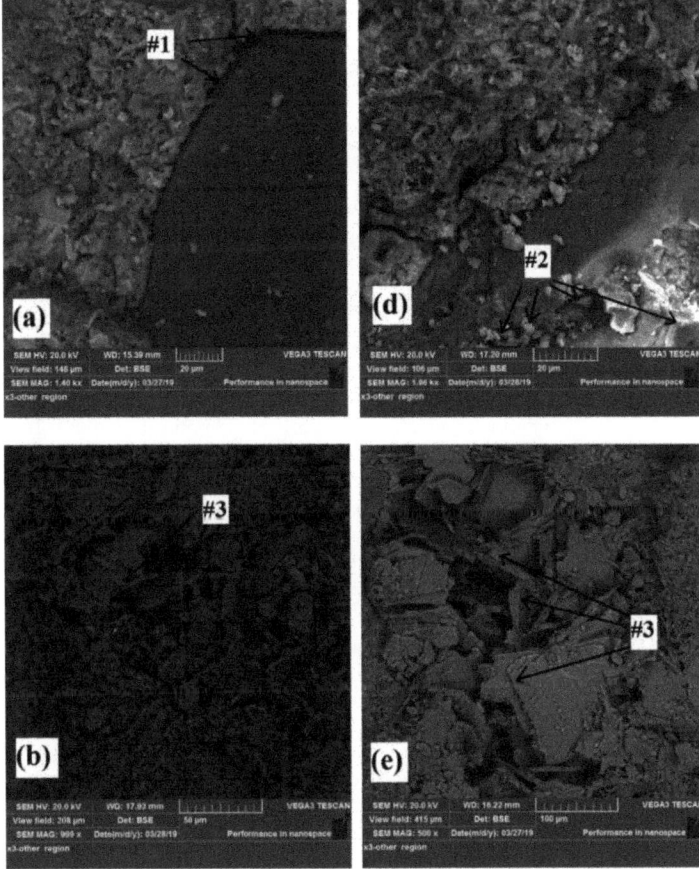

Figure 1. *Cont.*

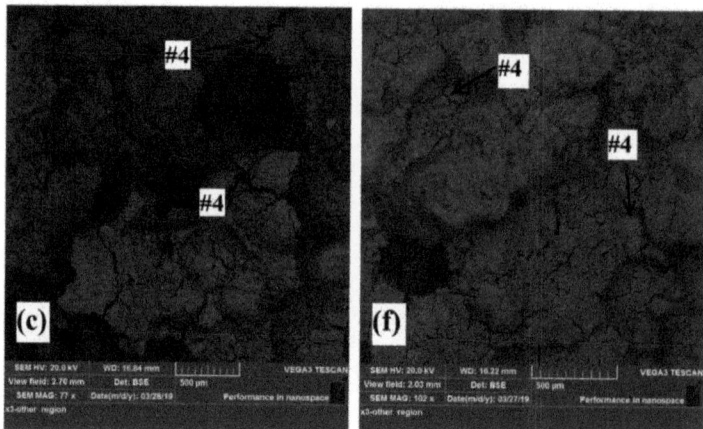

Figure 1. SEM images of fracture surfaces of control and modified mortar [28] (Reprinted/adapted with permission from Ref. [28]. Copyright 2018, ELSEVIER); (**a**) control at 25 °C, (**b**) control at 250 °C, (**c**) control at 900 °C, (**d**) NCMC mortar at 25 °C, (**e**) NCMC mortar at 250 °C, (**f**) NCMC mortar at 900 °C.

2.4.2. Thermogravimetric Analysis (TGA)

TG analysis is used to evaluate the influence of the temperature upon the considered material. Usually, the TGAs are conducted on cement paste samples because the aggregates significantly influence the final result. Generally, aggregates do not vary in weight when subjected to an increase in temperature so including them in the very small samples needed to run TG analysis would only lead to erroneous results [48,49].

Natural hydrophilic montmorillonite in 1% and 2% cement replacement was used to investigate the influence of temperature on the obtained cement paste. The heating rate for the TGA in this test was 30 °C/min. The study revealed that there were two main stages in the weight loss of cement paste: the dehydration of C-S-H at 105 °C and the CH decomposition at 470 °C [18]. The obtained results were in good agreement with the ones previously reported in general study on cement pastes [48]. At approximately 750 °C a third stage could be distinguished, corresponding to the decarbonation of calcium carbonate $CaCO_3$ [48]. It was shown that, as the percentage of nanoclay increases, the weight loss increases, indicating the property of natural montmorillonite to attract water molecules. Therefore, when heated, the samples began releasing the physically bounded water, its quantity being greater in case of samples with MMT compared to the control sample.

In contrast to the results presented in [18], the use of calcined natural montmorillonite clay resulted in a smaller weight loss compared to the control sample [28]. The stages of weight loss depending on the temperature were similar with the ones reported in [18]. However, during the first stage, i.e., 100–110 °C, there was no abrupt change in weight for the NCMC but a more gradual one due to the fact that the NCMC water absorption capacity has been significantly reduced by means of calcination. Between 430 °C and 470 °C the dehydration of weakly bound water from CH and CSH products occurred, followed by a reduction in weight between 580 and 680 °C when the strongly bound water was lost.

Similar results to the ones reported in [28] regarding the nano-calcined montmorillonite clay (CNCC) for 1, 2 and 3 wt% cement replacement were previously reported [29]. The earlier research work continued the study further on, including in the TG analysis samples of 1 wt% cement replacement by nanoclay modified with quaternary ammonium salt (NCC). It was observed that a smaller weight loss occurred when adding CNCC of 1 and 2 wt% than when using 1 wt% NCC. This result proved that nano-montmorillonite calcination has a greater impact on the decreasing the hydrophilicity than adding ammonium cations.

2.4.3. X-ray Power Diffraction (XRD) Analysis

When creating a new composite material, it is very important to know the influence that each component has upon the final result. In this regard, XRD analysis provides an insight regarding the mineralogical changes that occur inside the material, e.g., mortar/concrete, when adding nanomaterials, e.g., nanoclay [50]. Using 2 wt% nanoclay, untreated hydrophilic montmorillonite by weight of cement replacement and subjecting the samples to 25 °C, 200 °C, 400 °C and 600 °C, it was observed that the intensity of CSH is higher for all samples combined with NC, as shown in Figure 2, by comparison with the control samples, which leads to an increase in strength [18]. The increase in CSH crystals quantity became more prominent when increasing the temperature. Thus, for samples subjected to 400 °C and 600 °C, there was a strong peak for the CSH crystals (at $2\theta = 28°$), while up to 400 °C, the quantity was significantly smaller. Moreover, the intensity of CH ($2\theta = 18°$) decreased with the temperature, as the cement hydration accelerated [18].

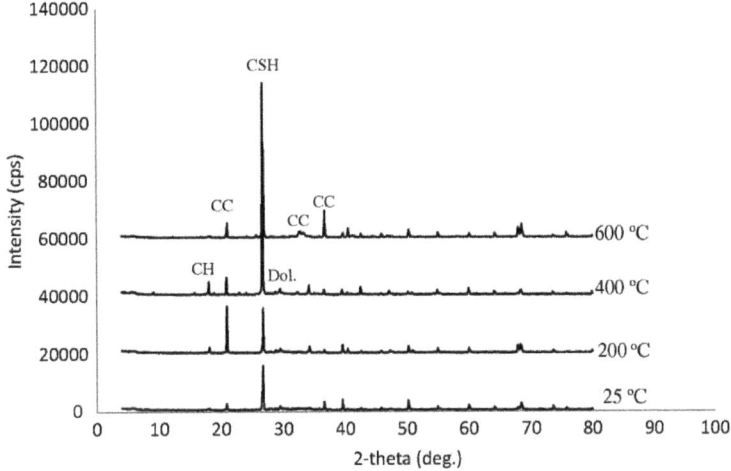

Figure 2. XRD patterns of mortar with 2% untreated hydrophilic montmorillonite nanoclay after exposure to different temperatures [18] (Reprinted/adapted with permission from Ref. [18]. Copyright 2020, ELSEVIER).

In another study, XRD analysis was performed on mortars combined with nanocalcined montmorillonite clay CNC, which lead to the conclusion that the calcined nanoclay had a positive effect upon the strength of the material, as it promoted the consumption of CH crystals and the formation of CSH gel [28]. The researchers used a 5 wt% nanoclay cement replacement and the analyzed samples were previously subjected to temperatures of 25 °C, 150 °C and 900 °C. The authors interpreted the presence of belite in the composite as an indicator for the level of both CH and CSH decomposition. Based on this remark, it can be stated that calcined nano-montmorillonite clay has a positive effect on the material properties, even at elevated temperatures as high as 900 °C. At that temperature, the control sample exhibited an important increase in belite, i.e., 8% higher than nanoclay mortar [28].

Similar results have been previously reported in a study in which the cement was replaced with 1, 2 and 3 wt% CNC and observed a reduction in the CH quantity (approximately 28% compared with control sample) [29]. This was attributed to the pozzolanic reaction which consumed the CH and led to the formation of CSH gel. Moreover, the increasing in the unreacted dicalcium silicate (C_2S) and tricalcium silicate (C_3S), compared with quantities from the control cement paste, confirmed the pozzolanic property of CNC. However, as the percentage of nanoclay introduced in the cement was increased, the reduction of CH crystals decreased. Moreover, a smaller quantity of C_2S and C_3S was obtained for 3 wt% CNC than for 1 wt% CNC, indicating a decrease in the pozzolanic activity. The XRD patterns for 1% CNC and the ones obtained for 1% cement replacement with montmorillonite nanoclay indicated that the calcined nanoclay led to a higher pozzolanic activity than the organo-modified nanoclay [29].

2.5. Material Strength Improvements

A definite remark upon the influence of nanoclay on cement mortar or concrete is difficult to advance, as the variables from one study to another are changing. An important change is the type of nanoclay used in the study, the number of curing days or the additives introduced in the mix, e.g., fly ash, polypropylene fibers, superplasticizer, silica fume. These influence the final measured values for the strengths of the composite. However, considering the reports from recent studies, the use of nano materials, nanoclay included, results in a denser microstructure of the material with net benefits in terms of strength and durability properties [51–53]. From this perspective, the current paper presents the literature results in a centralized manner identifying the possible influences on the reported results, where they are present. Moreover, the paper extracts from the literature only the results for the cement-based composites, where the main cement replacement material is nanoclay.

2.5.1. Splitting Tensile Strength

Taking into account that concrete/cement mortar elements have a very poor tensile strength and that their main strength develops under compression, the number of studies analyzing the tensile strength for a simple combination of cement-based material and nanoclay is very limited. However, based on the information presented in Sections 2.2 and 2.4 regarding the influence of nanoclay on the chemical and physical properties of the cement composite, an improvement in the tensile strength is expected. The porosity decrease, the prolonged hydration and the higher quantity of CSH crystals, all compared with the plain cement mortar, are all contributing towards a higher value of the tensile strength. Moreover, nanoclay particles attract cement particles and bond them together, thus obtaining a more durable composite [46,54].

It was also observed that the final tensile strength was also influenced by the curing method. Results on samples cured in water and samples cured by plastic wrapping were obtained and compared. Lower values of the tensile strength were obtained, especially after 90 days of curing, by applying the second method as the cement particles did not have enough water to fully hydrate. For 1% cement replacement with montmorillonite nanoclay, the difference between the two methods, in terms of strength, was 0.40 MPa at 90 days of curing [55].

The influence of temperature upon the cement mortar combined with 1 and 2 wt% natural hydrophilic montmorillonite sample was also investigated. It was observed that when the temperature increased up to 200 °C, the value of the tensile strength increased substantially for all samples, including the control sample. In case of 2% nanoclay cement replacement, a maximum difference of approximately 27% was reported between the values obtained for the tensile strength of the control sample and the nanoclay-cement composite, the nanoclay particles proving their efficiency. This increase in the value of the tensile strength was given by the chemical reactions which occurred during sample heating, namely the hydration acceleration, which promoted CSH formation around the nanoclay particles. As the temperature increased to 400 °C, the tensile strength dropped to 1.5–1.9 MPa for all samples. At 600 °C, a maximum value of approximately 0.2 MPa was obtained for the splitting tensile strength of the control mix. However, the nanoclay composite samples maintained consistently better values of the splitting tensile strength than the control one [18].

Table 1 summarizes the values of the tensile strength of mortar or concrete samples combined with different percentages of nanoclay, at room temperature, cured for 28 days in tap water or lime-saturated water.

Table 1. Splitting tensile strength values for different nanoclay-cement matrix samples.

Study	Type of Nanoclay	Type of Sample	Standard	w/b w/c	Additives	NC [%]	Tensile Strength [1] [MPa]
[18,56]	Natural hydrophilic montmorillonite	Cement mortar	ASTM C190	w/b 0.55	-	0 0.5 1 2	3.20 3.30 3.40 3.50
[46]	Metakaolin	Cement mortar	ASTM C307	w/b 0.50	-	0 2 4 6 8	3.60 4.05 4.48 4.92 5.36
[54]	Metakaolin	Concrete	ASTM C496	w/c 0.42	1.5% polypropylene	0 2 4 6 8 10	7.96 8.32 9.54 10.42 11.16 9.63
[55]	Montmorillonite (type not specified)	Concrete	ASTM C496	w/c 0.45	-	0 1 2 3	aprox. 4.30 aprox. 5.80 aprox. 5.50 aprox. 5.10

[1] Some values were approximated from the graphics, associated with the cited scientific research.

All samples registered a growth in the values of the tensile strength as the percentage of nanoclay increased, except for the sample analyzed in [54]. In that case, at 10 wt% NC addition, the value of the strength decreased by 13.70%, compared to those obtained for an 8 wt% NC addition. Taking into account that the nanoclay percentage was very high, obtaining a homogeneous nanoclay dispersion becomes very difficult, leading to flocculation. This automatically creates weak points within the matrix. The values of the splitting tensile strength reported in [54] were higher than those presented in [55] due to the use of polypropylene fibers, which succeed in controlling the occurrence and the propagation of cracks. In the former case, the authors used nanoclay in order to obtain a better bond between the polypropylene fibers and the matrix [54]. Although nanoclay is not able to completely prevent the occurrence and propagation of fractures, the increase in the value of the flexural tensile strength of the composite leads to smaller and fewer cracks inside the material. Therefore, the reinforcement is protected from the ingression of chemical agents and humidity increasing the durability of reinforced concrete elements.

The results obtained for mortar samples follow the same trend as the ones reported for concrete [18,46,56]. The difference of 0.55 MPa for 2 wt% NC can be explained by the smaller water to binder ratio used for the former.

2.5.2. Flexural Tensile Strength

Taking into account that the bending state of loading is present in every civil engineering structure, the researchers had to verify if and to what extent the content of nanoclay would influence the flexural tensile strength of cement mortar or concrete elements. There are several studies on the improvement of the flexural strength regarding the combination between nanoclay and cement-based materials. The focus was on preventing or limiting the occurrence of cracks, which, besides the fact that this leads to the exposure of reinforcement to the outside air, the reinforcement could not benefit from the protection of the concrete cover in case of a fire.

In case of exposure to high temperatures, the same trend observed for the splitting tensile strength was reported in [18] for the flexural tensile strength, for 1 and 2 wt% natural hydrophilic montmorillonite cement replacement, without any treatments. For all specimens and all temperatures, the 2 wt% NC composites exhibited the highest strength values compared with the control sample. The maximum flexural strength was reached at a temperature of 200 °C, at 11.60 MPa. When subjected to 400 °C, the flexural strength had a major drop in the value, of about 76% for the control sample, considering as reference the value for 200 °C, reaching a value of 2.60 MPa. The influence of nanoclay was evident at 400 °C, as the flexural strength value increased for 2 wt% NC by 138% compared to the control specimen. Unlike the tensile strength, the specimens preserved a certain value of flexural strength after their treatment at 600 °C, although very small—0.8 MPa for the control specimen. Adding 1 wt% NC, in this case, did not lead to flexural strength improvements, but for the 2 wt% NC combination the strength value reached 1.7 MPa [18].

A similar analysis was conducted for 1, 3 and 5 wt% nano-calcined montmorillonite clay [28]. The same trend of flexural strength increase was obtained at 250 °C, reaching the maximum value for 5 wt% nanoclay. At the next temperature stage, i.e., 450 °C, the strength decreased significantly for all specimens. It should be noted that at about 180 °C, polypropylene fibers started to melt and caused an increase in porosity which, in turn, led to smaller flexural strength values than the ones reported in [18].

A comparison was conducted between two types of curing methods applied to specimens with various percentages of nanoclay added in the cement matrix, i.e., water curing and plastic cover curing. The authors of the study observed that, similar to the tensile strength, the flexural strength value was smaller when curing the specimens by plastic wrapping than by immersing them in water. The major difference was obtained at 90 days of curing for 1 wt% NC sample, in which case the flexural strength decreased by approximately 14.9% for the second curing method compare to the traditional, water curing method [55]. This could be explained by the fact that the water cured specimens had enough moisture to continue the hydration process of the cement whereas the same process was significantly slowed down in case of plastic wrapping curing method once the mixing water was consumed.

Table 2 presents the values of the flexural tensile strength, for different inputs, regarding cement mortar and concrete composites with different nanoclay quantities. The values were selected for samples subjected only to room temperature and cured for 28 days in water or lime-saturated water.

Regarding what concerns the flexural strength for both cement mortar and concrete, the values vary significantly from one study to the other. The only common information resulted from these analyses would be the increase in the value of the flexural strength with the increasing nanoclay percentage within the cement matrix. The available data is sometimes conflicting, with some authors [28] reporting smaller values of the flexural tensile strength than others [18,56], although additives were used to improve workability and strength. One possible explanation could be the water to binder ratio. A w/b ratio of 0.55 could provide cement and nanoclay particles enough water to hydrate during mixing and curing period [18,56]. In another study [28], however, a lower w/b ratio was considered but a superplasticizer was used to improve workability. Generally, a small calculated w/b ratio provides a higher strength. However, due to the level of hydrophilicity of nanoclay, a higher amount of water could be needed. Moreover, there were two different types of montmorillonite used: a calcined one which was less hydrophilic and a natural one without treatment which had a high level of hydrophilicity.

Table 2. Flexural strength values for different nanoclay-cement matrix samples.

Study	Type of Nanoclay	Type of Sample	Standard	w/b w/c	Additives	NC %	Flexural Strength [1] [MPa]
[18,56]	Natural hydrophilic montmorillonite	Cement mortar	ASTM C348	w/b 0.55	-	0 0.5 1 2	7.40 7.60 7.60–7.70 7.80
[28]	Calcined hydrophilic montmorillonite clay	Cement mortar	ASTM C348	w/b 0.484	- non-absorbent monofilament polypropylene fibers - naphthalene-sulfonate-based superplasticizer	0 1 3 5	approx. 6.10 approx. 6.20 approx. 6.70 approx. 7.00
[54]	Metakaolin	Concrete	ASTM C293	w/c 0.42	1.95% polypropylene	0 2 4 6 8 10	9.36 9.92 11.24 12.02 12.76 11.50
[55]	Montmorillonite (type not specified)	Concrete	ASTM C78	w/c 0.45	-	0 1 2 3	approx. 5.00 approx. 6.60 approx. 6.30 approx. 5.90
[23]	Organo-montmorillonite clay	Cement mortar	ASTM C348	w/c 0.50	-	0 0.25 0.5 1	approx. 9.50 approx. 9.60 approx. 9.10 approx. 9.80

[1] Some values were approximated from the graphics, associated with the cited scientific research.

A decrease in the value of the flexural tensile strength for a higher percentage nanoclay was reported in [54,55], which may be due to the inhomogeneous specimens given by the large quantity of nanoclay present within the mixture. The values of the flexural tensile strength obtained by [54] on the concrete samples were higher for every nanoclay percentage, mainly because of the 1.5% polypropylene addition, which prevents crack formation and propagation.

The highest value of the flexural tensile strength from all the presented studies was reported by [23], although the authors did not use additives. This may be related to the organo-modified montmorillonite clay, which resulted after a treatment of the natural montmorillonite with dimethyl dehydrogenated tallow and quaternary ammonium chloride.

2.5.3. Compressive Strength

The most important material property of both cement mortar and concrete is their compressive strength.

Considering that there are several studies that investigated the variation in the compressive strength depending on temperature, the analysis within Section 2.5.3 pursues two directions. The first one will debate upon the compressive strength results obtained for specimens stored and tested at room temperature, while the second one will focus only on the studies where the specimens were subjected to elevated temperatures.

Strength Values at Room Temperature

From the data presented so far for splitting and flexural tensile strength, it was concluded that between the water curing and the plastic wrapping curing, the former is the best one to use for cement-based composites with nanoclay additions. The same trend was reported for the case of compressive strength. The latter curing method resulted in lower values of the compressive strength at 90 days, with on average a 14% decrease compared with water curing. As previously explained, cement and nanoclay particles need water to hydrate and if they do not have enough, the hydration reaction significantly decreases and even stop and no C-S-H gels will be produced anymore, thus limiting the strength increase. The highest difference was observed at 90 days of curing. On the other hand, at the age of 28 days, this difference was very small [55].

In another study, the enhancement of compressive strength due to 1, 2, 3 wt% nanoclay addition in self-consolidating concrete was investigated [42]. The considered curing ages were 3, 7, 14, 28 and 90 days. The obtained results showed that the highest increment in the values of the compressive strength was at 7 and 14 days of curing, compared to the control mix. In that time interval the hydration process reached a maximum level, as the CH particles were consumed and CSH gels were formed. As more nanoclay was added in the cement matrix, the difference between the compressive strength of the control sample and the compressive strength of the nanoclay enriched composite became larger. A maximum was reached for 3 wt% nanoclay addition, the difference between the value of the compressive strength for the control mix and the nanoclay composite at 7 and 14 days of curing being 31% and 14%, respectively. For the 2 wt% nanoclay addition, this difference was 20.70% and 4.70%, respectively [42].

The compressive strength of a nanoclay-concrete composite at 7, 14, 28, 49, 56 and 90 days of curing, for nanoclay percentages of 0.1, 0.3 and 0.5 wt% and two water to cement ratios, i.e., 0.40 and 0.50, was also investigated. A similar trend was observed with the one reported in [42], the largest increase in the value of the compressive strength being during the first 28 days. The study also investigated the influence of water to cement ratio on the values of the compressive strength. The nanoclay percentage for which the maximum compressive strength was obtained changed depending on the w/c ratio. For a w/c ratio of 0.40, the optimum nanoclay percentage was 0.50%, whereas for a w/c ratio of 0.50, it decreased to 0.30% [57].

Table 3 summarizes the findings on the values of the compressive strength for mortar/cement specimens with certain percentages of nanoclay additions, stored at room temperature and cured for 28 days in water or lime-saturated water.

The smallest values for cement mortar reported in [28] can be explained by a non-homogeneous distribution of nanoclay within the cement matrix, the authors resorting to manual mixing of the material. This method can cause nanoclay particles to agglomerate and, eventually, lead to a decrease in strength. The same can be observed in terms of values of the flexural tensile strength obtained in [28] compared to other studies. In case of concrete, the results presented in [57] were very low in comparison with the other studies.

According to [57], for low nanoclay percentages, there was a negative influence upon the compressive strength of the composite. An addition of only 0.1 wt% nanoclay led to a decrease of 14% in the value of the compressive strength, for a w/c ratio of 0.50. However, when increasing the nanoclay content, the compressive strength value exceeded the one of the control specimen but not significantly [57]. For percentages greater than 0.1 wt% nanoclay, the compressive strength value increased with the increase in nanoclay content within the cement matrix. However, a small decrease in the values of the compressive strength at the maximum analyzed nanoclay percentage, of about 5.15%, compared to the previous percentage, was reported in [54,55]. Taking into account the spread of the reported results, it is difficult to provide a reason applicable to all scenarios. On the other hand, as the percentage of nanoclay increases, the distance between nanoclay particles decreases and, as they tend to attract each other, flocculation of nanoparticles may occur, which leads to a decrease in the strength of the material.

Table 3. Compressive strength values for nanoclay—mortar/concrete specimens at room temperature.

Study	Type of Nanoclay	Type of Sample	Standard	w/b w/c	Additives	NC %	Compressive Strength [1] [MPa]
[23]	Organo-montmorillonite clay	Cement mortar	ASTM C109	w/c 0.50	-	0 0.25 0.5 1	approx. 45.50 approx. 49.00 approx. 55.00 approx. 52.50
[28]	Calcined hydrophilic montmorillonite clay	Cement mortar	ASTM C109	w/b 0.48	- non-absorbent monofilament polypropylene fibers - naphthalene-sulfonate-based superplasticizer	0 1 3 5	approx. 21.00 approx. 21.20 approx. 22.00 approx. 24.10
[18,56]	Natural hydrophilic montmorillonite clay	Cement mortar	ASTM C109	w/b 0.55	-	0 0.5 1 2	37.00–37.60 38.00 38.50–39.00 40.30–41.00
[46]	Metakaolin MKC	Cement mortar	ASTM C109	w/b 0.50	-	0 2 4 6 8	approx. 47.20 approx. 47.60 approx. 48.50 approx. 49.70 approx. 50.50
[54]	Metakaolin MKC	Concrete	BS 1881 -Part 116	w/c 0.42	1.5% polypropylene	0 2 4 6 8 10	52.32 53.80 55.90 57.50 58.80 55.60
[55]	Montmorillonite (type not specified)	Concrete	ASTM C470	w/c 0.45	-	0 1 2 3	approx. 45.00 approx. 61.00 approx. 58.00 approx. 55.00
[42]	Montmorillonite (type not specified)	Self-consolidated concrete	-	w/b 0.34	F-type poly-carboxylate-based superplasticizer	0 1 2 3	approx. 49.20 approx. 52.00 approx. 51.00 approx. 54.50
[57]	type not specified	Concrete	-	w/c 0.40	-	0 0.1 0.3 0.5	approx. 35.00 approx. 34.00 approx. 37.00 approx. 39.40
				w/c 0.50		0 0.1 0.3 0.5	approx. 34.00 approx. 29.00 approx. 36.00 approx. 35.00

[1] Some values were approximated from the graphics, associated with the cited scientific research.

Strength Values at Elevated Temperatures

When concrete or cement mortar specimens are subjected to high temperatures, a temperature gradient develops inside the elements. The chemical and physical phenomena which occur lead to spalling or fracture development. Therefore, a decrease in strength is registered, the material losing its ability to overtake the induced thermal generated stresses.

Table 4 presents the compressive strength variation for different cement mortar/concrete specimens with nanoclay addition, cured for 28 days in water or lime-saturated water, subjected to various temperatures. Taking into account that the studies which are presented in this section were also reported in Table 3, the given information will focus only on the temperature variation and the corresponding values of the compressive strength.

Table 4. Compressive strength values for nanoclay—mortar/concrete specimens at high temperature.

		[18]			[28]			
Type of Nanoclay		Natural Hydrophilic Montmorillonite			Calcinated Hydrophilic Montmorillonite			
Nanoclay Content [%]		0	1	2	0	1	3	5
Compressive strength [1] [MPa]	25 °C	37.6	38.5	40.3	21	21.2	22	24.1
	200 °C	54.4	56.1	59.7				
	250 °C				22.5	23	24	27
	400 °C	38.9	40.2	41.6				
	450 °C				11	12	14	17
	600 °C	9.2	9.8	10.6	10	10.5	12	13.5
	900 °C				2	2.5	5	5

[1] All values are approximated from the graphics, associated with the cited scientific research.

As temperature reaches 200 °C, the value of the compressive strength increases, according to the data reported in [18] and graphically presented in Figure 3a due to the acceleration of the hydration process and the CSH production. The highest value was obtained for 2% nanoclay cement replacement—59.7 MPa. On the other hand, the rate of increase in the values of the compressive strength for the three considered mixes was between 44.68% for the control mix and 48.14% for the mix containing 2 wt% nanoclay cement replacement. As the temperature increased up to 400 °C, the material lost its strength by about 30% but it still maintained a value higher than the control specimen kept at room temperature. It is interesting to observe that in this case, the highest gain in the value of the compressive strength was obtained by the mix with 1 wt% nanoclay cement replacement, at 4.42%. At 600 °C, all samples registered a very high strength loss of about 75% with respect to the 400 °C samples, the 2% nanoclay combination having the biggest residual compressive strength value, i.e., 10.6 MPa [18].

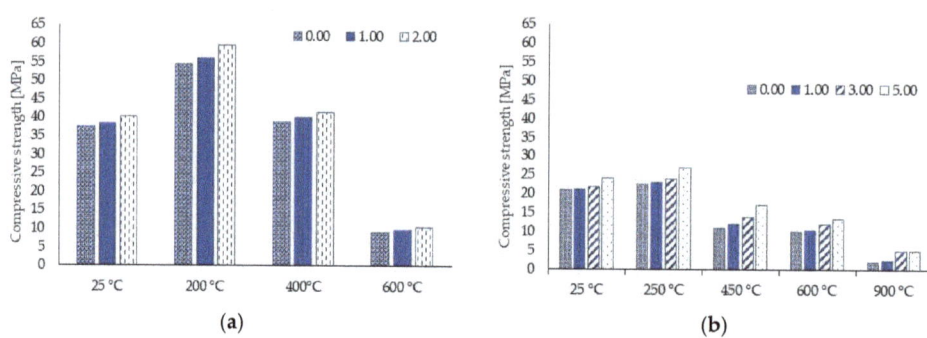

Figure 3. Variation in compressive strength as function of temperature and different percentages of nanoclay. (a) results reported in [18]; (b) results reported in [28].

Although the values were smaller than those presented in [18], the trend reported in [28] was similar for the cement mortar subjected to high temperatures, as seen in Figure 3b. The authors observed that as the temperature rose to 250 °C, the value of the compressive strength increased as well, as the hydration process is accelerated and more CSH is produced. For that temperature, the optimum nanoclay percentage was 5%. However, at 450 °C, the strength was already diminished significantly by about 37% compared to the 5% nanoclay cement replacement at 250 °C, when CSH was in the decomposing process. For 600 °C, there was an evident strength loss especially for the nanoclay-cement composite reaching a maximum of 20.6%. At 900 °C, the strength loss continued for all specimens but the nanoclay-cement composite still maintained a higher strength value than the control sample. It should be noted that there was no difference in residual compressive strength between the 3% and 5% nanoclay cement replacement [28].

2.6. Durability Tests

The durability of a material is essential in civil engineering where the structures must have a life cycle of decades and where some of them are subjected not only to external loads but also to water penetration which can lead to corrosion, to frost-thaw cycles, to air with various chemical pollutants, etc. There are several studies that investigated the influence of nanoclay on the long-term durability of cement mortar or concrete.

Gas permeability tests with methanol were used [14] on samples made of cement paste with different quantities of montmorillonite nanoclay. As the curing period increased, the permeability coefficient decreased. There was a pattern emerging from this analysis that is common to all curing periods. The smallest permeability coefficient was registered for 0.4% nanoclay cement replacement, whereas the highest corresponds to the control specimen. The relative decrease was significant, a value of 49.95% being obtained for the 56 days curing period [14]. This result demonstrates the filler effect that nanoclay has on the cement matrix, succeeding in decreasing porosity and blocking outside elements to penetrate inside the cement matrix.

However, the results of the tests on cement pastes could not be confirmed by the results obtained on concrete specimens [41]. Nano-metakaolin was used in the concrete mixes and oxygen permeability tests were conducted. A higher permeability was obtained for nanoclay samples compared to the control samples, with approximately 150% for the 1% nanoclay and 200% for the 2% nanoclay. The difference between the two studies may have several explanations: different types of permeability tests, increased porosity due to the presence of aggregates combined with an insufficiently dispersion of nanoclay particles within the cement matrix, type of nanoclay, etc.

Water penetration and water absorption test are relevant when assessing the probability of water reaching the reinforcement and eventually causing corrosion. Tests were conducted in accordance to the EN 12390-8 norm and it was observed that for 1–3 wt% nanoclay cement replacement, the quantity of water penetrating the concrete samples was less than in case of the control samples, for which a 31 mm depth was registered [55]. The smallest depth of water absorption, i.e., 20 mm, was obtained for the 1 wt% nanoclay. The reasons for the smaller water depth values for concrete combined with nanoclay are the filler effect that nanoclay has coupled with the higher development of CSH gels compared to the control sample resulting in a denser structure of the material. The above results were obtained at the age of 28 days of water curing. In case of plastic wrapping curing, all samples had a higher water absorption than the control sample, which confirms the results obtained for compressive, flexural and tensile strengths [55]. Similar results were obtained for self-consolidated concrete with 1, 2 and 3 wt% nanoclay addition, with 90 days of curing [42]. However, the smallest water penetration depth was registered for 3% nanoclay addition, which was 64.3% smaller than the control specimen [42].

A water absorption test based on ASTM C642 for concrete specimens with 1, 2 and 3% montmorillonite nanoclay was also used [55]. As in the case of water penetration test, all nanoclay specimens had a water absorption percentage smaller than the control specimen. The lowest water absorption percentage corresponded to the 1% nanoclay sample, i.e., 1.46%, 54.3% smaller compared to the control sample [55]. Similar results were reported for self-consolidated concrete samples with 1, 2 and 3% nanoclay addition [42]. In contrast with [55], the lowest water absorption percentage corresponded to the 3% nanoclay mix while the highest was for 2% nanoclay. A possible explanation could be attributed to the effect of shrinkage on the integrity of the mix [42]. For cement mortar on the other hand, the obtained results were different, as the water absorption percentage for a 0.5% nanoclay addition was smaller than the one for the control specimen, but for 1% and 2% nanoclay, the value was higher and has an increasing trend [56]. The authors associated their results with the capillary water absorption test, in which case the specimens were submerged in water only at 5–10 mm depth, according to BS EN 1015-18 norm. The lowest capillarity absorption coefficient results came from 1% and 2% nanoclay, reaching a value of approximately 0.03 kg/(m^2 min$^{0.5}$), compared with 0.04 kg/(m^2 min$^{0.5}$)—corresponding to the control sample [56].

When water trapped inside the cement matrix starts to evaporate, it creates pressure on the void walls which leads to the development of fine cracks. These cracks have a negative impact both on the strength and on the durability of the material. The influence of adding organo-montmorillonite nanoclay in the cement matrix upon the material plastic shrinkage was evaluated in [23]. It was concluded that nanoclay has a definite positive effect, as the plastic shrinkage value decreased by 70%, even for the lowest analyzed nanoclay quantity, i.e., 0.25 wt% cement replacement [23].

An impressed voltage test was employed to analyze the variation in corrosion current for self-consolidating concrete modified with 1, 2 and 3% nanoclay [42]. It resulted that, as the percentage of nanoclay increases, the deterioration time of the reinforcement extended, meaning that the use of nanoclay leads to a better protection of the steel reinforcement against corrosion mainly due to the denser structure of the cement matrix.

Table 5 summarizes the findings reported in this section of the paper. The information is structured based on the type of durability test, type of material it was conducted on (i.e., mortar or concrete) and nanoclay percentage. The main findings of each referenced study are also included.

Table 5. Durability tests on nanoclay—mortar/concrete specimens.

Measured Parameter	Scientific Paper	Test	Type of Sample	Type of Nanoclay	Nanoclay Percentage	Observation
Permeability coefficient	[14]	Gas permeability test (methanol)	Cement paste	Montmorillonite in liquid form (type not specified)	0; 0.2; 0.4; 0.6; 0.8	- As the curing period increased, the permeability coefficient decreased - The smallest permeability coefficient was registered for 0.4% nanoclay, and the highest for the control sample
	[41]	Oxygen permeability test	Concrete	Nano-metakaolin	0; 1; 2	- Higher permeability for nanoclay samples compared to control samples
Water absorption percentage	[55]	Water absorption test	Concrete	Montmorillonite (type not specified)	0; 1; 2; 3	- The test was carried out at 28 days of curing - Water absorption percentage smaller than the control specimen - The lowest value corresponds to 1 wt% nanoclay - Plastic wrapping curing samples had a higher water absorption than the water curing samples
	[42]	Water absorption test	Self-consolidated concrete	Montmorillonite (type not specified)	0; 1; 2; 3	- The test was carried out at 90 days of curing - Water absorption was reduced when nanoclay was added - The best result was for the 3 wt% nanoclay addition
	[56]	Water absorption test	Cement mortar	Natural hydrophilic montmorillonite clay	0; 0.5; 1; 2	- Water absorption percentage for 0.5% nanoclay was smaller compared to the control specimen - For 1 and 2% nanoclay, the value was higher than the control specimen

Table 5. Cont.

Measured Parameter	Scientific Paper	Test	Type of Sample	Type of Nanoclay	Nanoclay Percentage	Observation
Water penetration depth	[55]	Water penetration test	Concrete	Montmorillonite (type not specified)	0; 1; 2; 3	- The test was carried out at 28 days of curing - 1 wt% nanoclay—smallest depth of water absorption (20 mm) - Plastic wrapping curing samples had a higher water absorption than the water curing samples
	[42]	Water penetration test	Self-consolidated concrete	Montmorillonite (type not specified)	0; 1; 2; 3	- The test was carried out at 90 days of curing - The best result was for the 3 wt% nanoclay addition
Capillary water absorption coefficient	[56]	Water absorption test	Cement mortar	Natural hydrophilic montmorillonite clay	0; 0.5; 1; 2	- The lowest capillarity absorption coefficients resulted for 1 and 2% nanoclay
Plastic shrinkage	[23]	Plastic shrinkage test	Cement mortar	Organo-montmorillonite clay	0; 0.25; 0.5; 1	- Nanoclay has a definite positive effect - Plastic shrinkage decreased by 70% for 0.25 wt% nanoclay cement replacement
Corrosion current	[42]	Impressed voltage test	Self-consolidated concrete	Montmorillonite (type not specified)	0; 1; 2; 3	- As the percentage of nanoclay increases, the deterioration time of the reinforcement ex-tended

2.7. Remarks on the Impact of Using Nanoclay in Cement Composites

Nanoclay particles can be used in a modified or unmodified physical state, depending on their properties. From the scientific literature analysis, it can be stated that nanoclay has a positive impact as it succeeds in diminishing some of the weak points characteristic to mortar and concrete, the newly composite being characterized by:

- higher values of tensile, flexural and compressive strength, for both specimens kept at room temperature and subjected to high temperatures
- lower water absorption percentage and water penetration depth
- lower plastic shrinkage
- extended reinforcement deterioration time

The combined physical and chemical properties of nanoclay make this nanomaterial suitable for use in mortar/concrete mixes. Although all results concur to this idea, the values differ from one study to another, especially the ones related to the mechanical properties. At this moment, most of the related studies are focused on the cement paste. Therefore, there is a great gap of knowledge regarding mortar and, especially, concrete modified with nanoclay. Moreover, taking into account the promising values of compressive strength for specimens subjected to high temperatures and the lack of research in this area, more studies should be conducted on this subject.

3. Influence of Titanium Dioxide on Cement-Based Materials

The use of titanium dioxide (TiO_2) in the construction industry did not have a structural purpose in the beginning, but a more architectural and ecological one. Torre de Especialidades from Mexico City and the Jubilee Church in Rome Italy are two of the buildings for which titanium dioxide was used in the concrete formula with the purpose of decreasing the level of pollutants, i.e., nitrogen oxide and nitrogen dioxide. Except the practical application, the esthetics of these buildings stands out not only due to the architecture but also due to their bright whiteness [58–61].

As a nanomaterial, TiO_2 captured the interest of researchers in the fields of building services and electrotechnics. It started to be used in photovoltaic cells, in the composition of semiconductors and even in bio-medical applications and cancer therapy [62]. Moreover, as its properties have the potential to improve the strength and durability of cement mortar and concrete elements, the research in civil engineering is still on-going. The positive influence on both ecology and civil engineering is a material property which is constantly searched for, in view of the new stricter regulations in terms of greenhouse gas emissions. Titanium dioxide combines, at a certain level, these two areas of interest. In addition, TiO_2 is a naturally occurring oxide, being found in the Earth's crust [63], its addition to concrete enhancing the sustainability index of the new material [64].

3.1. Chemical Structure and Properties—Types of TiO_2 Used in Cement Mortar and Concrete

The crystalline structure of titanium dioxide, TiO_2, is found in three main different forms: anatase, rutile and brookite. Both anatase and rutile have a tetragonal crystal structure, while brookite has an orthorhombic crystal structure [63,65,66]. From these three crystal structures, the most commonly used in civil engineering are anatase and rutile. Both are wide band gap semiconductors, meaning that they can resist higher temperatures, unlike brookite [65,67].

Rutile is considered as the most stable form of titanium dioxide but only for a particle size greater than 35 nm. Below that size, the thermodynamic stability decreases. Another characteristic of rutile is its behavior at high temperatures. When the calcination temperature increases, its particle size increases also with a growth rate higher than in case of anatase [63]. Anatase, on the other hand, develops a higher photocatalytic activity than rutile and, thus, was preferred for various element coatings. Moreover, it was demonstrated that it has better success in breaking both inorganic and organic pollutants. On the other hand, it was shown that combining the anatase and rutile phases, leads to an increase in the photocatalytic activity [65,68]. This breaking of pollutants during the photocatalytic process begins when a light with enough energy strikes the material containing TiO_2, i.e., the catalyst, and an oxidation-reduction reaction takes place. During the process, the pollutants are mineralized, but the quantity of TiO_2 is not consumed. However, between 550 °C and 1000 °C anatase transforms into rutile, the transformation temperature depending on the existent impurities and the morphology of the sample [65,66,68].

Several general chemical properties of TiO_2 are listed in [66], with some of them making it adequate for use in mortar and concrete, such as its chemical stability, biocompatibility, low toxicity, and low cost compared with other nanomaterials.

3.2. Input of TiO_2 Particles in Cementitious Materials—Technological Flow

There are two possibilities of using TiO_2 combined with mortar or concrete. The first one is by introducing a certain quantity of nanomaterial in the cement matrix and mixing it. The other one is by coating the element with a special formula in order to protect the element from the exterior polluted environment. This study will focus on the first method, taking into account that the main interest is the material strength. Moreover, there is an important probability that the coating could be damaged during execution or service life [65]. Unlike nanoclay, titanium dioxide does not require any additional special mixing processes before being added to the cement matrix, as it does not have the tendency to flocculate. There are two methods of adding TiO_2 in the cement matrix, namely the wet and the dry mixing procedures.

In the first method, TiO_2 is introduced in water and mixed for several minutes. The cement and aggregates are mixed in dry conditions, separately. Afterwards, the water-nanomaterials solution is added to the dry mix, the rest of the water is added and a final mixing is performed. If there was fiber reinforcement to be added, e.g., PP fibers, it is introduced into the matrix in the final step and mixed again [58,69–71].

In the second method, the aggregates and all the powder materials, including TiO_2, are dry mixed together. The water is gradually introduced, along with the superplasticizer and mixed for 3–5 min until a proper consistency is reached [72–75].

3.3. Structural Influence of TiO_2

Similar to other nanoparticles, titanium dioxide nanoparticles have been used for their high specific surface area which can result in promoting the hydration reaction and for their pore-filling effect. In addition to these benefits, common to this size, TiO_2 has specific properties that make it attractive in its use in cementitious composites, such as its photocatalytic properties and thermal stability.

Titanium dioxide does not possess pozzolanic activity, as shown in several studies [59,76]. Nevertheless, it increases the rate of hydration due to the nanoparticles acting as nucleation sites. This process, together with the small size induced filling effect, creates a denser structure. It has been found that CH is influenced by the presence of these nanoparticles, which decrease the size of CH crystals either by limiting their growth space or by promoting accelerated CSH gels formation [59,76].

Studies on cement paste incorporating TiO_2 nanoparticles used TG analysis in order to render evident the hydration reaction acceleration [77]. It can be inferred that the pure cement specimen contains less non-evaporable water, chemically bound water, than the cement-titanium blend, which is a result of the presence of more hydration products [77].

3.4. Material Strength Improvements

Titanium dioxide nanoparticles succeed in improving the mechanical properties of cement mortar and concrete due to their chemical and physical properties. They are usually used in mortar/concrete mixes combined with superplasticizers in order to obtain a higher workability. Taking into account that TiO_2 is a non-reactive powder, there are studies in which pozzolans were added with the purpose of promoting the cement hydration reaction, e.g., fly ash, silica fume. In addition, in comparison with nanoclay studies, in the case of TiO_2, the number of scientific experiments made on mortars and concrete is significantly larger. This section of the state-of-the-art article presents the strength values of mortar and concrete modified with certain quantities of TiO_2 and other additives. The results will focus on the effect that TiO_2 has on the mortar and cement, with or without supplementary materials.

3.4.1. Splitting Tensile Strength

As in the case of nanoclay, the number of research works focusing on the tensile strength is less than in case of compressive strength. However, this mechanical property is also important, mainly from the point of view of cracks occurrence which must be prevented. Even micro-cracks can lead to an exponential decrease in the concrete and/or reinforced concrete durability. Due to the large specific surface area of TiO_2 particle there is more area available for the cement hydration reaction to occur and to produce more CSH, which leads to an increase in strength [78].

The value of the tensile strength of concrete when the cement was replaced with 1 wt% nano-TiO_2 was compared with the one obtained when the cement was replaced with 1 wt% nano-Fe_3O_4 [79]. The results showed that the TiO_2-modified specimen developed 18% higher values for the tensile strength. It should be noted that adding Fe_3O_4 to the mix results in values of the tensile strength smaller than the ones corresponding to the control specimen.

Other studies focused on the synergistic effect of using two nano-materials, e.g., both TiO_2 nanoparticles and carbon nanofibers (CNF) [80]. By adding only 0.2% and 0.4% of CNF in the cement matrix, the tensile strength values increased at about 3.90 MPa and 4.20 MPa, respectively. Compared with specimens containing only TiO_2, these values were much higher, proving that CNF have a higher efficiency in strength improvement than TiO_2. When introducing both CNF and TiO_2 nanoparticles in the composition, although the values of the tensile strength were higher compared to the composite with a single type of nanomaterial addition, the dispersion deficiencies and particle agglomeration started to emerge. Thus, for the highest nanoparticle quantity combination considered, i.e., 0.4% CNF and 5% TiO_2, the tensile strength value was smaller than for the 0.4% CNF and 3% TiO_2 at 28 days of curing. Moreover, at 180 days of curing, the tensile strength value for the maximum nanoparticle combination became smaller than the composite having only 0.4% CNF and no TiO_2. Another remark that could be made regarding the values of the tensile strength for the 3% and 5% TiO_2 combination, after 90 days of curing, was that the difference between these values started to decrease slowly and reached almost the same value of 4.8 MPa at 180 days of curing [80].

In another study, ZnO and TiO_2 nanoparticles were used as supplementary materials to improve the mechanical properties of concrete containing 0.6% polypropylene fibers. The focus of the research work was to find the best combination for which a maximum strength was achieved. The cement was replaced with ZnO and TiO_2 in the following percentages: (1%; 0.5%), (2%; 1%), (3%; 1.5%), (4%; 2%), (5%; 2.5%). For the splitting tensile strength, the best combination was (4%; 2%), which proved that both ZnO and TiO_2 have a positive effect on the material properties. However, when the nanomaterials quantities were increased to (5%; 2.5%), a decrease in the values of the tensile strength was registered due to the problems in nanoparticles dispersion [71].

Table 6 lists a series of research works in which various cement mortar or concrete specimens modified with TiO_2 nanoparticles were tested to obtain the values of the tensile strength at 28 days of curing. It should be noted that the list focuses on the research works that used TiO_2 as principal nanomaterial.

Table 6. Splitting tensile strength values mortar/concrete specimens modified with TiO_2 nanoparticles, at room temperature.

Study	Type of TiO_2	Type of Sample	Standard	w/b w/c	Additives	TiO_2 %	Splitting Tensile Strength [3] [MPa]
[78]	Type not specified (probably anatase)	concrete	-	w/c 0.45	Sulphonated naphthalene formaldehyde (superplasticizer)	0 0.5 1 1.5	3.45 3.54 3.76 3.65
[79]	Type not specified (probably anatase/rutile combination)	concrete	-	w/c 0.57	-	1	3.36
[80]	Type not specified (probably anatase)	concrete	ASTM C496	w/c 0.45	Polycarboxylate-based HRWR [1] agent (superplasticizer)	0 3 5	approx. 3.00 approx. 3.60 approx. 3.30
[70]	White powder (probably anatase)	concrete	-	w/c 0.38	15% Silica fume Sika ViscoCrete-3425 superplasticizer	0 0.5 1 1.5	3.37 4.39 4.93 4.76
[81]	Type not specified (probably anatase)	concrete	ASTM C496	w/b 0.40	-	0 0.5 1 1.5 2	1.80 2.60 3.00 2.70 1.90
[82]	Type not specified (probably anatase)	SCC [2]	ASTM C496	w/b 0.40	Polycarboxylate (superplasticizer)	0 1 2 3 4 5	1.60 1.60 2.00 2.50 2.90 2.60

[1] High-range water reducer. [2] Self-compacting concrete. [3] Some values were approximated from the graphics, associated with the cited scientific research.

The main common observation of these studies is that for high nanoparticle quantities, the value of the tensile strength decreased. Two main reasons are given for this phenomenon in the scientific literature. The first one is the dispersion difficulties that emerge when adding high percentages of nanoparticles as it increases the probability of agglomeration occurring. The second one is that if the quantity of TiO_2 was greater than the quantity needed for CH hydration, an excess of silica would be found inside the cement matrix, leading to a decrease in strength [70,81]. This percentage varies, but the majority of the listed research works showed that 1% cement replacement with TiO_2 is the quantity for which a maximum value of the tensile strength was obtained in case of concrete specimens [70,78,81].

The highest values of the tensile strength compared to the other studies was reported in [70]. It this case, the addition of 15% silica fume proved its efficiency [70]. The lowest values were obtained for the self-consolidated concrete specimens and plain concrete, respectively, both modified with various TiO_2 percentages [81,82].

Altogether, the TiO_2 nanoparticle addition in the cement matrix leads to a definite increase in the values of the splitting tensile strength, as the specific surface area increases and CSH particles formation accelerates. However, the increment is not well defined as it depends on a series of variables which are different, usually, from study to study, e.g., additives, water/cement or water/binder ratio, type of cement, type of mixing, particle size etc.

3.4.2. Flexural Tensile Strength

The flexural strength for various combinations of CNF and TiO_2 nanoparticles added in concrete was also investigated [80]. All combinations had a higher strength value than the control sample. However, the specimens containing CNF had superior strength, whereas the 5% TiO_2 mix had the smallest strength value from the modified formula samples. When combining 0.4% CNF with 3% and 5% TiO_2, the highest flexural strength values were obtained. In case of flexural tensile strength, unlike the splitting tensile test results, all CNF mixes, including combinations, led to superior strength values, especially after 180 days of curing. Therefore, the combination 0.2% CNF and 5% TiO_2 registered, at 180 days of curing, a value of the flexural strength smaller than that of the sample with only 3% TiO_2. These results confirm the observations made in the case of splitting tensile strength, that as the quantity of nanoparticle increases above a certain critical level, the strength begins to decrease because silica accumulates within the matrix and particle agglomerations occur [80]. For the flexural strength tests carried out on ZnO and TiO_2 specimens, the best combination was (4%; 2%) [71].

Table 7 summarizes the values for the flexural tensile strength for a series of studies on concrete specimens modified with a certain quantity of TiO_2 nanoparticles, stored at room temperature, after 28 days of curing.

For the TiO_2 modified cement mortar, the values of the flexural tensile strength were higher than in the case of concrete specimens. When adding coarse aggregates, although the compressive strength increased, voids occurred at the interfacial transition zone, which tended to develop under tensile stresses. Therefore, these voids led to smaller values for splitting and flexural tensile strength. However, the nanoscale particles have a filler effect on the matrix due to their reduced size. They can enter these voids and reconstruct the ITZ so that the weak areas are reduced in size and the development of microcracks is either slowed down or completely arrested. This phenomenon was observed by comparing the strength increment when adding TiO_2 nanoparticles in the matrix, in both cement mortar and concrete, respectively. Higher strength differences were registered for concrete specimens when they were modified with TiO_2 compared with the control samples, i.e., greater than 1 MPa [80–82]. However, in case of cement mortar, these improvements in the flexural tensile strength remained smaller than 1 MPa [83–85].

Table 7. Flexural strength values mortar/concrete specimens modified with TiO_2 nanoparticles, at room temperature.

Study	Type of TiO_2	Type of Sample	Standard	w/b w/c	Additives	TiO_2 %	Flexural Strength [3] [MPa]
[80]	Type not specified (probably anatase)	concrete	ASTM C78	w/c 0.45	Polycarboxylate-based HRWR [1] agent (superplasticizer)	0 3 5	approx. 4.00 approx. 5.20 approx. 4.50
[83]	Type not specified (probably anatase)	Cement mortar	ASTM C293	w/b 0.485	30% Fly ash by weight of cement	0 1 3 5	approx. 5.15 approx. 5.15 approx. 5.70 approx. 5.15
[81]	Type not specified (probably anatase)	concrete	ASTM C293	w/b 0.40	-	0 0.5 1 1.5 2	4.40 5.10 5.50 5.40 5.10
[82]	Type not specified (probably anatase)	SCC [2]	ASTM C293	w/b 0.40	Polycarboxylate (superplasticizer)	0 1 2 3 4 5	4.20 4.00 4.90 5.60 6.30 6.00
[84]	anatase	Cement mortar	Chinese standard	w/c 0.32	-	0 0.1 1	10.10 10.80 9.60
[85]	anatase	Cement mortar	ASTM C348	w/b 0.40	GGBFS [2] Polycarboxylate (superplasticizer)	0 3 6 9 12	approx. 5.30 approx. 5.50 approx. 4.40 approx. 4.00 approx. 3.80

[1] High-range water reducer. [2] Ground granulated blast furnace slag. [3] Some values were approximated from the graphics, associated with the cited scientific research.

Experiments conducted based on the Chinese standard resulted in the highest flexural strength values, reaching up to 10 MPa for cement mortar [84]. However, the researchers who used ASTM norms, whether or not additives were used, obtained values in the range of 4.0–6.0 MPa [80–83,85].

Taking into account the variability in TiO_2 percentages from one study to another and the difference in results, there is no optimum percentage which can be clearly defined. Smaller percentage increments, i.e., smaller than 1%, should be selected and the range of their variability should be larger in order to be able to thoroughly analyze the dependency between strength and the quantity of TiO_2 introduced. However, there is a certain trend which is also respected in the case of flexural tensile strength, namely, as the quantity of TiO_2 exceeded a certain level, the value of the flexural strength decreased.

3.4.3. Compressive Strength

The most important material property of both cement mortar and concrete is their compressive strength. As in the case of nanoclay, some studies subjected the specimens to high temperatures while others analyzed the samples only at room temperature. Taking into account that the high temperature tests are conducted in steps, for different temperature levels, corresponding to various physical and chemical phenomena, they are separately presented in the present paper.

Compressive Strength Values Obtained at Room Temperature

In case of compressive strength, a similar trend of the results was obtained as for flexural strength [80]. The CNF modified concrete exhibited the highest values of the compressive strengths compared with the specimens modified only with TiO_2. However, the maximum strength from the analyzed samples corresponds to a concrete specimen modified with both CNF and TiO_2 nanoparticles, i.e., 0.4% CNF and 3% TiO_2. Similar to the flexural tensile strength, the samples of CNF+ TiO_2 having 5% TiO_2 exhibited lower compressive strength value than the corresponding CNF modified concrete samples but without the TiO_2 addition. The trend was maintained at least up to 180 days of curing. Moreover, for the 0.2% CNF and 5% TiO_2, the registered values of the compressive strength was smaller than the 3% TiO_2 mix, at 28 days of curing [80].

Unlike the tensile strength case, the compressive strength of the 1% TiO_2 concrete specimens was lower than for the 1% Fe_3O_4 concrete specimens [79]. TiO_2 was used in combination with ZnO and 0.6% polypropylene fibers PPF to improve the concrete formula [71]. Samples were tested in compression according to the Indian standard IS: 516–1959. For both mixes, with or without PPF, the best nanomaterial combination was (4%; 2%) by weight of cement, the result being applicable for all curing ages (7, 14, 28 and 90 days). For the largest nanomaterial quantity, i.e., 5% ZnO and 2.5% TiO_2, the compressive strength value decreased at all ages below the values corresponding to the control sample [71].

Table 8 summarizes the values of compressive strength for concrete specimens modified with TiO_2 nanoparticles, kept at room temperature, and cured for 28 days.

The importance of TiO_2 nanoparticle dimension on the compressive strength value of cement mortar was demonstrated in [77]. The study employed two types on TiO_2, with dimensions of 21 nm and 350 nm. The mortar modified with larger TiO_2 particles had smaller values for the compressive strengths than the one with 21 nm. As the particle dimension increased, the specific surface area decreased, being promoted less CH for hydration and thus, less CSH in the cement matrix.

All samples registered a decrease in the values of mechanical properties after a certain TiO_2 quantity when the high number of particles increased the probability of flocculation and thus stress concentrations could occur. Moreover, as in the nanoclay case, when introducing a higher quantity of nanomaterial than the one needed for CH hydration, an excess of silica is available, resulting in a deficiency in strength [70]. The TiO_2 optimum percentage, from which this decrease initiated, and emphasized in Table 8, is not well defined when considering all the studies. It varies from 1% to 6% by weight of cement. High compressive strength was also obtained at 10%, according to [77], but there were no higher TiO_2 percentages analyzed, so there is no specific optimum TiO_2 percentage.

On the other hand, a previous study [60] reported that the values of the compressive strength decreased with the increase of TiO_2 percentage. Taking into account that the TiO_2 percentages analyzed in that study were relatively high, i.e., 5% and 10%, there is a high probability that the main cause of the obtained results was the agglomeration of nano particles during the mixing procedure.

To sum up, the compressive strength results presented in Table 8 demonstrate the positive effect that TiO_2 particles have on the cement mortar/concrete compressive strength. Unlike the case of the flexural tensile strength, there is no significant difference between the values of the compressive strength concrete and mortar, due to the fact that, during compression, the voids present in the ITZ tend to close.

Table 8. Compressive strength values mortar/concrete specimens modified with TiO_2 nanoparticles, at room temperature.

Study	Type of TiO_2	Type of Sample	Standard	w/b w/c	Additives	TiO_2 %	Compressive Strength [3] [MPa]
[80]	-	Concrete	ASTM C39	w/c 0.45	Polycarboxylate-based HRWR [1] agent (superplasticizer)	0 3 5	approx. 33.00 approx. 42.00 approx. 37.00
[83]	-	Cement mortar	ASTM C109M-16a	w/b 0.485	30% Fly ash by weight of cement	0 1 3 5	approx. 26.50 approx. 33.00 approx. 36.50 approx. 30.70
[81]	-	Concrete	ASTM C39	w/b 0.40	-	0 0.5 1 1.5 2	36.80 41.90 43.40 42.50 39.30
[82]	-	SCC[2]	ASTM C39	w/b 0.40	Polycarboxylate (superplasticizer)	0 1 2 3 4 5	31.60 35.20 38.30 44.50 50.10 48.70
[79]	-	Concrete	-	w/c 0.57	-	1	approx. 31.00
[85]	anatase	Cement mortar	ASTM C109	w/b 0.40	GGBFS [2] Polycarboxylate (superplasticizer)	0 3 6 9 12	approx. 40.00 approx. 44.00 approx. 46.00 approx. 42.00 approx. 36.00
[77]	75% anatase and 25% rutile (21 nm)	Cement mortar	ASTM C109	w/c 0.485	-	0 5 10	approx. 39.50 approx. 47.50 approx. 49.00
	99% anatase (350 nm)					5 10	approx. 43.00 approx. 44.50
[69]	Rutile and anatase	Cement mortar	-	w/b 0.45	-	0 1 3 5	43.70 47.60 48.20 48.80
[86]	Anatase and rutile—Aeroxide P25	High strength mortar	ASTM C109	w/b 0.35	2% naphthalene sulfonate base superplasticizer 5% silica fume	0 1 2 3	approx. 55.00 approx. 63.00 approx. 64.00 approx. 61.00
[70]	White powder (probably anatase)	Concrete	-	w/c 0.38	15% Silica fume Sika ViscoCrete-3425 superplasticizer	0 0.5 1 15	49.86 55.74 58.79 57.42
[60]	80% anatase and 20% rutile	ECC [1]	ASTM C109	w/c 0.30	PVA [2]	0 5 10	66.53 60.05 58.49
[78]	-	Concrete	-	w/c 0.45	Sulphonated naphthalene formaldehyde (superplasticizer)	0 05 1 1.5	33.00 35.00 38.00 30.00
[75]	-	Concrete	BS 1881-part 116	w/c 0.40	Superplasticizer	0 2 4 6 8	34.00 41.40 44.20 48.40 46.50
[76]	-	Cement mortar	-	w/c 0.50	-	0 1 2 5	approx. 50.00 approx. 50.00 approx. 52.00 approx. 48.00

[1] Engineered cementitious composite. [2] Polyvinyl alcohol fibers. [3] Some values were approximated from the graphics, associated with the cited scientific research.

Compressive Strength Values at Elevated Temperatures

Considering the filler effect that titanium dioxide has upon the cement matrix, along with promoting of the development of hydration compounds, it has been proven that the TiO_2 introduction in the cement matrix succeeded in improving the material strength to elevated temperatures. Table 9 presents some of the compressive strength results on cement mortar/concrete when the specimen was subjected to elevated temperatures.

Table 9. Compressive strength values for TiO$_2$—mortar/concrete specimens at high temperature.

		[86]				[87]			
	TiO$_2$ [%]	0	1	2	3	0	2	4	6
Compressive strength [1] [MPa]	25 °C	55	63	64	61	33	36.3	46.2	48
	100 °C	47	58	62	54				
	200 °C	46	52	58	51	38	43.5	47	54
	300 °C	35	46	51	48				
	400 °C	35	47	49	44	35	42	46	51.5
	600 °C	31	37	35	31	25	31	33.5	38.5
	800 °C	17	17	22	17				
	1000 °C	6	6	7	5				

[1] All values are approximated from the graphics, associated with the cited scientific research.

A high strength mortar was modified with 1%, 2% and 3% Aeroxide P25, i.e., a multiphasic titanium dioxide containing both anatase and rutile. A 5% silica fume and a superplasticizer were added in the mix. The water to binder ratio was maintained at 0.35 [86]. On the other hand, another study focused on heavy concrete samples, having a density greater than 2600 kg/m^3 with magnetite aggregates of 25 mm maximum size. The authors modified the samples by adding 2%, 4% and 6% TiO$_2$ as a cement replacement [87]. Although the difference between the values obtained in the two studies was expected to occur, the positive influence of TiO$_2$ is evident in both cases.

However, while a 2% TiO$_2$ was obtained as the optimum percentage for the compressive strength of, for all temperatures [86], as shown in Figure 4a, the maximum strength was recorded at the maximum nanoparticle addition of 6% [87], as seen in Figure 4b. Taking into account the use of magnetite aggregates, they will have a different behavior to high temperatures than the normal aggregates used in previous studies. Moreover, only fine aggregates were used in [86], thus obtaining a more homogeneous sample with less voids and less disturbances than in [87]. Therefore, a comparison between these two studies can be made only by analyzing the positive variation that TiO$_2$ had on the samples. Although the compressive strength decreased for the 3% TiO$_2$ sample, it still remained higher than the one corresponding to the control sample up to 400 °C [86]. After this temperature limit, the compressive strength maintained a value approximately equal with the one associated to the control specimen.

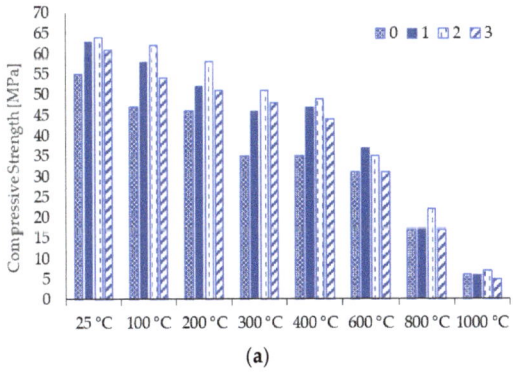
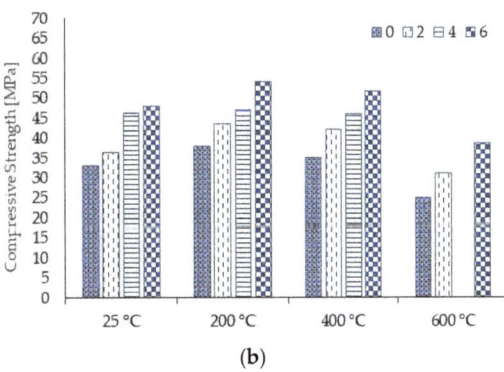

Figure 4. Variation in compressive strength as function of temperature and different percentages of TiO$_2$. (**a**) results reported in [86]; (**b**) results reported in [87].

3.5. Durability Tests

When the TiO_2 nanoparticles are introduced in the cement matrix, due to their small size, they start filling the existent voids, gathering around them CH compounds and promoting the CH hydration. Thus, the density of the specimen increases and water absorption is expected to decrease. For the anatase-based modified cement mortar used in [85], a constant decrease in water absorption up to the maximum considered TiO_2 percentage, from approximately 9%—for 0 wt% TiO_2, to 7%—for 12 wt% TiO_2 was recorded. These results were confirmed by [82]. A comparison between the influence of 1% Fe_3O_4 and 1% TiO_2 on concrete water absorption was conducted in [79]. From the analysis, nano-titania particles succeeded in restricting the water absorption better than Fe_3O_4, also having a smaller void ratio.

As the water absorption and void ratio decreased, the probability of liquid infiltration was significantly reduced, as the chloride and sulphate exposure analyses proved. A higher level of steel reinforcement corrosion protection was gained by using nano-TiO_2 as supplementary material in concrete, according to [79]. The potentio-dynamic polarization test confirmed the benefits of using TiO_2 in the cement matrix as the reinforcement corrosion rates were lower for all adverse exposure environments studied (tap water, saline water, acidic solution), compared to the control specimen [69].

However, it was recommended that all cementitious materials using nano-titania as additive should be specially designed to sulphate attack, as the expansion rate of the mortar increased with the quantity of TiO_2 and important cracks occur, compared to the control sample [85].

Table 10 summarizes the findings in terms of durability tests on mortar and concrete samples enhanced with TiO_2.

Table 10. Durability tests on TiO_2—mortar/concrete specimens.

Measured Parameter	Scientific Paper	Type of Sample	TiO_2 Percentage	Observation
Water absorption percentage	[85]	Cement mortar	0; 3; 6; 9; 12	- Decrease in water absorption with the increase in TiO_2 content
	[82]	SCC	0; 1; 2; 3; 4; 5	- During the first 2 days, the water absorption percentage increased compared to the control sample - For 7 and 28 days of curing, the water absorption percentage decreased compared to the control sample
	[79]	Concrete	1	- Nano-TiO_2 particles restrict water absorption better than Fe_3O_4
Durability to chloride and sulphate ions (corrosion potential)	[79]	Concrete	1	- Nano-TiO_2 leads to a higher level of steel reinforcement corrosion protection than Fe_3O_4
	[69]	Cement mortar	0; 1; 3; 5	- Reinforcement corrosion rates were lower for all adverse exposure environments studied compared to control samples - The optimum TiO_2 percentage was 5 wt%
Expansion rate	[85]	Cement mortar	0; 3; 6; 9; 12	- The expansion rate of the mortar increases with the quantity of TiO_2 and important cracks occur, compared to control sample

3.6. Remarks on the Impact of Using TiO_2 in Cement Composites

Titanium dioxide is a nanomaterial with multiple benefits, in various areas of interest, which has proven its significance in the field of civil engineering. Its nano-sized structure characterized by a large specific surface area promotes the acceleration of the CH hydration activity. Its filling effect, together with a high quantity of hydration products, i.e., CSH, results in higher values of compressive, flexural and splitting tensile strength compared to the control specimens. The filling effect also decreases the void ratio and the water absorption, increasing the reinforcement protection against tap/saline water or acidic solutions.

4. Conclusions

Nanomaterials have proven their benefits when used in cement mortar or concrete. Their size and chemical properties largely improve the strength and physical characteristics of cementitious composites. Nanoclay acts as a pozzolanic and filler nanomaterial, while titanium dioxide is an inert filler and promotor of nucleation sites for CSH development. Both nanoclay and titanium dioxide nano-particles succeed in promoting CH hydration when added to the cement matrix, thus leading to strength improvements. The water absorption decreases as well as the void ratio, increasing the reinforcement protection to corrosion. For both of them, the behavior is enhanced when subjected to high temperatures, compared with the control samples. Practically, they both improve the mechanical and physical characteristics of cement mortar and/or concrete. However, the manner in which this improvement is manifests varies.

The use of NC definitely influences the values of the splitting tensile strength for both mortar and concrete in a positive manner. However, the magnitude of this improvement greatly varies from one study to another because of the type of NC used and water/binder ratio. However, the use of montmorillonite-based NC shows its influence at lower percentages by weight of cement, for both mortar (2 wt%) and concrete (1 wt%), than metakaolin based NC (8 wt%). Based on the analyzed data, the rate of improvement also greatly depends on the water/binder ratio.

Similar conclusions can be drawn in terms of flexural tensile strength of cement-based mortar and concrete. For mortars, both natural hydrophilic montmorillonite and organo-montmorillonite based NC shown similar rates of improvement, compared to the reference mix, for 2 wt% replacement of cement for natural hydrophilic montmorillonite NC and 1 wt% replacement for organo-montmorillonite based NC. On the other hand, the use of calcinated hydrophilic montmorillonite NC leads to highest values in terms of flexural tensile strength of mortars when used in 5 wt% replacement of cement. For concrete mixes, metakaolin based NC should be used in higher percentages (8 wt% of cement) to obtain the highest values for the flexural tensile strength as compared to montmorillonite NC which leads to the best results at only 1 wt% of cement.

The compressive strength of both mortar and concrete are positively influenced by the addition of nanoclay, but the rate of improvement is different from mortar to concrete. Lower percentages of montmorillonite-based NC are required to obtain the highest values of compressive strength compared to metakaolin based NC. While for the former the highest values of the compressive strength of mortar are obtained at 0.5 wt% in case of organo-montmorillonite and up to 5 wt% in case of calcinated hydrophilic montmorillonite NC, in case of metakaolin based NC a higher percentage is required, 8 wt% of cement. Similar observations can be made for concrete, although the exact type of NC is not clearly specified. For montmorillonite-based NC the percentage varies between 1 wt% and 3 wt%, while for metakaolin based NC the percentage stays unchanged, namely at 8 wt% of cement.

The loss of compressive strength at elevated temperatures is inversely proportional to the percentage of NC in concrete. Based on the available scientific data, the peak performance is reached for 2% natural hydrophilic montmorillonite NC at 200 °C, while a similar trend is observed at 5% calcinated hydrophilic montmorillonite NC at 250 °C.

In case of durability of cement mortar and concrete, the best values are reached for much lower values of NC percentages than in the case of mechanical properties. Most of the available scientific literature reporting results on durability performance of nanoclay modified mortar and concrete focuses on the use of montmorillonite-based NC. Significant improvements are obtained for percentages as low as 0.4 wt% of cement.

On the other hand, titanium dioxide, with a smaller particle dimension, is only an inert filler which does not have the possibility to promote pozzolanic activity as nanoclay does. The CH hydration, in this case, is accelerated only by the nucleation effect.

Most of the studies related to the use of TiO_2 in cement-based mortar and concrete used anatase based nano TiO_2. Therefore, the scattering of the reported results in terms of the optimum content of TiO_2 is smaller compared to their counterparts using NC.

From the point of view of the splitting tensile strength most of the studies recommend an optimum percentage of 1 wt% of cement in order to obtain the highest increase in performance for regular concrete. Self-compacted concrete, on the other hand, requires a higher percentage of nano TiO_2, up to 4%.

For the flexural tensile strength, the use of nano TiO_2 in 3 wt% of cement results in the highest values both for mortar and concrete. In this case however, there seems to be a large discrepancy of the obtained results depending on the standard applied.

When it comes to the compressive strength, although improvements were reported for both mortar and concrete with nano TiO_2 addition, the reported results in terms of optimum nano-TiO_2 content varies significantly from one study to another. While for concrete the optimum percentage varies from 1 wt% of cement up to 6 wt% of cement, with most of the studies recommending 1% addition of nano TiO_2 for best performance gains, the interval is much larger in case of mortar. In this case, the optimum percentage varies from 2 wt% of cement and up to 10 wt% of cement. It should be pointed out that the larger recommended interval is due to the fact that the input parameters significantly changed from one study to another: supplementary cementitious materials (fly ash, GGBS, silica fume), type of TiO_2 (anatase and combinations of anatase and rutile in different percentages).

The decrease in the compressive strength values of concrete with nano TiO_2 subjected to elevated temperatures followed a similar trend with the one reported for NC. The use of higher percentages of nano TiO_2 resulted in significantly lower strength losses with the increase in temperature compared to the reference mix. The temperature threshold beyond which strength loss become significant is similar to the one reported for NC, namely 400 °C. This suggests that, in case of elevated temperatures, the limitation is related to the behavior of cement rather than the other constituent materials.

In terms of durability performance of mortar and concrete, the optimum percentage varies from one study to another. According to the available data, a 5 wt% of cement leads to the best performance in durability tests. Higher percentages result in slightly better performance but the benefit is not as significant, percentage-wise.

Each of the two nanomaterials presented in this study lack some properties from the other, while having other similar effects on the cementitious materials. Therefore, a combination of these two types of nanoparticles has the potential of improving the mortar and/or concrete properties beyond the level set by each nanomaterial on its own due to their synergistic effect. While the synergistic effect of using TiO_2 in combination with silica fume [70], CNF [80], fly ash [83] and GGBS [85] was rendered evident in some previous studies, the use of TiO_2 in conjunction with nanoclay was not so intensively studied and only recently a very limited number of studies emerged in this direction.

A recent study proved the potential of this combination by adding to concrete 1%, 2%, 3% and 4% nanoclay as fine aggregate and 1%, 2%, 3% and 4% TiO_2 as cement replacement. The best results were obtained for the 2%-TiO_2 and 3%-nanoclay combination, with a compressive strength increment of 48.64% compared to the control sample and 21.83% increment compared to the concrete modified with only 2% TiO_2 [88].

A similar research work investigated the combined used of TiO_2 and NC in fly ash geopolymer concrete [89]. The main difference between this study and the previous one resides in the fact that lower percentages were used; only 1% NC and 1.25% TiO_2, by mass of fly ash. Improvements were observed in the values of splitting tensile and compressive strengths of investigated mixes. The SEM images revealed the absence of interfacial voids and crack in the ITZ and the formation of needle-like structures at the interface regions between paste and aggregates. Those formations were attributed to pozzolanc interaction between the nanomaterials and the promotion of the nucleation sites.

Both aforementioned studies showed that better material properties can be obtained when TiO_2 and NC are used together, compared to the case when they are individually used and that lower percentages of the two nanomaterials are needed to obtain those improved material properties compared to each nanomaterial used individually. However, there is a lack of research on this combination of nanomaterials, which, taking into account their beneficial potential, should be studied further.

Author Contributions: Conceptualization, I.-O.T. and G.B.; methodology, G.B., S.-M.A.-S., P.M. and I.-O.T.; formal analysis, G.B., S.-M.A.-S. and P.M.; investigation, G.B., S.-M.A.-S., P.M. and I.-O.T.; writing—original draft preparation, G.B. and S.-M.A.-S.; writing—review and editing, P.M. and I.-O.T.; supervision, P.M. and I.-O.T.; project administration, G.B.; funding acquisition, G.B. All authors have read and agreed to the published version of the manuscript.

Funding: This research was funded by the Romanian Government through the Ministry of Research, Innovation and Digitalization, grant number PN III 27PFE/2021; The APC was funded by The "Gheorghe Asachi" Technical University of Iasi.

Acknowledgments: This paper was realized with the support of COMPETE 2.0 Project nr. 27PFE/2021, financed by the Romanian Government, Ministry of Research, Innovation and Digitalization.

Conflicts of Interest: The authors declare no conflict of interest.

References

1. United Nations Environment Programme. *Global Status Report for Buildings and Construction 2021*; United Nations Environment Programme: Nairobi, Kenya, 2021.
2. European Union. EU Commission recommendation of 18 October 2011 on the definition of nanomaterial (2011/696/EU). *Off. J. Eur. Union* **2011**, *275*, 38–40.
3. Sikora, P.; Abd Elrahman, M.; Stephan, D. The Influence of Nanomaterials on the Thermal Resistance of Cement-Based Composites—A Review. *Nanomaterials* **2018**, *8*, 465. [CrossRef] [PubMed]
4. Saleem, H.; Zaidi, S.J.; Alnuaimi, N.A. Recent Advancements in the Nanomaterial Application in Concrete and Its Ecological Impact. *Materials* **2021**, *14*, 6387. [CrossRef] [PubMed]
5. Monteiro, H.; Moura, B.; Soares, N. Advancements in nano-enabled cement and concrete: Innovative properties and environmental implications. *J. Build. Eng.* **2022**, *56*, 104736. [CrossRef]
6. Duquesne, S.; Jama, C.; Le Bras, M.; Delobel, R.; Recourt, P.; Gloaguen, J. Elaboration of EVA–nanoclay systems—Characterization, thermal behaviour and fire performance. *Compos. Sci. Technol.* **2003**, *63*, 1141–1148. [CrossRef]
7. Park, J.H.; Jana, S.C. The relationship between nano- and micro-structures and mechanical properties in PMMA–epoxy–nanoclay composites. *Polymer* **2003**, *44*, 2091–2100. [CrossRef]
8. Xu, L.; Lee, L.J. Effect of nanoclay on shrinkage control of low profile unsaturated polyester (UP) resin cured at room temperature. *Polymer* **2004**, *45*, 7325–7334. [CrossRef]
9. Lee, H.-C.; Lee, T.-W.; Kim, T.-H.; Park, O.O. Fabrication and characterization of polymer/nanoclay hybrid ultrathin multilayer film by spin self-assembly method. *Thin Solid Films* **2004**, *458*, 9–14. [CrossRef]
10. LAU, K.; LU, M.; QI, J.; ZHAO, D.; CHEUNG, H.; LAM, C.; LI, H. Cobalt hydroxide colloidal particles precipitation on nanoclay layers for the formation of novel nanocomposites of carbon nanotubes/nanoclay. *Compos. Sci. Technol.* **2006**, *66*, 450–458. [CrossRef]
11. Lau, K.; Gu, C.; Hui, D. A critical review on nanotube and nanotube/nanoclay related polymer composite materials. *Compos. Part B Eng.* **2006**, *37*, 425–436. [CrossRef]
12. Kuo, W.Y.; Huang, J.S.; Lin, C.-H. Effects of organo-modified montmorillonite on strengths and permeability of cement mortars. *Cem. Concr. Res.* **2006**, *36*, 886–895. [CrossRef]
13. He, X.; Shi, X. Chloride Permeability and Microstructure of Portland Cement Mortars Incorporating Nanomaterials. *Transp. Res. Rec. J. Transp. Res. Board* **2008**, *2070*, 13–21. [CrossRef]
14. Chang, T.-P.; Shih, J.-Y.; Yang, K.-M.; Hsiao, T.-C. Material properties of portland cement paste with nano-montmorillonite. *J. Mater. Sci.* **2007**, *42*, 7478–7487. [CrossRef]
15. Dhiman, N.K.; Sidhu, N.; Agnihotri, S.; Mukherjee, A.; Reddy, M.S. Role of nanomaterials in protecting building materials from degradation and deterioration. *Biodegrad. Biodeterior. Nanoscale* **2022**, 405–475. [CrossRef]

16. Ali, S.W.; Basak, S.; Shukla, A. Nanoparticles: A potential alternative to classical fire retardants for textile substrates. *Handb. Nanomater. Manuf. Appl.* **2020**, 265–278. [CrossRef]
17. Yu, P.; Wang, Z.; Lai, P.; Zhang, P.; Wang, J. Evaluation of mechanic damping properties of montmorillonite/organo-modified montmorillonite-reinforced cement paste. *Constr. Build. Mater.* **2019**, *203*, 356–365. [CrossRef]
18. Irshidat, M.R.; Al-Saleh, M.H. Thermal performance and fire resistance of nanoclay modified cementitious materials. *Constr. Build. Mater.* **2018**, *159*, 213–219. [CrossRef]
19. Pavlidou, S.; Papaspyrides, C.D. A review on polymer–layered silicate nanocomposites. *Prog. Polym. Sci.* **2008**, *33*, 1119–1198. [CrossRef]
20. Nehdi, M.L. Clay in cement-based materials: Critical overview of state-of-the-art. *Constr. Build. Mater.* **2014**, *51*, 372–382. [CrossRef]
21. Kawashima, S.; Wang, K.; Ferron, R.D.; Kim, J.H.; Tregger, N.; Shah, S. A review of the effect of nanoclays on the fresh and hardened properties of cement-based materials. *Cem. Concr. Res.* **2021**, *147*, 106502. [CrossRef]
22. Moraes, M.K.; Maria da Costa, E. Effect of adding organo-modified montmorillonite nanoclay on the performance of oil-well cement paste in CO2-rich environments. *Cem. Concr. Compos.* **2022**, *127*, 104400. [CrossRef]
23. Lee, S.J.; Kawashima, S.; Kim, K.J.; Woo, S.K.; Won, J.P. Shrinkage characteristics and strength recovery of nanomaterials-cement composites. *Compos. Struct.* **2018**, *202*, 559–565. [CrossRef]
24. Papatzani, S.; Badogiannis, E.G.; Paine, K. The pozzolanic properties of inorganic and organomodified nano-montmorillonite dispersions. *Constr. Build. Mater.* **2018**, *167*, 299–316. [CrossRef]
25. Kalpokait-Dičkuvien, R.; Lukošiūt, I.; Čcsnien, J.; Baltušnikas, A.; Brinkien, K. Influence of the Organically Modified Nanoclay on Properties of Cement Paste. Available online: https://www.laviosa.com/wp-content/uploads/2015/01/LAVIOSA_Plastic_InfluenceOfTheOrganicallyModifiedNanoclayOnPropertiesOfCementPaste.pdf (accessed on 7 November 2022).
26. Kuo, W.-Y.; Huang, J.-S.; Yu, B.-Y. Evaluation of strengthening through stress relaxation testing of organo-modified montmorillonite reinforced cement mortars. *Constr. Build. Mater.* **2011**, *25*, 2771–2776. [CrossRef]
27. Papatzani, S.; Paine, K. *Nanotechnology in Construction*; Sobolev, K., Shah, S.P., Eds.; Springer International Publishing: Cham, Switzerland, 2015; ISBN 978-3-319-17087-9.
28. Alani, S.; Hassan, M.S.; Jaber, A.A.; Ali, I.M. Effects of elevated temperatures on strength and microstructure of mortar containing nano-calcined montmorillonite clay. *Constr. Build. Mater.* **2020**, *263*, 120895. [CrossRef]
29. Hakamy, A.; Shaikh, F.U.A.; Low, I.M. Characteristics of nanoclay and calcined nanoclay-cement nanocomposites. *Compos. Part B Eng.* **2015**, *78*, 174–184. [CrossRef]
30. Rehman, S.U.; Yaqub, M.; Ali, T.; Shahzada, K.; Khan, S.W.; Noman, M. Durability of Mortars Modified with Calcined Montmorillonite Clay. *Civ. Eng. J.* **2019**, *5*, 1490–1505. [CrossRef]
31. Wakoya, G.; Quezon, E.; Ararsa, W.; Chimdi, J.; Lemu, D.G. Effect and Suitability of Calcined Montmorillonite Clay Powder and Waste Khat Husk Ash in the Strength and Durability of C-25 Concrete and its Benefits Cost Analysis. *J. Sustain. Constr. Mater. Proj. Manag.* **2021**, *1*.
32. Said-Mansour, M.; Kadri, E.-H.; Kenai, S.; Ghrici, M.; Bennaceur, R. Influence of calcined kaolin on mortar properties. *Constr. Build. Mater.* **2011**, *25*, 2275–2282. [CrossRef]
33. Chandrasekhar, S.; Ramaswamy, S. Influence of mineral impurities on the properties of kaolin and its thermally treated products. *Appl. Clay Sci.* **2002**, *21*, 133–142. [CrossRef]
34. Badogiannis, E.; Kakali, G.; Tsivilis, S. Metakaolin as supplementary cementitious material. *J. Therm. Anal. Calorim.* **2005**, *81*, 457–462. [CrossRef]
35. Fadzil, M.A.; Nurhasri, M.S.M.; Norliyati, M.A.; Hamidah, M.S.; Ibrahim, M.H.W.; Assrul, R.Z. Characterization of Kaolin as Nano Material for High Quality Construction. *MATEC Web Conf.* **2017**, *103*, 09019. [CrossRef]
36. Siddique, R.; Klaus, J. Influence of metakaolin on the properties of mortar and concrete: A review. *Appl. Clay Sci.* **2009**, *43*, 392–400. [CrossRef]
37. Mansi, A.; Sor, N.H.; Hilal, N.; Qaidi, S.M.A. The Impact of Nano Clay on Normal and High-Performance Concrete Characteristics: A Review. *IOP Conf. Ser. Earth Environ. Sci.* **2022**, *961*, 012085. [CrossRef]
38. Firoozi, A.A.; Firoozi, A.A.; Baghini, M.S. A Review of Clayey Soils. *Asian J. Appl. Sci.* **2016**, *4*, 1319–1330.
39. Nisticò, R. A Comprehensive Study on the Applications of Clays into Advanced Technologies, with a Particular Attention on Biomedicine and Environmental Remediation. *Inorganics* **2022**, *10*, 40. [CrossRef]
40. Salam, H.; Dong, Y.; Davies, I. Development of biobased polymer/clay nanocomposites. In *Fillers and Reinforcements for Advanced Nanocomposites*; Elsevier: Amsterdam, The Netherlands, 2015; pp. 101–132; ISBN 9780081000823.
41. Patel, K. The Use of Nanoclay as a Constructional Material. *Int. J. Eng. Res. Appl.* **2012**, *2*, 1382–1386.
42. Mirgozar Langaroudi, M.A.; Mohammadi, Y. Effect of nano-clay on workability, mechanical, and durability properties of self-consolidating concrete containing mineral admixtures. *Constr. Build. Mater.* **2018**, *191*, 619–634. [CrossRef]
43. Irshidat, M.R.; Al-Saleh, M.H.; Sanad, S. Effect of Nanoclay on the Expansive Potential of Cement Mortar due to Alkali-Silica Reaction. *ACI Mater. J.* **2015**, *112*, 801–808. [CrossRef]
44. Liu, G.; Zhang, S.; Fan, Y.; Shah, S.P. Study on Shrinkage Cracking Morphology of Cement Mortar with Different Nanoclay Particles under Restraint. *Buildings* **2022**, *12*, 1459. [CrossRef]

45. Noori, A.; Yubin, L.; Saffari, P.; Zhang, Y.; Wang, M. The optimum percentage of nano clay (NC) in both direct-additive and sonicated modes to improve the mechanical properties of self-compacting concrete (SCC). *Case Stud. Constr. Mater.* **2022**, *17*, e01493. [CrossRef]
46. Morsy, M.S.; Alsayed, S.H.; Aqel, M. Effect of Nano-clay on Mechanical Properties and Microstructure of Ordinary Portland Cement Mortar. *Int. J. Civ. Environ. Eng. IJCEE-IJENS* **2010**, *10*, 21–25.
47. Tantawy, M.A. Effect of High Temperatures on the Microstructure of Cement Paste. *J. Mater. Sci. Chem. Eng.* **2017**, *05*, 33–48. [CrossRef]
48. Alarcon-Ruiz, L.; Platret, G.; Massieu, E.; Ehrlacher, A. The use of thermal analysis in assessing the effect of temperature on a cement paste. *Cem. Concr. Res.* **2005**, *35*, 609–613. [CrossRef]
49. Han, B.; Li, Z.; Zhang, L.; Zeng, S.; Yu, X.; Han, B.; Ou, J. Reactive powder concrete reinforced with nano SiO_2-coated TiO_2. *Constr. Build. Mater.* **2017**, *148*, 104–112. [CrossRef]
50. Mohammed, A.; Rafiq, S.; Mahmood, W.; Al-Darkazalir, H.; Noaman, R.; Qadir, W.; Ghafor, K. Artificial Neural Network and NLR techniques to predict the rheological properties and compression strength of cement past modified with nanoclay. *Ain Shams Eng. J.* **2021**, *12*, 1313–1328. [CrossRef]
51. Allalou, S.; Kheribet, R.; Benmounah, A. Effects of calcined halloysite nano-clay on the mechanical properties and microstructure of low-clinker cement mortar. *Case Stud. Constr. Mater.* **2019**, *10*, e00213. [CrossRef]
52. Emamian, S.A.; Eskandari-Naddaf, H. Effect of porosity on predicting compressive and flexural strength of cement mortar containing micro and nano-silica by ANN and GEP. *Constr. Build. Mater.* **2019**, *218*, 8–27. [CrossRef]
53. Zhang, A.; Yang, W.; Ge, Y.; Liu, P. Effect of nanomaterials on the mechanical properties and microstructure of cement mortar under low air pressure curing. *Constr. Build. Mater.* **2020**, *249*, 118787. [CrossRef]
54. Naji, H.F.; Khalid, N.N.; Alsaraj, W.K. Influence of Nanoclay on the Behavior of Reinforced Concrete Slabs. *IOP Conf. Ser. Mater. Sci. Eng.* **2020**, *870*, 012107. [CrossRef]
55. ali Shafabakhsh, G.; Janaki, A.M.; Ani, O.J. Laboratory Investigation on Durability of Nano Clay Modified Concrete Pavement. *Eng. J.* **2020**, *24*, 35–44. [CrossRef]
56. Irshidat, M.R.; Al-Saleh, M.H. Influence of Nanoclay on the Properties and Morphology of Cement Mortar. *KSCE J. Civ. Eng.* **2018**, *22*, 4056–4063. [CrossRef]
57. Wang, W.-C. Compressive strength and thermal conductivity of concrete with nanoclay under Various High-Temperatures. *Constr. Build. Mater.* **2017**, *147*, 305–311. [CrossRef]
58. Shafaei, D.; Yang, S.; Berlouis, L.; Minto, J. Multiscale pore structure analysis of nano titanium dioxide cement mortar composite. *Mater. Today Commun.* **2020**, *22*, 100779. [CrossRef]
59. Francioso, V.; Moro, C.; Martinez-Lage, I.; Velay-Lizancos, M. Curing temperature: A key factor that changes the effect of TiO_2 nanoparticles on mechanical properties, calcium hydroxide formation and pore structure of cement mortars. *Cem. Concr. Compos.* **2019**, *104*, 103374. [CrossRef]
60. Zhao, A.; Yang, J.; Yang, E.-H. Self-cleaning engineered cementitious composites. *Cem. Concr. Compos.* **2015**, *64*, 74–83. [CrossRef]
61. Tokuç, A.; Özkaban, F.F.; Çakır, Ö.A. Biomimetic Facade Applications for a More Sustainable Future. In *Interdisciplinary Expansions in Engineering and Design With the Power of Biomimicry*; InTech: London, UK, 2018.
62. Moradeeya, P.G.; Sharma, A.; Kumar, M.A.; Basha, S. Titanium dioxide based nanocomposites—Current trends and emerging strategies for the photocatalytic degradation of ruinous environmental pollutants. *Environ. Res.* **2022**, *204*, 112384. [CrossRef]
63. Janczarek, M.; Klapiszewski, Ł.; Jędrzejczak, P.; Klapiszewska, I.; Ślosarczyk, A.; Jesionowski, T. Progress of functionalized TiO_2-based nanomaterials in the construction industry: A comprehensive review. *Chem. Eng. J.* **2022**, *430*, 132062. [CrossRef]
64. Jayapalan, A.R.; Lee, B.Y.; Kurtis, K.E. Can nanotechnology be 'green'? Comparing efficacy of nano and microparticles in cementitious materials. *Cem. Concr. Compos.* **2013**, *36*, 16–24. [CrossRef]
65. Hamidi, F.; Aslani, F. TiO_2-based Photocatalytic Cementitious Composites: Materials, Properties, Influential Parameters, and Assessment Techniques. *Nanomaterials* **2019**, *9*, 1444. [CrossRef]
66. Castro-Hoyos, A.M.; Manzano, M.A.R.; Maury-Ramírez, A. Challenges and Opportunities of Using Titanium Dioxide Photocatalysis on Cement-Based Materials. *Coatings* **2022**, *12*, 968. [CrossRef]
67. Albetran, H.M. Thermal expansion coefficient determination of pure, doped, and co-doped anatase nanoparticles heated in sealed quartz capillaries using in-situ high-temperature synchrotron radiation diffraction. *Heliyon* **2020**, *6*, e04501. [CrossRef] [PubMed]
68. Mohamed, H.; Deutou, J.G.N.; Kaze, C.R.; Beleuk à Moungam, L.M.; Kamseu, E.; Melo, U.C.; Leonelli, C. Mechanical and microstructural properties of geopolymer mortars from meta-halloysite: Effect of titanium dioxide TiO_2 (anatase and rutile) content. *SN Appl. Sci.* **2020**, *2*, 1573. [CrossRef]
69. Daniyal, M.; Akhtar, S.; Azam, A. Effect of nano-TiO_2 on the properties of cementitious composites under different exposure environments. *J. Mater. Res. Technol.* **2019**, *8*, 6158–6172. [CrossRef]
70. Mustafa, T.S.; El Hariri, M.O.R.; Nader, M.A.; Montaser, W.M. Enhanced shear behaviour of reinforced concrete beams containing Nano-Titanium. *Eng. Struct.* **2022**, *257*, 114082. [CrossRef]
71. Reshma, T.V.; Manjunatha, M.; Bharath, A.; Tangadagi, R.B.; Vengala, J.; Manjunatha, L. Influence of ZnO and TiO_2 on mechanical and durability properties of concrete prepared with and without polypropylene fibers. *Materialia* **2021**, *18*, 101138. [CrossRef]
72. Moro, C.; Francioso, V.; Velay-Lizancos, M. Nano-TiO_2 effects on high temperature resistance of recycled mortars. *J. Clean. Prod.* **2020**, *263*, 121581. [CrossRef]

73. Nazari, A.; Riahi, S.; Shamekhi, S.F.; Khademno, A. Assessment of the effects of the cement paste composite in presence TiO_2 nanoparticles. *J. Am. Sci.* **2010**, *6*, 43–46.
74. Jalal, M. Durability enhancement of concrete by incorporating titanium dioxide nanopowder into binder. *J. Am. Sci.* **2012**, *8*, 289–294.
75. Nikbin, I.M.; Mohebbi, R.; Dezhampanah, S.; Mehdipour, S.; Mohammadi, R.; Nejat, T. Gamma ray shielding properties of heavy-weight concrete containing Nano-TiO_2. *Radiat. Phys. Chem.* **2019**, *162*, 157–167. [CrossRef]
76. Nochaiya, T.; Chaipanich, A. The effect of nano-TiO_2 addition on Portland cement properties. In Proceedings of the 2010 3rd International Nanoelectronics Conference (INEC), Hong Kong, China, 3–8 January 2010; Volume 2, pp. 1479–1480.
77. Chen, J.; Kou, S.; Poon, C. Hydration and properties of nano-TiO_2 blended cement composites. *Cem. Concr. Compos.* **2012**, *34*, 642–649. [CrossRef]
78. Aravind, R.; Devasena, M.; Sreevidya, V.; Vadivel, M. Dispersion Characteristics and Flexural Behavior of Concrete Using Nano Titanium Dioxide. *Int. J. Earth Sci. Eng.* **2016**, *9*, 443–447.
79. d'Orey Gaivão Portella Bragança, M.; Portella, K.F.; Gobi, C.M.; de Mesquita Silva, E.; Alberti, E. The Use of 1% Nano-Fe_3O_4 and 1% Nano-TiO_2 as Partial Replacement of Cement to Enhance the Chemical Performance of Reinforced Concrete Structures. *Athens J. Technology Eng.* **2017**, *4*, 97–108. [CrossRef]
80. Joshaghani, A. Evaluating the effects of titanium dioxide (TiO_2) and carbon-nanofibers (CNF) as cement partial replacement on concrete properties. *MOJ Civ. Eng.* **2018**, *4*, 29–38. [CrossRef]
81. Nazari, A.; Riahi, S.; Riahi, S.; Shamekhi, S.F.; Khademno, A. Improvement the mechanical properties of the cementitious composite by using TiO_2 nanoparticles. *J. Am. Sci.* **2010**, *6*, 98–101.
82. Nazari, A.; Riahi, S. The effect of TiO_2 nanoparticles on water permeability and thermal and mechanical properties of high strength self-compacting concrete. *Mater. Sci. Eng. A* **2010**, *528*, 756–763. [CrossRef]
83. Siang Ng, D.; Paul, S.C.; Anggraini, V.; Kong, S.Y.; Qureshi, T.S.; Rodriguez, C.R.; Liu, Q.; Šavija, B. Influence of SiO_2, TiO_2 and Fe_2O_3 nanoparticles on the properties of fly ash blended cement mortars. *Constr. Build. Mater.* **2020**, *258*, 119627. [CrossRef]
84. Doko, V.K.; Hounkpe, S.P.; Kotchoni, S.O.; Hui, L.; Datchossa, A.T. Changing Mechanicals Characteristiques of Cementitious Materials Using Titanium Dioxide. *Mater. Sci. Appl.* **2021**, *12*, 297–313. [CrossRef]
85. Qudoos, A.; Kim, H.; Ryou, J.-S. Influence of Titanium Dioxide Nanoparticles on the Sulfate Attack upon Ordinary Portland Cement and Slag-Blended Mortars. *Materials* **2018**, *11*, 356. [CrossRef]
86. Farzadnia, N.; Abang Ali, A.A.; Demirboga, R.; Anwar, M.P. Characterization of high strength mortars with nano Titania at elevated temperatures. *Constr. Build. Mater.* **2013**, *43*, 469–479. [CrossRef]
87. Nikbin, I.M.; Mehdipour, S.; Dezhampanah, S.; Mohammadi, R.; Mohebbi, R.; Moghadam, H.H.; Sadrmomtazi, A. Effect of high temperature on mechanical and gamma ray shielding properties of concrete containing nano-TiO_2. *Radiat. Phys. Chem.* **2020**, *174*, 108967. [CrossRef]
88. Selvasofia, S.D.A.; Sarojini, E.; Moulica, G.; Thomas, S.; Tharani, M.; Saravanakumar, P.T.; Kumar, P.M. Study on the mechanical properties of the nanoconcrete using nano-TiO_2 and nanoclay. *Mater. Today Proc.* **2022**, *50*, 1319–1325. [CrossRef]
89. Jumaa, N.H.; Ali, I.M.; Nasr, M.S.; Falah, M.W. Strength and microstructural properties of binary and ternary blends in fly ash-based geopolymer concrete. *Case Stud. Constr. Mater.* **2022**, *17*, e01317. [CrossRef]

Disclaimer/Publisher's Note: The statements, opinions and data contained in all publications are solely those of the individual author(s) and contributor(s) and not of MDPI and/or the editor(s). MDPI and/or the editor(s) disclaim responsibility for any injury to people or property resulting from any ideas, methods, instructions or products referred to in the content.

MDPI
St. Alban-Anlage 66
4052 Basel
Switzerland
www.mdpi.com

Coatings Editorial Office
E-mail: coatings@mdpi.com
www.mdpi.com/journal/coatings

Disclaimer/Publisher's Note: The statements, opinions and data contained in all publications are solely those of the individual author(s) and contributor(s) and not of MDPI and/or the editor(s). MDPI and/or the editor(s) disclaim responsibility for any injury to people or property resulting from any ideas, methods, instructions or products referred to in the content.

www.ingramcontent.com/pod-product-compliance
Lightning Source LLC
LaVergne TN
LVHW070200100526
838202LV00015B/1975